Differential Antennas

IEEE Press
445 Hoes Lane
Piscataway, NJ 08854

Differential Antennas

Theory and Practice

Yueping Zhang
School of Electrical and Electronic Engineering,
Nanyang Technological University, Singapore

IEEE Press

WILEY

Published by John Wiley & Sons, Inc., Hoboken, New Jersey.
Published simultaneously in Canada.

Library of Congress Cataloging-in-Publication Data
Names: Zhang, Yueping, author. | John Wiley & Sons, publisher.
Title: Differential antennas : theory and practice / Yueping Zhang.
Description: Hoboken, New Jersey : Wiley, [2024] | Includes index.
Identifiers: LCCN 2024008700 (print) | LCCN 2024008701 (ebook) | ISBN 9781394273089 (hardback) | ISBN 9781394273102 (adobe pdf) | ISBN 9781394273096 (epub)
Subjects: LCSH: Microstrip antennas–Design and construction. | Microstrip antennas–Data processing.
Classification: LCC TK7871.67.M5 Z48 2024 (print) | LCC TK7871.67.M5 (ebook) | DDC 621.382/4–dc23/eng/20240318
LC record available at https://lccn.loc.gov/2024008700
LC ebook record available at https://lccn.loc.gov/2024008701

Cover Design: Wiley
Cover Image: © Andrey Suslov/Shutterstock

Set in 9.5/12.5pt STIXTwoText by Straive, Pondicherry, India

To the memory of my parents

Contents

About the Author

Yueping Zhang is currently Professor with the School of Electrical and Electronic Engineering at Nanyang Technological University, Singapore. He was Distinguished Lecturer of the IEEE Antennas and Propagation Society (IEEE AP-S) (2018–2022) and Associate Editor of the IEEE Transactions on Antennas and Propagation (IEEE TAP) (2010–2016).

Prof. Zhang has published numerous papers, including two invited and one regular papers in the Proceedings of the IEEE and one invited paper in the IEEE TAP. He is the Chinese radio scientist who has published a historical article in an English learned journal such as the IEEE Antennas and Propagation Magazine. He received the 2012 IEEE AP-S Sergei A. Schelkunoff Prize Paper Award. He also received the Best Paper Award of the *Journal of Microwaves of Chinese Institute of Electronics* (CIE) in 2023.

Prof. Zhang has been invited to deliver plenary, keynote, and invited speeches at the flagship conferences organized by national and international learned societies since 2000. He received a few Best Paper Awards from the International Conferences held in the United Kingdom and China.

Prof. Zhang holds eight US patents. He has made pioneering and significant contributions to the development of antenna-in-package (AiP) technology that have had a profound impact on antennas and electronic packaging. He received the 2020 IEEE AP-S John Kraus Antenna Award. He also received the 2022 Exceptional Technical Achievement Award from the IEEE Electronics Packaging Society.

His current research interests include the development of antenna-on-chip (AoC) technology for very large-scale antenna integration and characterization of chip-scale propagation channels at terahertz for wireless chip area network (WCAN).

Preface

A differential antenna is an antenna with two terminals connecting a differential signal source. A single-ended antenna is an antenna with a single terminal connecting a single-ended signal source. A dipole is such a differential antenna, while a monopole is such a single-ended antenna. The differential antenna has no ground plane, but the single-ended antenna has the ground plane.

A differential antenna can also be an antenna having two single-ended ports. Each single-ended port must connect a single-ended signal source. The two single-ended signal sources applied on the two single-ended ports should have the same amplitude but opposite phases. The single-ended counterpart of such a differential antenna possesses one single-ended port excited by a single-ended signal source, while the other single-ended port maintains an open-circuit configuration. Both such differential and single-ended antennas contain ground planes.

Single-ended antennas show higher gain values and smaller radiator sizes than differential antennas. Therefore, single-ended antennas have dominated the design of antennas for wireless systems. However, it must be stressed that the concept of lower gains and bulkier sizes of differential antennas compared to single-ended counterparts is not always true. It has been demonstrated that differential microstrip patch antennas can possess comparable or even smaller sizes and higher gain values than single-ended microstrip patch antennas.

Differential circuits permit higher linearity and lower offset, leaving them more immune to power supply variations, temperature changes, and substrate noise than single-ended circuits. Consequently, differential circuits are more popular than single-ended circuits in integrated circuit design. Differential circuits naturally call for differential antennas, which is particularly essential in the design of a wireless system-on-chip (SoC) or system-in-package (SiP) device. Differential antennas perfectly interconnect with differential circuits. No lossy balanced/unbalanced conversion circuit (balun) is needed. As a result, the receiver noise performance and transmitter power efficiency are improved. Differential antennas reduce cross-polarized radiation, remove pattern distortion, and produce improved axial ratio for circular polarization.

This book presents the theory and practice of differential antennas for the first time. An effort has been made to give a theoretical treatment of differential antennas while keeping in mind the aspects of practical applications from simple discrete wire to sophisticated integrated designs in antenna-in-package (AiP) or antenna-on-chip (AoC) technologies.

Chapter 1 introduces some basic concepts associated with differential antennas including balanced and unbalanced antennas, even and odd modes, differential and single-ended circuits, mixed-mode S-parameters, and typical baluns. The chapter also highlights that there exists the important ratio of 2 : 1 for many differential and single-ended structures.

Differential wire antennas such as the dipole antenna and the loop antenna are the earliest and most basic antennas. Hence, Chapter 2 begins with the dipole antenna and the loop antenna to analyze differential wire antennas. The chapter ends with the Yagi-Uda antenna to highlight directional differential wire antennas. In addition, the chapter briefly considers the single-ended counterparts of differential dipole, loop, and Yagi-Uda antennas to show the relationship between differential and single-ended wire antennas.

Chapter 3 deals with the slot antenna. It will be treated as complementary to a dipole antenna based on Booker's seminal paper. An emphasis will be given to extend Booker's relation for differential complementary slot and dipole to single-ended counterparts. Then, the self-complementary antennas will be described in terms of the Mushiake's work. Finally, the Yin-Yang antennas will be discussed.

The first microstrip patch antenna was realized by printing a circular conductor patch on a dielectric substrate with a conductor ground plane and feeding it through two coaxial cables in 1970. It is interesting to mention that the first microstrip patch antenna was also a differential microstrip patch antenna, which is the topic of Chapter 4. As a practical concept for solving many antenna system problems, microstrip patch antennas not only gave birth to a new antenna industry but also enriched antenna theory. It was found for the first time that the ratio of the impedance of a differential microstrip patch antenna to its single-ended counterpart is 4 : 1 rather than 2 : 1. This finding has important implications for impedance matching.

A half-wave microstrip patch antenna operates in the fundamental mode. A virtual electric wall exists along the center line between the two radiating edges. Hence, a physical metal wall was proposed to replace the virtual electric wall to create a quarter-wave microstrip patch antenna or a shorted patch antenna. Although the shorted patch antenna preserves the advantages of a patch antenna with a reduced size, it has a major issue of lower H-plane cross-polarization level, which limits the shorted patch antenna from many applications. Chapter 5 focuses on the differential shorted patch antenna, and their variations, aiming to improve H-plane cross-polarization level.

Microstrip slot antennas naturally produce bidirectional radiation patterns, while unidirectional radiation patterns can also be obtained through adding reflectors. Microstrip slot antennas are less sensitive to fabrication tolerances and so can be manufactured at lower cost. In addition, microstrip slot antennas do not add weight and size to the system. Hence, they are suitable for applications where cost, size, and weight are of significance. Most of the reported microstrip slot antennas are single-ended designs. Chapter 6 presents differential microstrip slot antennas and compares them with their single-ended counterparts.

The first microstrip grid array antenna was published in 1981. Chapter 7 deals with the analysis and design of differential microstrip grid array antennas. It explains the principle of operation through careful examination of the current distribution and radiation patterns of the microstrip grid array antenna. It compares the novel with the usual differential excitation technique. It describes the design procedure and various design techniques.

A printed antenna is an antenna printed on one surface of a dielectric substrate where no ground plane is printed directly underneath the radiator on the other surface of the dielectric substrate. There are many types of printed antennas. Chapter 8 covers quasi-Yagi, fractal, and spiral antennas for end-fire radiation, multiband, and wideband operations, respectively.

Chapter 9 describes differential antenna measurements with an emphasis on the balun calibration techniques.

Acknowledgments

The author expresses the highest appreciation to a number of his associates and students who assisted him in completing this book, providing data, figures, tables, and materials related to differential antennas. Particular acknowledgment should go to Drs. Junjun Wang, Mei Sun, Zihong Tu, Bing Zhang, Zihao Chen, Guanghui Xu, Zijian Shao, Jun Shu, Yaowei Hou, Ziyang Zheng, Yulin Fang, Binshan Zhao, Junming Ding, Mrs. Wei Lin, Yang Xue, Zhihao Zhu, and Ms. Lin Zhang.

The author thanks Prof. Jianhuan Sheng of Taiyuan University of Technology, Shanxi, China; Dr. Yeongming Hwang and Prof. Pak Chung Ching of the Chinese University of Hong Kong, Hong Kong, China; Prof. Tung Sang Ng of the University of Hong Kong, Hong Kong, China; and Prof. Manh Anh Do of Nanyang Technological University, Singapore, for their guidance.

The author greatly appreciates the long-term collaboration with Profs. Junfa Mao, Linsheng Wu, Min Tang, Liang Zhou, and Changzhan Gu of Shanghai Jiao Tong University, Shanghai, China; Prof. Wenmei Zhang of Shanxi University, Taiyuan, Shanxi, China; Prof. Guoxin Zheng of Shanghai University, Shanghai, China; Prof. Qingxin Chu of South China University of Technology, Guangzhou, Guangdong, China; Prof. Yilong Lu of Nanyang Technological University, Singapore; and Dr. Duixian Liu of IBM Thomas J. Watson Research Center, New York, United States.

The author is grateful to Prof. Xiaowei Zhu and Ms. Yinyue Zhao of Southeast University, Nanjing, Jiangsu, China, for their contribution to the analysis and design of differential magnetoelectric dipole and array.

Finally, the author pays tribute to his wife Aihua Wei, daughter Yu Zhang, and son Yaohua Zhang for their love, support, patience, sacrifice, and understanding.

1

Introduction

1.1 Background

During 1864 and 1867, James Clerk Maxwell established his theory of electromagnetism, which predicted that electric and magnetic fields travel through space as waves moving at the speed of light.[1] Between 1886 and 1889, Heinrich Rudolf Hertz conducted a series of experiments that demonstrated the existence of electromagnetic waves and validated Maxwell's theory.[2] By the mid-1890s, the scientific and technical foundation had been laid for Guglielmo Giovanni Maria Marconi to develop wireless telegraphy systems.[3] At the turn of the 20th century, wireless telegraphy began to be used commercially and wireless telephony was also demonstrated, indicating that the wireless age came.[4] Since then, antennas that made wireless communication possible have opened up many other possibilities.

1.2 Balanced and Unbalanced Antennas

A dipole antenna commonly consists of two identical conductive elements such as metal wires or rods. A loop antenna is usually made of a coil of metal wire or another electrical conductor. A dipole antenna was used as a transmitting antenna and a loop antenna as a receiving antenna by Hertz for the discovery of electromagnetic waves. Hence, the dipole and loop antennas are the earliest antennas. Figure 1.1 shows photos of Hertz's sphere-loaded dipole and loop antennas [1].

1 https://en.wikipedia.org/wiki/James_Clerk_Maxwell
2 https://en.wikipedia.org/wiki/Heinrich_Hertz
3 https://en.wikipedia.org/wiki/Guglielmo_Marconi
4 https://en.wikipedia.org/wiki/Reginald_Fessenden

Differential Antennas: Theory and Practice, First Edition. Yueping Zhang.

(b)

(a)

Figure 1.1 Photos of (a) Hertz's dipole and (b) loop antennas. *Sources:* (a) Heinrich Hertz/Public Domain. (b) Rollo Appleyard/Wikimedia Commons/Public domain.

A monopole antenna normally consists of a straight metal wire or rod, often mounted perpendicularly over a conductive surface, called a ground plane. The monopole antenna was invented in 1895 and patented in 1896 by Marconi [2]. He found that the monopole antenna could cover longer distances than the dipole antenna in his radio transmission experiments. Figure 1.2 shows a photo of the monopole antenna setup by Marconi himself at Shanghai Jiaotong University (SJTU) on 8 December 1933.[5]

Antennas can be balanced or unbalanced. A dipole antenna is balanced for its structural symmetry about the feed point, while a monopole antenna is unbalanced for its structural asymmetry about the feed point. It is noted that the old terms of balanced and unbalanced antennas cause confusion. For example, if a dipole antenna is installed parallel to the Earth's surface, it is indeed balanced. However, if the dipole antenna is installed vertically to the Earth's surface, it is unbalanced because one arm of the dipole antenna is closer to the Earth's surface than the other. In addition, if a dipole antenna is fed off-center, the dipole antenna is obviously unbalanced, but the feeding source is balanced. To avoid confusion,

5 https://museum.sjtu.edu.cn/

Figure 1.2 Photo of Marconi and his monopole antenna taken at SJTU on 8 December 1933. *Source:* Shanghai Jiao Tong University/Wikimedia Commons/Public domain.

the new terms of differential and single-ended antennas are adopted in this book. The new terminology is based on feeding sources rather than antenna structures. A differential antenna is an antenna fed with a differential or an equivalently differential signal source. A dipole antenna is a differential antenna. A single-ended antenna is an antenna fed with a single-ended signal source. A monopole antenna is a single-ended antenna.

1.3 Even and Odd Modes

Three principal types of transmission lines are microstrip, stripline, and coplanar waveguide (CPW). They are widely used in modern wireless systems. A transmission line pair can be formed with any of them. Figure 1.3 shows the transmission line pair in a microstrip structure. We use it as an example to describe the even and odd modes of propagation.

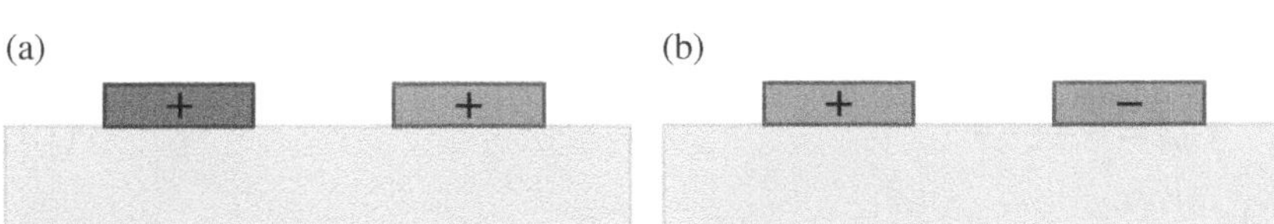

Figure 1.3 Illustration of the transmission line pair in a microstrip structure. (a) Even and (b) odd modes.

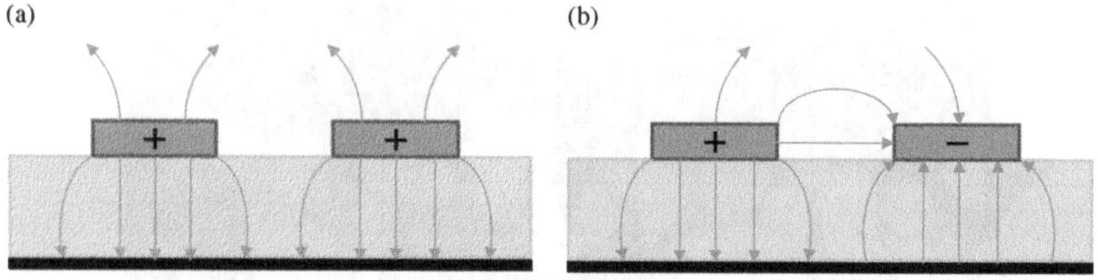

Figure 1.4 Illustration of the electric field distributions (a) for the even and (b) for the odd modes.

The even mode is the mode corresponding to both lines having the same potential V to the ground plane and carrying the identical current in the same direction. The odd mode is the mode corresponding to both lines having opposite potentials V and $-V$, relative to the ground plane and carrying the identical current in the opposite directions. Figure 1.4 shows a sketch of electric field lines for the two modes. Note a magnetic wall exists in the even mode whereas an electric wall exists in the odd mode. The wall separates the whole structure into two identical half structures.

The even mode has an associated characteristic impedance Z_{0e}, which can be calculated from the half structure of the even mode. Similarly, the odd mode corresponds to a characteristic impedance Z_{0o}, which can be calculated from the half structure of the odd mode. It should be able to figure out from the field distributions in Figure 1.4 that Z_{0e} is larger than Z_{0o}.

1.4 Differential and Single-Ended Circuits

A differential circuit deals with the difference between two input signals, while a single-ended circuit accepts a single input signal. Figure 1.5 shows the schematic diagrams of differential and single-ended bipolar junction transistor (BJT) amplifiers. We will use them as examples to illustrate the responses of differential and single-ended circuits to differential and common-mode signals.

Note that both V_1 and V_2 are two input signals. The differential-mode input signal is defined as

$$V_{dm} = \frac{V_1 - V_2}{2} \tag{1.1}$$

and the common-mode input signal as

$$V_{cm} = \frac{V_1 + V_2}{2}. \tag{1.2}$$

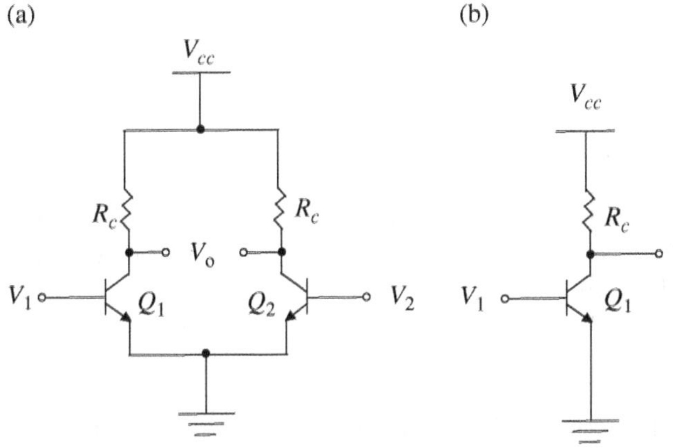

Figure 1.5 Schematic diagrams of (a) differential and (b) single-ended BJT amplifiers.

Thus, the two input signals V_1 and V_2 can be expressed as

$$V_1 = V_{cm} + V_{dm} \tag{1.3}$$

and

$$V_2 = V_{cm} - V_{dm}. \tag{1.4}$$

Assuming that the common-mode signal is an interference, and the differential-mode signal is desirable, the single-ended amplifier will amplify both the interference and desirable signal, while the differential amplifier has the advantage of amplifying only the desirable signal and rejecting the interference.

1.5 An Important Ratio

There exists an important ratio between many differential and single-ended structures. For a differential antenna such as a half-wavelength dipole, it has an input impedance

$$Z_d = 73.1 + j42.5\,\Omega, \tag{1.5}$$

while for a single-ended antenna such as a quarter-wavelength monopole, it has an input impedance

$$Z_s = 36.5 + j21.3\,\Omega. \tag{1.6}$$

Hence, the impedance ratio of the differential dipole to the single-ended monopole is 2 : 1, and so is the size ratio.

Let us turn to the differential and single-ended amplifiers as shown in Figure 1.5. The differential amplifier consists of two transistors and two resistors. The area is assumed to be A_d. The input resistance R_d is given by

$$R_d = \frac{2V_T}{I_B}\ \Omega \tag{1.7}$$

where V_T is the thermal voltage and I_B is the base bias current. The single-ended amplifier consists of one transistor and one resistor. The area is assumed to be A_s. The input resistance R_s is given by

$$R_s = \frac{V_T}{I_B}\ \Omega \tag{1.8}$$

Hence, the ratio of 2 : 1 is also true for the circuits. In Chapter 4, we will show that this important ratio is not always correct.

1.6 Mixed-Mode *S*-Parameters

S-parameters refer to the scattering parameters of a microwave network. Taking a three-port network as an example, the *S*-parameters in a matrix form are expressed as

$$\begin{bmatrix} b_1 \\ b_2 \\ b_3 \end{bmatrix} = \begin{bmatrix} S_{11} & S_{12} & S_{13} \\ S_{21} & S_{22} & S_{23} \\ S_{31} & S_{32} & S_{33} \end{bmatrix} \begin{bmatrix} a_1 \\ a_2 \\ a_3 \end{bmatrix} \tag{1.9}$$

where the variable a_i represents a power wave incident to port i and the variable b_j a power wave reflected from port j. If each port is terminated in the reference impedance Z_0, the *S*-parameters of the three-port vector network are defined as

$$S_{ij} = \frac{b_i}{a_j} \tag{1.10}$$

where i and j are from 1 to 3 and a_{j-1} and a_{j-2} should be set to zero. *S*-parameters are measured with a two-port vector network analyzer.

Mixed-mode *S*-parameters are used for the characterization of differential structures [3]. Let us reconfigure the three-port network as one single-ended port (port 1) and one differential port (ports 2 and 3). Assuming that a single-ended signal exists at the single-ended port, a differential-mode signal, and a common-mode signal exist at the differential port, we use the nomenclature *s*, *d*, and *c* to represent the single-ended, differential, and common modes, respectively, and we obtain the mixed-mode S-parameters in a matrix form as

$$\begin{bmatrix} b_{s1} \\ b_{d2} \\ b_{c2} \end{bmatrix} = \begin{bmatrix} S_{ss11} & S_{sd12} & S_{sc12} \\ S_{ds21} & S_{dd22} & S_{dc22} \\ S_{cs21} & S_{cd22} & S_{cc22} \end{bmatrix} \begin{bmatrix} a_{s1} \\ a_{d2} \\ a_{c2} \end{bmatrix}. \quad (1.11)$$

Mixed-mode S-parameters are measured with a four-port vector network analyzer. They can also be calculated using the single-ended S-parameters measured by a two-port network analyzer [3], for example,

$$S_{ss11} = S_{11}, \quad (1.12)$$

$$S_{sd12} = S_{ds21} = \frac{1}{\sqrt{2}}(S_{21} - S_{31}), \quad (1.13)$$

$$S_{sc12} = S_{cs21} = \frac{1}{\sqrt{2}}(S_{21} + S_{31}), \quad (1.14)$$

$$S_{dd22} = \frac{1}{2}(S_{22} - S_{23} - S_{32} + S_{33}), \quad (1.15)$$

$$S_{dc22} = \frac{1}{2}(S_{22} + S_{23} - S_{32} - S_{33}), \quad (1.16)$$

$$S_{cd22} = \frac{1}{2}(S_{22} - S_{23} + S_{32} - S_{33}), \quad (1.17)$$

$$S_{cc22} = \frac{1}{2}(S_{22} + S_{23} + S_{32} + S_{33}). \quad (1.18)$$

For a differential antenna, there are two single-ended ports, which can only form one differential or common-mode port. So, the mixed-mode S-parameters for the differential antenna are

$$S_{dd11} = (S_{11} - S_{12}) = S_{d11} \quad (1.19)$$

and

$$S_{cc11} = (S_{11} + S_{12}) = S_{c11}. \quad (1.20)$$

For an array with two differential antenna elements, there are four single-ended ports [4], which can form two differential ports or two common-mode ports. So, the mixed-mode S-parameters for the array are

$$S_{mm} = \begin{bmatrix} S_{dd} & S_{dc} \\ S_{cd} & S_{cc} \end{bmatrix} = \begin{bmatrix} S_{dd11} & S_{dd12} & S_{dc11} & S_{dc12} \\ S_{dd21} & S_{dd22} & S_{dc21} & S_{dc22} \\ S_{cd11} & S_{cd12} & S_{cc11} & S_{cc12} \\ S_{cd21} & S_{cd22} & S_{cc21} & S_{cc22} \end{bmatrix} \quad (1.21)$$

where

$$S_{dd} = \begin{bmatrix} S_{dd11} & S_{dd12} \\ S_{dd21} & S_{dd22} \end{bmatrix} = \frac{1}{2}\begin{bmatrix} S_{11} - S_{13} - S_{31} + S_{33} & S_{12} - S_{14} - S_{32} + S_{34} \\ S_{21} - S_{23} - S_{41} + S_{43} & S_{22} - S_{24} - S_{42} + S_{44} \end{bmatrix}, \tag{1.22}$$

$$S_{dc} = \begin{bmatrix} S_{dc11} & S_{dc12} \\ S_{dc21} & S_{dc22} \end{bmatrix} = \frac{1}{2}\begin{bmatrix} S_{11} + S_{13} - S_{31} - S_{33} & S_{12} + S_{14} - S_{32} - S_{34} \\ S_{21} + S_{23} - S_{41} - S_{43} & S_{22} + S_{24} - S_{42} - S_{44} \end{bmatrix}, \tag{1.23}$$

$$S_{cd} = \begin{bmatrix} S_{cd11} & S_{cd12} \\ S_{cd21} & S_{cd22} \end{bmatrix} = \frac{1}{2}\begin{bmatrix} S_{11} - S_{13} + S_{31} - S_{33} & S_{12} - S_{14} + S_{32} - S_{34} \\ S_{21} - S_{23} + S_{41} - S_{43} & S_{22} - S_{24} + S_{42} - S_{44} \end{bmatrix}, \tag{1.24}$$

$$S_{cc} = \begin{bmatrix} S_{cc11} & S_{cc12} \\ S_{cc21} & S_{cc22} \end{bmatrix} = \frac{1}{2}\begin{bmatrix} S_{11} + S_{13} + S_{31} + S_{33} & S_{12} + S_{14} + S_{32} + S_{34} \\ S_{21} + S_{23} + S_{41} + S_{43} & S_{22} + S_{24} + S_{42} + S_{44} \end{bmatrix}. \tag{1.25}$$

1.7 Balun

A balun is a three-port device, which bridges between the single-ended and the differential structures. It is succinctly defined by the required (ideal) S-parameters as

$$S_{11} = 0, \tag{1.26}$$

$$S_{12} = -S_{13} = S_{21} = -S_{31}. \tag{1.27}$$

Note that the single-ended input port is matched to the line characteristic impedance (usually 50 Ω) and the two single-ended output ports will provide the signals of equal amplitude and opposite phase. Also note that the two single-ended output ports are not necessarily matched, may or may not be isolated, and there may be a different return loss for differential and common mode signals.

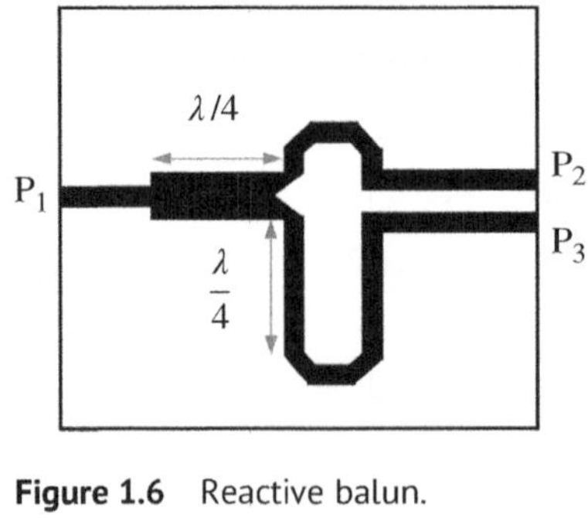

Figure 1.6 Reactive balun.

A balun can take many forms. Three typical baluns that will be used in the design examples of this book are discussed as follows. Figure 1.6 shows a reactive balun. It consists of a reactive 3-dB power divider, a line with the length of l, and another line with the length of $l + \lambda/2$. A third line that has the length of $\lambda/4$ and the characteristic impedance of $\sqrt{2}Z_0$ acts as an impedance transformer to provide impedance matching between the input and output lines of the same characteristic impedance of Z_0. There are three single-ended ports by three lines. Port 1 (P_1) is the input port. Ports 2 (P_2) and 3 (P_3) are the output ports, which can be configured as a differential port or a common-mode port. For simplicity, the line with the length of $l + \lambda/2$ is assumed to have a constant π phase shift over the frequency band, and the phase shift passing through the $\lambda/4$ matching section is ignored. Thus, the standard S matrix for an ideal reactive balun can be found as

$$S = \begin{bmatrix} 0 & \frac{e^{-j\frac{2\pi l}{\lambda}}}{\sqrt{2}} & -\frac{e^{-j\frac{2\pi l}{\lambda}}}{\sqrt{2}} \\ \frac{e^{-j\frac{2\pi l}{\lambda}}}{\sqrt{2}} & -\frac{e^{-j\frac{4\pi l}{\lambda}}}{2} & -\frac{e^{-j\frac{4\pi l}{\lambda}}}{2} \\ -\frac{e^{-j\frac{2\pi l}{\lambda}}}{\sqrt{2}} & -\frac{e^{-j\frac{4\pi l}{\lambda}}}{2} & -\frac{e^{-j\frac{4\pi l}{\lambda}}}{2} \end{bmatrix}, \tag{1.28}$$

which reveals that the input port is matched but the output ports are not matched and isolated. Now let us find the mixed-mode S matrix for the ideal reactive balun. This can be easily done by substituting the standard S-parameters in (1.28) into (1.12)–(1.18). The resultant mixed-mode S matrix is

$$S_{mm} = \begin{bmatrix} 0 & e^{-j\frac{2\pi l}{\lambda}} & 0 \\ e^{-j\frac{2\pi l}{\lambda}} & 0 & 0 \\ 0 & 0 & e^{-j\frac{4\pi l}{\lambda}} \end{bmatrix}. \tag{1.29}$$

It is clear from (1.29) that the single-ended input port and the differential output port are matched but the common-mode output port is not matched, the single-ended input port and the differential output port are not isolated, the single-ended input port and the common-mode output port are isolated, and the differential output port and the common-mode output port are isolated.

Figure 1.7 shows a resistive balun. It consists of a resistive 3-dB power divider, a line with the length of l, and another line with the length of $l + \lambda/2$. With the same assumption and ignorance as in the reactive balun, the standard scattering matrix for an ideal resistive balun can be found as

$$S = \frac{-j}{\sqrt{2}} \begin{bmatrix} 0 & e^{-j\frac{2\pi l}{\lambda}} & -e^{-j\frac{2\pi l}{\lambda}} \\ e^{-j\frac{2\pi l}{\lambda}} & 0 & 0 \\ -e^{-j\frac{2\pi l}{\lambda}} & 0 & 0 \end{bmatrix}, \tag{1.30}$$

which shows ports 1, 2, and 3 are all matched and ports 2 and 3 are isolated. When a signal inputs to port 1, half of the

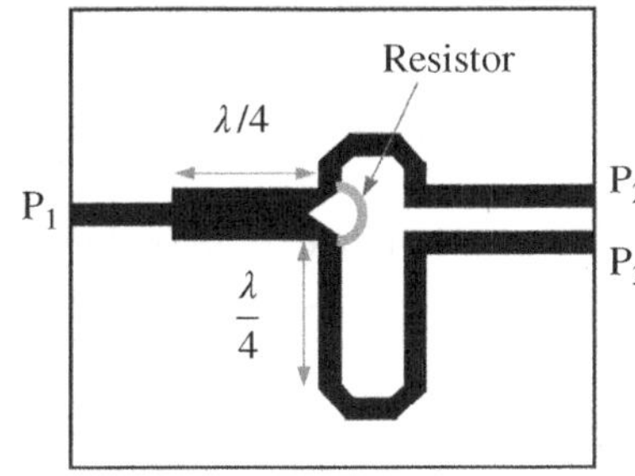

Figure 1.7 Resistive balun.

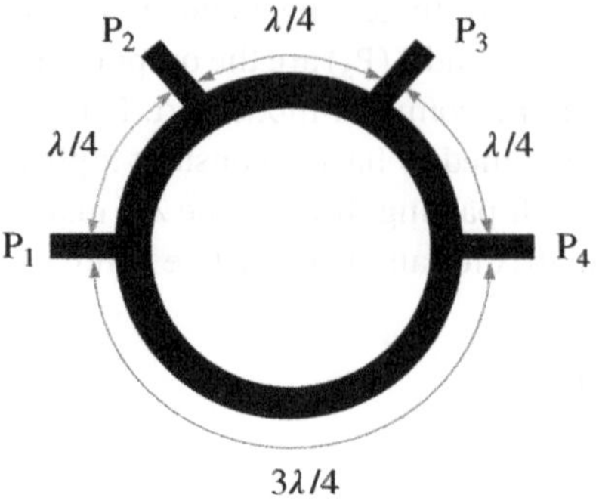

Figure 1.8 Rat-race coupler.

signal gets transmitted to ports 2 and 3 out of phase, respectively. The mixed-mode S matrix for the ideal resistive balun is

$$S_{mm} = -j\begin{bmatrix} 0 & e^{-j\frac{2\pi l}{\lambda}} & 0 \\ e^{-j\frac{2\pi l}{\lambda}} & 0 & 0 \\ 0 & 0 & 0 \end{bmatrix}. \tag{1.31}$$

It is obvious from (1.31) that the single-ended input port, the differential output port, and the common-mode output port are all matched. The single-ended input port and the common-mode output port are isolated, the differential output port and the common-mode output port are isolated, but the single-ended input port and the differential output port are not isolated.

Figure 1.8 shows a rat-race coupler. It has four ports, each placed $\lambda/4$ away from each other around the top half of the ring. The bottom half of the ring is $3\lambda/4$ in length. The ring has a characteristic impedance of factor $\sqrt{2}$ compared to port impedance. A signal input on port 1 will be split equally between ports 2 and 4, and port 3 will be isolated. As the signals on ports 2 and 4 are equal in amplitude and 180° out of phase, the rat-race coupler can be configured as a balun. The standard scattering matrix for an ideal 3 dB rat-race coupler is

$$S = \frac{-j}{\sqrt{2}}\begin{bmatrix} 0 & 1 & 0 & -1 \\ 1 & 0 & 1 & 0 \\ 0 & 1 & 0 & 1 \\ -1 & 0 & 1 & 0 \end{bmatrix}. \tag{1.32}$$

Deleting the S-parameters associated with port 3 in (1.32), one can get the standard scattering matrix for ports 1, 2, and 4 as

$$S = \frac{-j}{\sqrt{2}}\begin{bmatrix} 0 & 1 & -1 \\ 1 & 0 & 0 \\ -1 & 0 & 0 \end{bmatrix}. \tag{1.33}$$

The mixed-mode S matrix for (1.32) is

$$S_{mm} = -j\begin{bmatrix} 0 & 1 & 0 \\ 1 & 0 & 0 \\ 0 & 0 & 0 \end{bmatrix}, \tag{1.34}$$

which indicates that the single-ended input port, the differential output port, and the common-mode output port are all matched, the single-ended input port and the differential output port are not isolated, but the differential output port and the common-mode output port are isolated.

The key figures of merit of a balun are balance and common-mode rejection ratio (CMRR) over the operating bandwidth. The balance tells how close the balanced outputs are to having equal power and opposite phases. The balance is typically characterized in both magnitude and phase. It can be calculated using the standard S-parameters as

$$\Delta = -\frac{S_{21}}{S_{31}}. \tag{1.35}$$

Amplitude balance is usually specified in dB and actually gives the match between output power magnitude. Low-performance baluns have amplitude balance of ± 1.5 dB max and ± 1 dB typical, while high-performance baluns approach ± 0.5 dB max and ± 0.2 dB typical. Phase balance is the measure of how closely the inverted output is to 180° out of phase with the non-inverted output, usually given in degrees. Typical phase balance for standard microwave baluns is $\pm 15°$ max and $\pm 10°$ typical, while high-performance baluns approach $\pm 5°$ max and $\pm 2°$ typical.[6] The CMRR informs how well the rejection of undesired common-mode signal with minimal impact on the desired differential-mode signal. It is defined as the ratio of powers of the differential signal to the common-mode signal and can be calculated using the mixed-mode S-parameters as

$$CMRR = 20 \log \frac{S_{ds21}}{S_{cs21}}. \tag{1.36}$$

A low-performance balun will have 15–20 dB of CMRR, while high-performance baluns can achieve 25–55 dB of CMRR (see Footnote 5).

1.8 Concluding Remarks

In this chapter, we briefly introduced balanced and unbalanced antennas, even and odd modes, differential and single-ended circuits, mixed-mode S-parameters, and typical baluns. We should point out that terms such as balanced, odd, and differential modes have similar technical meanings as do even and common modes as well as single-ended and unbalanced ports. We also highlighted that there exists an important ratio of 2 : 1 for many differential and single-ended structures. Furthermore, we would like to make the following remarks: Compared with single-ended

6 https://www.markimicrowave.com/home/

circuits, differential circuits permit higher linearity and lower offset and make them immune to power supply variations, temperature changes, and substrate noise. As a result, differential circuits have dominated in integrated circuit designs. Compared with differential antennas, single-ended antennas usually have smaller size and therefore single-ended antennas have dominated in antenna designs. Differential circuits call for differential antennas. This is particularly essential in highly integrated RF system-on-chip (SoC) and system-in-package (SiP) solutions where the system ground plane may be much smaller than one free-space wavelength. Differential antennas perfectly marry (match) with differential circuits. No lossy balanced/unbalanced conversion circuit (balun) is needed. As a result, the receiver noise performance and transmitter power efficiency are improved.

References

1 R. Appleyard, "Pioneers of electrical communication part 5 – Heinrich Rudolph Hertz," *Electrical Communication*. New York: International Standard Electric Corp., vol. 6, no. 2, pp. 63–77, Oct. 1927.

2 G. Marconi, "Improvements in transmitting electrical impulses and signals, and in apparatus therefor," British patent No. 12,039 (1897).

3 D. E. Bockelman, W. R. Eisenstadt, "Combined differential and common-mode scattering parameters: Theory and simulation," *IEEE Trans. Microw. Theory Technol.*, vol. 43, no.7, pp. 1530–1539, Jul. 1995.

4 W. Fan, A . Lu, L. L. Wai, B. K. Lok, "Mixed-mode S-parameter characterization of differential structures," in *Proc. Electronics Packaging Technol. Conf.*, Singapore, 10–12 Dec. 2003.

2

Differential Wire Antennas

2.1 Introduction

Differential wire antennas such as the dipole antenna and the loop antenna are the earliest and most basic antennas. Hence, we begin with the dipole antenna and the loop antenna to analyze differential wire antennas [1–5]. For simplicity's sake, we assume that the dipole antenna and the loop antenna are made of electrically thin wires. Thus, we can express the current distribution on the dipole antenna by a sinusoidal function and on the loop antenna by a Fourier series. For the dipole antenna, we repeat the important results from classical textbooks [1, 2], while for the loop antenna, we validate Storer's theory for impedance [3] and Rao's derivation for radiation [4]. With the knowledge of the dipole antenna and the loop antenna, we can then proceed to a study of the loop-dipole composite antenna. We end with the Yagi-Uda antenna to highlight directional differential wire antennas. In addition, we briefly consider the single-ended counterparts of differential dipole, loop, and Yagi-Uda antennas to show the relationship between differential and single-ended wire antennas.

2.2 Dipole and Monopole Antennas

The dipole antenna, as shown in Figure 2.1, is placed along the z-axis and with its center at the origin in free space. It has a length of l and a radius of a. It is energized by a differential signal source V_d through a twin-wire transmission line. The monopole antenna, as shown in Figure 2.1, is mounted above a ground plane on the $z = 0$ surface. It has a length of $l/2$ and a radius of a. It is excited by a single-ended signal source V_s through a single-wire transmission line.

Differential Antennas: Theory and Practice, First Edition. Yueping Zhang.

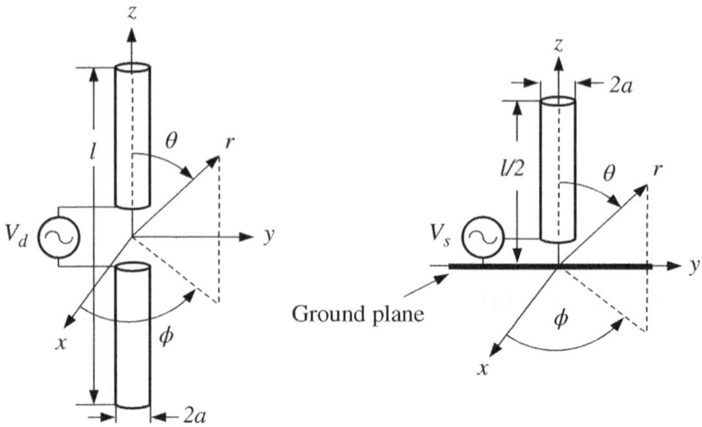

Figure 2.1 Dipole and monopole antennas and coordinates.

For the dipole antenna, we assume that the current distribution is sinusoidal as

$$I(z) = I_m \sin\left[k_0\left(\frac{l}{2} - |z|\right)\right] \tag{2.1}$$

where I_m is the maximum current and k_0 is the wave number in free space [1]. It has been verified experimentally that the assumption is valid for electrically thin dipole antennas with $a < 0.01\lambda$ [2].

For the current distribution of (2.1), the nonzero electric and magnetic field components at any distance from the dipole antenna in cylindrical coordinates can be expressed as

$$E_\rho = j\frac{\eta_0 I_m}{4\pi\rho}\left[\left(\frac{z-\frac{l}{2}}{R_1}\right)e^{-jk_0R_1} + \left(\frac{z+\frac{l}{2}}{R_2}\right)e^{-jk_0R_2} - 2z\cos\left(\frac{k_0 l}{2}\right)\frac{e^{-jk_0 r}}{r}\right], \tag{2.2}$$

$$E_z = -j\frac{\eta_0 I_m}{4\pi}\left[\frac{e^{-jk_0R_1}}{R_1} + \frac{e^{-jk_0R_2}}{R_2} - 2\cos\left(\frac{k_0 l}{2}\right)\frac{e^{-jk_0 r}}{r}\right], \tag{2.3}$$

$$H_\phi = j\frac{I_m}{4\pi\rho}\left[e^{-jk_0R_1} + e^{-jk_0R_2} - 2\cos\left(\frac{k_0 l}{2}\right)e^{-jk_0 r}\right] \tag{2.4}$$

where η_0 is the free-space characteristic impedance of 377 Ω and

$$r = \sqrt{\rho^2 + z^2}, \tag{2.5}$$

$$R_1 = \sqrt{\rho^2 + \left(z - \frac{l}{2}\right)^2}, \tag{2.6}$$

$$R_2 = \sqrt{\rho^2 + \left(z + \frac{l}{2}\right)^2}. \tag{2.7}$$

It is clear from (2.2) to (2.4) that the field at a point in free space is the sum of three terms originating from the ends and the center of the dipole antenna.

The impedance of the dipole antenna referred to at the current maximum can be calculated as

$$Z = R + jX = -\frac{1}{I_m^2}\int_{-l/2}^{l/2} I_z(\rho = a, z = z')E_z(\rho = a, z = z')dz'. \tag{2.8}$$

Substituting (2.1) and (2.3) into (2.8), it can be shown, after tedious but straightforward derivations, that the resistance and reactance of the dipole antennas can be expressed as

$$R = \frac{\eta_0}{2\pi}\left\{\begin{array}{c} 0.577 + \ln(k_0 l) - C_i(k_0 l) + \frac{1}{2}\sin(k_0 l)[S_i(2k_0 l) - 2S_i(k_0 l)] \\ + \frac{1}{2}\cos(k_0 l)[0.577 + \ln(k_0 l/2) + C_i(2k_0 l) - 2C_i(k_0 l)] \end{array}\right\} \tag{2.9}$$

and

$$X = \frac{\eta_0}{4\pi}\left\{\begin{array}{c} 2S_i(k_0 l) + \cos(k_0 l)[2S_i(k_0 l) - S_i(2k_0 l)] \\ -\sin(k_0 l)\left[2C_i(k_0 l) - C_i(2k_0 l) - C_i\left(\frac{2k_0 a^2}{l}\right)\right] \end{array}\right\} \tag{2.10}$$

where $C_i(x)$ and $S_i(x)$ are the cosine and sine integrals [1]. The input impedance of the dipole antenna referred to as the current at the input terminals can be obtained as

$$Z_d = \frac{Z}{\sin^2\left(\frac{k_0 l}{2}\right)}. \tag{2.11}$$

For the monopole antenna, we assume that the ground plane is an infinite perfect electric conductor, and, therefore, the image theory can be used here. As a result, the field expressions of (2.2)–(2.4) for the dipole antenna can be directly applied to calculate the field distribution of the monopole antenna above the ground plane. However, the impedance expression of (2.8) for the dipole antenna cannot be directly employed to calculate the impedance of the monopole antenna. It must be modified to account for the radiation of the monopole antenna only to the upper hemisphere, which yields that the input impedance of the monopole antenna is equal to one-half that of the dipole antenna given by

$$Z_m = \frac{Z_d}{2}. \tag{2.12}$$

A half-wavelength ($l = \lambda/2$) dipole antenna is the most common dipole antenna. The electric and magnetic field components of the half-wavelength dipole antenna can be obtained by replacing l in (2.2)–(2.4) with $\lambda/2$. Doing so, they reduce to

$$E_\rho = j\frac{\eta_0 I_m}{4\pi r \sin\theta}\left[\left(\frac{z-\frac{\lambda}{4}}{R_1}\right)e^{-jk_0R_1} + \left(\frac{z+\frac{\lambda}{4}}{R_2}\right)e^{-jk_0R_2}\right], \tag{2.13}$$

$$E_z = -j\frac{\eta_0 I_m}{4\pi}\left[\frac{e^{-jk_0R_1}}{R_1} + \frac{e^{-jk_0R_2}}{R_2}\right], \tag{2.14}$$

$$H_\phi = j\frac{I_m}{4\pi r \sin\theta}\left[e^{-jk_0R_1} + e^{-jk_0R_2}\right]. \tag{2.15}$$

For the far-field region where $k_0 r \gg 1$, (2.13)–(2.15) can be approximated in spherical coordinates as

$$E_\theta = E_\rho \cos\theta - E_z \sin\theta \approx j\eta_0 \frac{I_m e^{-jk_0 r}}{2\pi r}\left[\frac{\cos\left(\frac{\pi}{2}\cos\theta\right)}{\sin\theta}\right] \tag{2.16}$$

and

$$H_\phi \approx j\frac{I_m e^{-jk_0 r}}{2\pi r}\left[\frac{\cos\left(\frac{\pi}{2}\cos\theta\right)}{\sin\theta}\right]. \tag{2.17}$$

One can find a field pattern factor in (2.16) and (2.17), which is

$$F(\theta) = \frac{\cos\left(\frac{\pi}{2}\cos\theta\right)}{\sin\theta}. \tag{2.18}$$

It is seen from (2.18) that the field pattern factor is independent of the azimuth angle of ϕ, implying equal radiation in all directions on the azimuth plane. Figure 2.2 plots the normalized H- and E-plane patterns. It is noted from the figure that the radiation is omnidirectional in the H plane but directional in the E plane. The directivity is given by

$$D = 1.64\left[\frac{\cos\left(\frac{\pi}{2}\cos\theta\right)}{\sin\theta}\right]^2. \tag{2.19}$$

The maximum directivity is 1.64 or 2.15 dBi at $\theta = 90°$ and the 3-dB beam width is 78°.

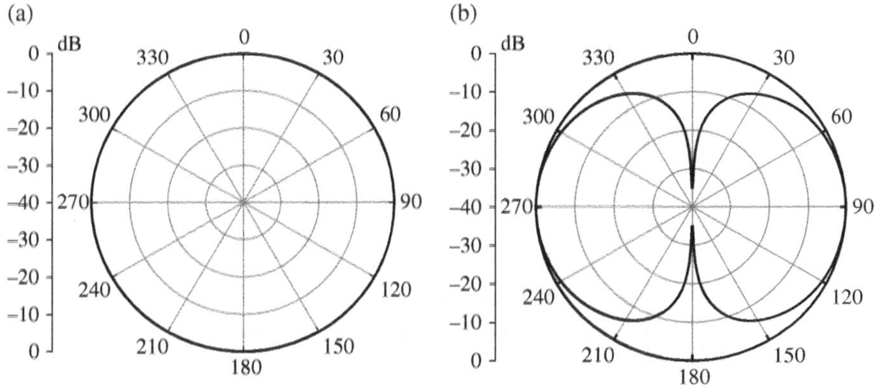

Figure 2.2 Normalized radiation patterns of the half-wavelength dipole antenna. (a) The *H* plane and (b) the *E* plane.

Similarly, the input impedance of the half-wavelength dipole antenna can be found by letting $l = \lambda/2$ in (2.8)–(2.11) as

$$Z_d = R + jX = 73.1 + j42.5\,\Omega. \tag{2.20}$$

It is an inductive impedance. Thus, one needs to resonate the inductive reactance of 42.5 Ω with the equal capacitive reactance. In practice, the stray capacitance between the driven ends of the half-wavelength dipole antenna is explored together with slightly reducing the length to achieve resonance.

A quarter-wavelength ($l = \lambda/4$) monopole antenna is the most popular monopole antenna. The input impedance of the quarter-wavelength monopole antenna can be found from (2.12) and (2.20) as

$$Z_m = \frac{Z_d}{2} = 36.5\,\Omega + j21.3\,\Omega. \tag{2.21}$$

The monopole antenna only radiates to the upper hemisphere. So, its directivity should be twice the corresponding dipole antenna. Thus, the directivity of the quarter-wavelength monopole antenna is twice that of the half-wavelength dipole antenna. The maximum directivity of the quarter-wavelength monopole antenna is 5.15 dBi at $\theta = 90°$ and the 3-dB beam width is 39°. The higher directivity also partially explains the wide use of the quarter-wavelength monopole antenna.

2.3 Folded Dipole and Monopole Antennas

The folded dipole antenna, as shown in Figure 2.3, is formed as a narrow rectangular loop with the long side of l and the short side of s. It is energized by a differential signal source V_d through a twin-wire transmission line. The folded

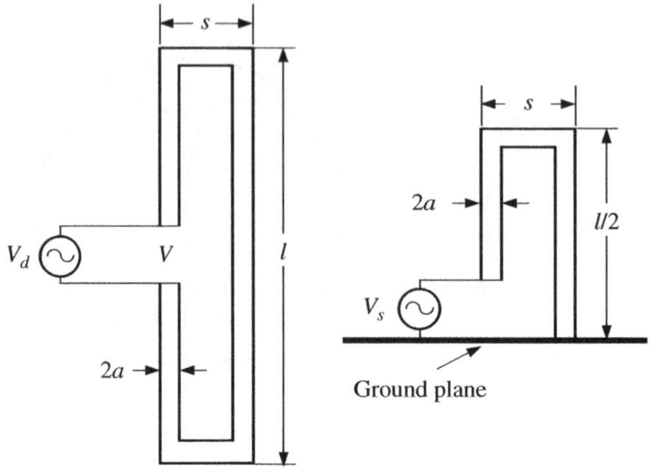

Figure 2.3 Folded dipole and monopole antennas.

monopole antenna, as shown in Figure 2.3, is half the size of the folded dipole antenna. It is mounted above a ground plane and excited by a single-ended signal source V_s through a single-wire transmission line. For the folded dipole and monopole antennas, s is typically chosen to be less than 0.05λ. As a result, the folded dipole and monopole antennas have essentially the same radiation patterns but increased input impedances as the dipole and monopole antennas.

The input impedance of the folded dipole antenna can be derived by decomposing the current into a transmission-line mode and an antenna mode [6]. For the transmission-line mode of Figure 2.4, the input impedance of each shorted transmission line is

$$Z_t = jZ_0 \tan\left(\frac{k_0 l}{2}\right) \tag{2.22}$$

where Z_0 is the characteristic impedance of a twin-wire transmission line given by

$$Z_0 = \frac{\eta_0}{\pi} \operatorname{arccosh}\left(\frac{s}{2a}\right). \tag{2.23}$$

Thus, the current in the transmission-line mode is

$$I_t = \frac{V}{2Z_t}. \tag{2.24}$$

For the antenna mode of Figure 2.4, it operates the same as the dipole antenna with an equivalent radius of

$$a_e = \sqrt{as}. \tag{2.25}$$

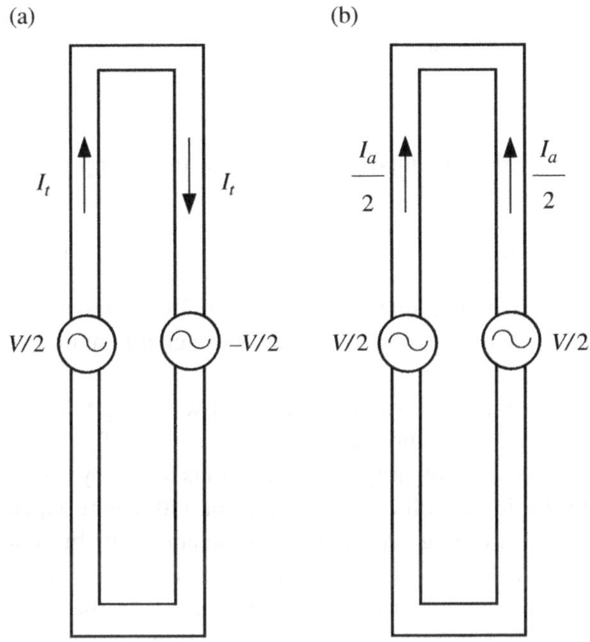

Figure 2.4 Transmission-line mode (a) and antenna mode (b) for folded dipole antenna.

Hence, the current in the antenna mode is

$$I_a = \frac{V}{2Z_d}. \tag{2.26}$$

The total current on the feed leg of the folded dipole can be written as

$$I = I_t + \frac{I_a}{2} = \frac{V}{2Z_t} + \frac{V}{4Z_d}. \tag{2.27}$$

The input impedance at the feed is given by

$$Z_{fd} = \frac{V}{I} = \frac{4Z_t Z_d}{2Z_d + Z_t} \tag{2.28}$$

Since $l = \lambda/2$, Z_t goes to infinitely large, which leads to

$$Z_{fd} = 4Z_d = 292.4 + j170\,\Omega \tag{2.29}$$

or that the input impedance of the half-wavelength folded dipole antenna is four times greater than that of the half-wavelength dipole antenna.

Similarly, referring to the modelling in Figure 2.4a and b, the input impedance of the folded monopole antenna can be easily obtained as

$$Z_{fm} = \frac{4Z_t Z_m}{2Z_m + Z_t}. \tag{2.30}$$

For the quarter-wavelength folded monopole antenna, the input impedance reduces to

$$Z_{fm} = 4Z_m = 146.2 + j85\,\Omega. \tag{2.31}$$

Or that the input impedance of the quarter-wavelength folded monopole antenna is four times greater than that of the quarter-wavelength monopole antenna.

It should be mentioned that the input resistance of the half-wavelength folded dipole antenna is quite close to the characteristic impedance of 300 Ω of a typical twin-wire transmission line. Thus, the matching between them is relatively easier. One only needs to resonate the input inductive reactance of 170 Ω with equal capacitive reactance. In practice, a tuning capacitance can be added in shunt at the feed to achieve resonance. The input resistance of the quarter-wavelength folded monopole antenna is approximately two or three times greater than that of the characteristic impedance of 75 or 50 Ω of a typical coaxial transmission line. A quarter-wavelength impedance transformer is usually inserted between the quarter-wave folded monopole antenna and the coaxial transmission line to achieve impedance matching.

2.4 Loop and Half-Loop Antennas

The loop antenna can be configured in different forms. The circular loop antenna is the most popular for its simplicity in analysis and construction. Figure 2.5 shows the circular loop antenna located with its center at the origin of the coordinates in free space. It has a radius of b and is energized by a differential signal source V_d through a twin-wire transmission line. Figure 2.5 shows the circular half-loop antenna mounted above a ground plane on the $z = 0$ surface. It has a radius of b and is excited by a single-ended signal source V_s through a single-wire transmission line.

For the circular loop antenna, the analysis follows that given by Storer [3]. The current distribution on the loop is predominately in the azimuthal direction and is conveniently expressed as a Fourier series

$$I(\phi) = I_0 + 2\sum_{n=1}^{\infty} I_n \cos(n\phi) \tag{2.32}$$

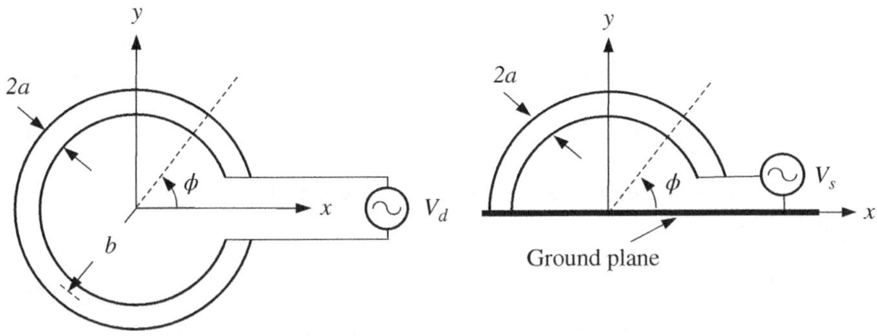

Figure 2.5 Loop and half-loop antennas and coordinates.

where

$$I_n = \frac{V}{j\pi\eta_0 a_n} \tag{2.33}$$

with

$$a_n = \frac{k_0 b}{2}(K_{n+1} + K_{n-1}) - \frac{n^2}{k_0 b}K_n \tag{2.34}$$

and

$$K_n = \frac{1}{2\pi}\int_{-\pi}^{\pi} \frac{e^{-jk_0 b r(\phi)}}{r(\phi)} e^{-jn\phi} d\phi. \tag{2.35}$$

Here V is the voltage exciting the loop at the feed and r is the distance from the feed point to the observation point given by

$$r(\phi) = \left[4\sin^2\left(\frac{\phi}{2}\right) + \frac{a^2}{b^2}\right]^{1/2}. \tag{2.36}$$

Substituting (2.33) into (2.32) yields

$$I(\phi) = \frac{V}{j\pi\eta_0}\left(\frac{1}{a_0} + 2\sum_{n=1}^{\infty}\frac{\cos(n\phi)}{a_n}\right). \tag{2.37}$$

From this, the input impedance of the circular loop antenna is found to be

$$Z_l = \frac{V}{I(0)} = j\pi\eta_0\left(\frac{1}{a_0} + 2\sum_{n=1}^{\infty}\frac{1}{a_n}\right)^{-1}. \tag{2.38}$$

Note that (2.37) constitutes only a formal solution of the circular loop antenna. In order to make it useful, some way must be found to evaluate the series. Hallen proved that, for large n, the coefficients in (2.37) approach asymptotically the value

$$a_n \cong -\frac{n^2}{\pi k_0 b}\left[\ln\left(\frac{2b}{a}\right) - \gamma - \ln(n)\right] \tag{2.39}$$

where $\gamma = 0.5772$ is the Euler's constant. It is apparent that the value goes to zero for such n as

$$n \cong n_0 = \frac{2b}{a}e^{-\gamma}. \tag{2.40}$$

Thus, the series (2.37) has a singularity [4]. Storer treated the singularity problem by including the dominant complex term as

$$a_n \cong \frac{1}{\pi}\left(k_0 b - \frac{n^2}{k_0 b}\right)\left[\ln(n_0) - \ln(n) - j\frac{(k_0 b)^{2n+1}}{\Gamma(2n+2)}\right] \tag{2.41}$$

for $n \gg k_0 b$ and $n \gg 1$ where $\Gamma(x)$ is the Gamma function. It will be clear later that the inclusion of the complex term permits a replacement of the series by an integral. The series (2.37) can then be written in the form

$$I(\phi) = \frac{V}{j\pi\eta_0}\left(\frac{1}{a_0} + 2\sum_{n=1}^{4}\frac{\cos(n\phi)}{a_n} + \psi(\phi)\right) \tag{2.42}$$

where

$$\psi(\phi) = 2\sum_{n=5}^{\infty}\frac{\cos(n\phi)}{a_n}. \tag{2.43}$$

It can be shown that an excellent approximation exists for (2.43), which is

$$\psi(\phi) = -2\pi k_0 b\sum_{n=5}^{\infty}\frac{\cos(n\phi)}{(n^2 - k_0^2 b^2)\left[\ln(n_0) - \ln(n) - j(k_0 b)^{2n+1}/\Gamma(2n+2)\right]} \tag{2.44}$$

for $k_0 b \leq 2.5$ and $n \geq 5$. The condition of $k_0 b < 2.5$ can be satisfied with the circumference of the loop antenna less than 2.5 wavelengths. Almost all loop antennas of practical interest are contained in this range. The series (2.44) will now be replaced by an integral

$$\begin{aligned}
\psi(\phi) &= -2\pi k_0 b\left[\int_{4.5}^{\infty}\frac{\cos(x\phi)}{\ln(n_0) - \ln(x)}\frac{dx}{x^2} + \int_{4.5}^{\infty}\frac{\cos(x\phi)}{\ln(n_0) - \ln(x)}\frac{k_0^2 b^2 dx}{x^2(x^2 - k^2 b^2)}\right]\\
&\cong -2\pi k_0 b\left[\int_{4.5}^{\infty}\frac{\cos(x\phi)}{\ln(n_0) - \ln(x)}\frac{dx}{x^2} + \frac{1}{\ln(n_0) - \ln(4.5)}\int_{4.5}^{\infty}\frac{k_0^2 b^2\cos(x\phi)}{x^4}dx\right]\\
&= -\frac{2\pi k_0 b}{\ln(n_0/4.5)}\left[\int_{1}^{\infty}\frac{\ln(n_0/4.5)}{\ln(n_0/4.5) - \ln(x)}\frac{\cos(4.5x\phi)}{4.5x^2}dx + \int_{1}^{\infty}\frac{k_0^2 b^2\cos(4.5x\phi)}{4.5^3 x^4}dx\right]\\
&= -\frac{2\pi}{\ln(n_0/4.5)}\left[\frac{k_0 b}{4.5}\int_{1}^{\infty}\frac{\ln(n_0/4.5)}{\ln(n_0/4.5) - \ln(x)}\frac{\cos(4.5x\phi)}{x^2}dx + \left(\frac{k_0 b}{4.5}\right)^3\int_{1}^{\infty}\frac{\cos(4.5x\phi)}{x^4}dx\right].
\end{aligned} \tag{2.45}$$

Substituting (2.45) into (2.42) yields an explicit expression for the current distribution as

$$I(\phi) = \frac{V}{j\pi\eta_0}\left\{\frac{1}{a_0} + 2\sum_{n=1}^{4}\frac{\cos(n\phi)}{a_n} - \frac{2\pi}{\ln(n_0/4.5)}\left[\frac{k_0 b}{4.5}\chi_1(\phi) + \left(\frac{k_0 b}{4.5}\right)^3 \chi_2(\phi)\right]\right\}. \tag{2.46}$$

The expression for the input impedance of the circular loop antenna becomes

$$Z_l = j\pi\eta_0\left\{\frac{1}{a_0} + 2\sum_{n=1}^{4}\frac{1}{a_n} - \frac{2\pi}{\ln(n_0/4.5)}\left[\frac{k_0 b}{4.5}\chi_1(0) + \left(\frac{k_0 b}{4.5}\right)^3 \chi_2(0)\right]\right\} \tag{2.47}$$

where $\chi_{1,2}(0)$ are explicitly given by

$$\chi_1(0) = \frac{\ln(n_0/4.5)}{(n_0/4.5)}\int_{-\infty}^{\ln\left(\frac{n_0}{4.5}\right)}\frac{e^x}{x}dx \tag{2.48}$$

and

$$\chi_2(0) = \frac{1}{3}. \tag{2.49}$$

In (2.46) and (2.47), the calculation of a_n where n is from 0 to 4 needs to be made from (2.34) and (2.35). It turns out that there exists a recursion formula for (2.35), i.e.,

$$\Delta_n = K_{n+1} - K_n = \Omega_{2n+1}(2k_0 b) + J_{2n+1}(2k_0 b) \tag{2.50}$$

where $\Omega_{2n+1}(x)$ is the Lommel–Weber function of order $2n+1$ and $J_{2n+1}(x)$ is the Bessel function of order $2n+1$. Therefore, all that remains to evaluate is

$$K_0 = \frac{1}{2\pi}\int_{-\pi}^{\pi}\frac{e^{-jk_0 b r(\phi)}}{r(\phi)}d\phi = \frac{1}{\pi}\ln\frac{8b}{a} - \frac{1}{2}\int_0^{2k_0 b}\Omega_0(x)dx - \frac{j}{2}\int_0^{2k_0 b}J_0(x)dx. \tag{2.51}$$

For the current distribution of (2.32), the general field expressions can be found in [7]. However, they are somewhat complex and thus not considered here. Instead, we will adopt the field expressions derived by Rao with the current distribution of (2.46) for the far-field region as

$$E_\phi = -\frac{V k_0 b}{4\pi}\frac{e^{-jk_0 r}}{r}\left\{2a_0 J_1(k_0 b\sin\theta) + \sum_{n=1}^{4}\frac{a_n\cos(n\phi)}{2}(j)^n[J_{n+1}(k_0 b\sin\theta) - J_{n-1}(k_0 b\sin\theta)]\right\}, \tag{2.52}$$

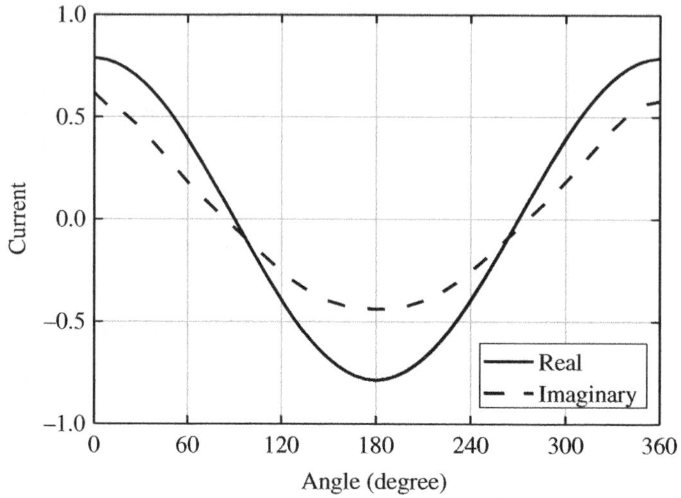

Figure 2.6 Calculated and normalized current distribution on the loop antenna with one-wavelength circumference.

$$E_\theta = j\frac{Vk_0b}{4\pi}\frac{e^{-jk_0r}}{r}\cos\theta\sum_{n=1}^{4}\frac{a_n\sin(n\phi)}{2}(j)^n[J_{n+1}(k_0b\sin\theta)+J_{n-1}(k_0b\sin\theta)] \tag{2.53}$$

where J_n is the Bessel function of the first kind and order n [5]. It should be noted that Rao only took the first five terms of (2.46) into account and thought that the contribution from the higher-order terms was negligible.

Recently, Fang and Zhang have revisited Storer's theory and confirmed its accuracy for the thin loop antenna with a circumference less than 2.5 wavelengths [8]. They also validated that the higher-order terms are indeed negligible in the far-field calculation. As the one-wavelength ($C = \lambda$) loop antenna is the most common loop antenna, we discuss its calculated results.

Figure 2.6 shows the calculated and normalized current distribution of the one-wavelength loop antenna at 3 GHz for a wire radius of 0.001λ. Note that the current is not uniform. It first reduces from the maximum at the input $\phi = 0^\circ$ to the minimum at $\phi = 180^\circ$, and then increases from the minimum to the maximum at $\phi = 360^\circ$.

Figure 2.7 shows the calculated input impedance of the circular loop antenna as a function of frequency for the wire radius of 0.001λ. The input impedance of the circular loop antenna can be found from the figure to be

$$Z_l = R + jX = 145\,\Omega - j102\,\Omega, \tag{2.54}$$

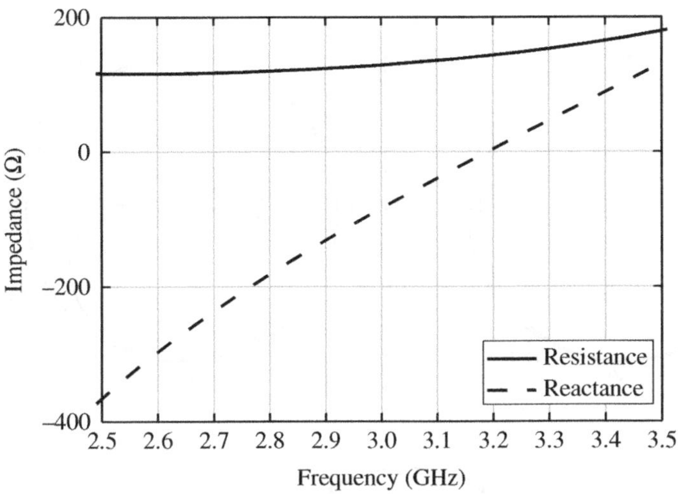

Figure 2.7 Calculated input impedance of the circular loop antenna.

which corresponds to one wavelength at 3.0 GHz. It is a capacitive impedance. Thus, one needs to resonate the capacitive reactance of 102 Ω with the equal inductive reactance. In practice, the one-wavelength loop antenna slightly increases the circumference to achieve resonance. As can be seen from the figure, resonance occurs at 3.2 GHz (or $C = 1.07\lambda$) where the resistance is 150 Ω and the reactance is zero.

Figure 2.8 plots the calculated and normalized H- and E-plane patterns for the one-wavelength loop antenna at 3 GHz. It is noted from the figure that the radiation is maximum normal to the plane of the loop.

Figure 2.9 shows the calculated directivities of a circular loop antenna as a function of circumference for a wire radius of 0.001λ. It is seen that the directivity is 3.4 dBi for the one-wavelength circular loop antenna at $\theta = 0°$ or 180°. The directivity increases to 3.67 dBi at a circumference of 1.07λ. The maximum directivity is 4.46 dBi at a circumference of 1.42λ.

The input impedance of the circular half-loop antenna is equal to one-half that of the circular loop antenna given by

$$Z_{hl} = \frac{Z_l}{2}. \tag{2.55}$$

The circular half-loop antenna only radiates to the upper hemisphere. So, its directivity should be twice of the corresponding circular loop antenna [9]. The directivity of the one-wavelength circular half-loop antenna is 6.4 dBi at $\theta = 0°$ or 180°.

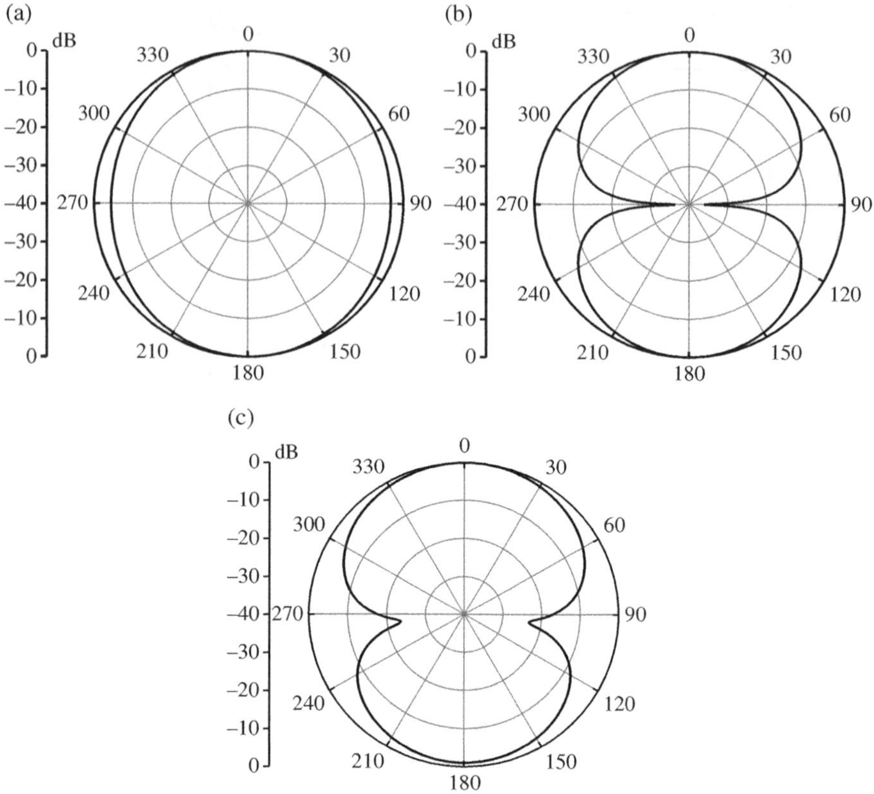

Figure 2.8 Normalized radiation pattern of the one-wavelength circular loop antenna. (a) The H plane or the xoz plane, (b) the E plane or the yoz plane, and (c) the E plane or the xoy plane.

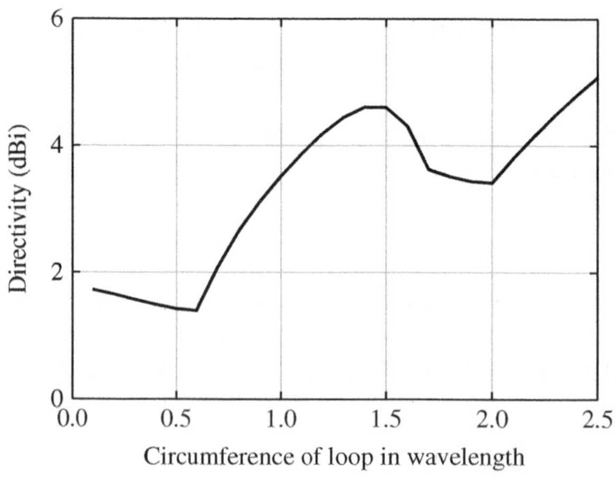

Figure 2.9 Calculated directivities of circular loop antennas.

2.5 Loop-Dipole and Half-Loop-Monopole Antennas

The loop antenna can be connected in either parallel or series with the dipole antenna to form the loop-dipole composite antenna. Figure 2.10 shows the composite antenna where the loop is connected in parallel with the dipole. The circular loop antenna is located with its center at the origin of the coordinates and is connected in parallel with the dipole antenna in free space. The circular loop antenna has the radius of b, while the dipole antenna has the length of l. They are made of wire with a radius of a. The loop-dipole composite antenna is energized by a differential signal source V_d through a twin-wire transmission line. Figure 2.10 also shows the circular half-loop-monopole composite antenna mounted above a ground plane on the $z = 0$ surface and is excited by a single-ended signal source V_s through a single-wire transmission line.

If the currents in the dipole and loop are in the same forms as that (2.1) for a dipole of the same length and (2.32) for a loop of the same circumference, respectively, the input impedance of the loop-dipole composite antenna can be found as

$$Z_{ld} = \frac{Z_d Z_l}{Z_d + Z_l} \tag{2.56}$$

where Z_d and Z_l were given by (2.11) and (2.47), respectively. Figure 2.11 shows the calculated input impedances of the loop-dipole composite antennas as a function of frequency for the three cases with the same wire radius of 0.001λ [8] For case one, the circumference of the loop and the length of the dipole are 0.5λ at 3 GHz. The calculated input impedance is $Z_{ld} = 73 - j50\ \Omega$ at 3 GHz. The resonance occurs at 3.1 GHz. For case two, the circumference of the loop and the length of the dipole are 0.5λ and 1.0λ at 3 GHz, respectively. The calculated input impedance is

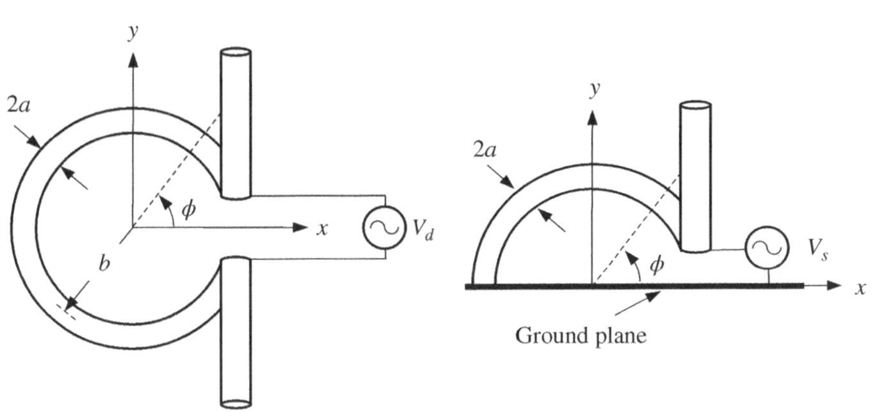

Figure 2.10 Loop-dipole and half-loop-monopole composite antennas and coordinates.

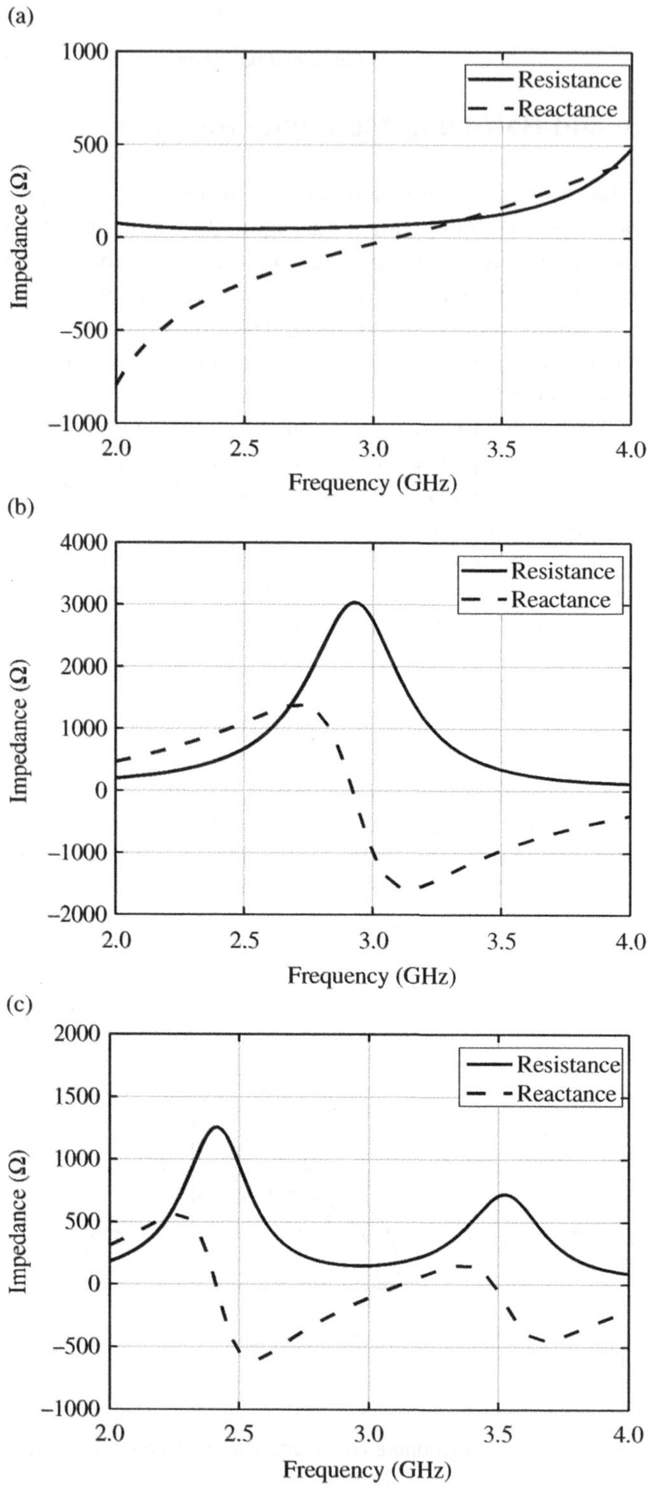

Figure 2.11 Calculated input impedances of loop-dipole composite antennas with different dimensions. (a) $C = 0.5\lambda$ and $l = 0.5\lambda$, (b) $C = 0.5\lambda$ and $l = 1.0\lambda$, and (c) $C = 2.0\lambda$ and $l = 1.0\lambda$.

Z_{ld} = 2800 − j1000 Ω. The resonance occurs at 2.87 GHz. For case three, the circumference of the loop and the length of the dipole are 2.0λ and 1.0λ at 3 GHz, respectively. The calculated input impedance is Z_{ld} = 150 − j150 Ω. The multiple resonances occur at 2.38, 3.18, and 3.48 GHz, respectively.

The electromagnetic field components in the far-field region for the loop-dipole antenna can be obtained by superposition of the separate fields for the loop antenna and the dipole antenna. The expressions for the far-field components of the loop antenna given in (2.52) and (2.53) can be directly used, while the expressions for the far-field components of the dipole antenna need to be modified to account for its shifted location as

$$E_\theta^d = -\frac{j\eta_0 I_m e^{-jk_0 r}}{4\pi r} e^{jk_0 b \sin\theta\cos\phi} \cos\theta \sin\phi \left\{ \frac{\cos\left(k_0 \frac{l}{2}\sin\theta\sin\phi\right) - \cos\left[k_0\frac{l}{2} - k_0 \upsilon(1+\sin\theta\sin\phi)\right]}{1+\sin\theta\sin\phi} + \frac{\cos\left(k_0 \frac{l}{2}\sin\theta\sin\phi\right) - \cos\left[k_0\frac{l}{2} - k_0 \upsilon(1-\sin\theta\sin\phi)\right]}{1-\sin\theta\sin\phi} \right\} \tag{2.57}$$

and

$$E_\phi^d = -\frac{j\eta_0 I_m e^{-jk_0 r}}{4\pi r} e^{jk_0 b \sin\theta\cos\phi} \cos\phi \left\{ \frac{\cos\left(k_0 \frac{l}{2}\sin\theta\sin\phi\right) - \cos\left[k_0\frac{l}{2} - k_0 \upsilon(1+\sin\theta\sin\phi)\right]}{1+\sin\theta\sin\phi} + \frac{\cos\left(k_0 \frac{l}{2}\sin\theta\sin\phi\right) - \cos\left[k_0\frac{l}{2} - k_0 \upsilon(1-\sin\theta\sin\phi)\right]}{1-\sin\theta\sin\phi} \right\} \tag{2.58}$$

where υ is the half gap for the delta-function source.

Figures 2.12–2.14 plot the calculated and normalized patterns on the three planes for the loop-dipole composite antenna of C = 0.5λ and l = 0.5λ at 3 GHz with the wire radius of 0.001λ [8]. Note that the patterns are similar to those of the half-wavelength dipole antenna because a negligible current flows to the loop antenna for its huge impedance. Figure 2.13 plots the calculated and normalized patterns on the three planes for the loop-dipole composite antenna of C = 0.5λ and l = 1.0λ at 3 GHz with a wire radius of 0.001λ [8]. Note that the patterns are similar to those of the one-wavelength dipole antenna but are distorted

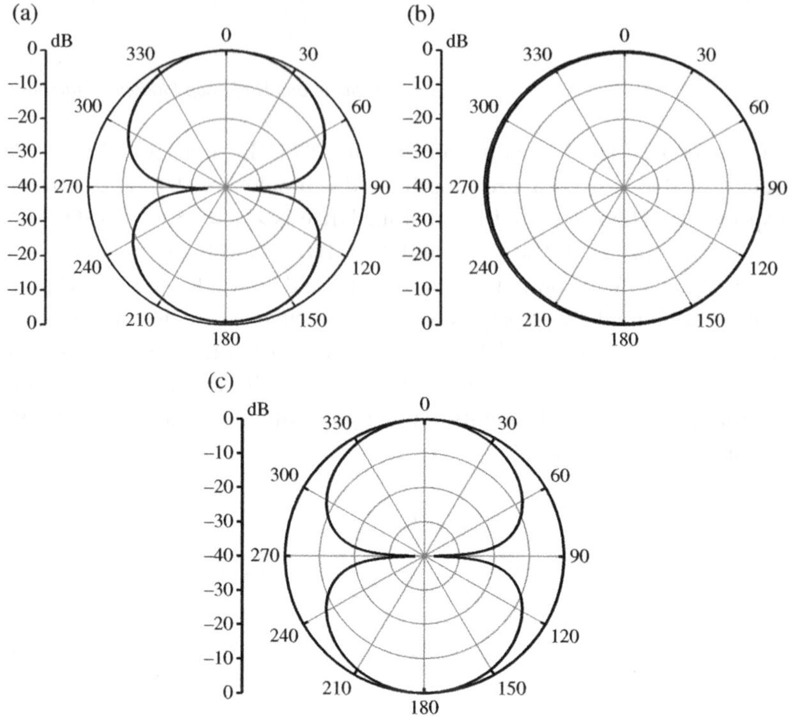

Figure 2.12 Calculated normalized patterns of loop-dipole composite antenna with $C = 0.5\lambda$ and $l = 0.5\lambda$ at 3 GHz. (a) On the *xoy*, (b) *xoz*, and (c) *xoz* planes.

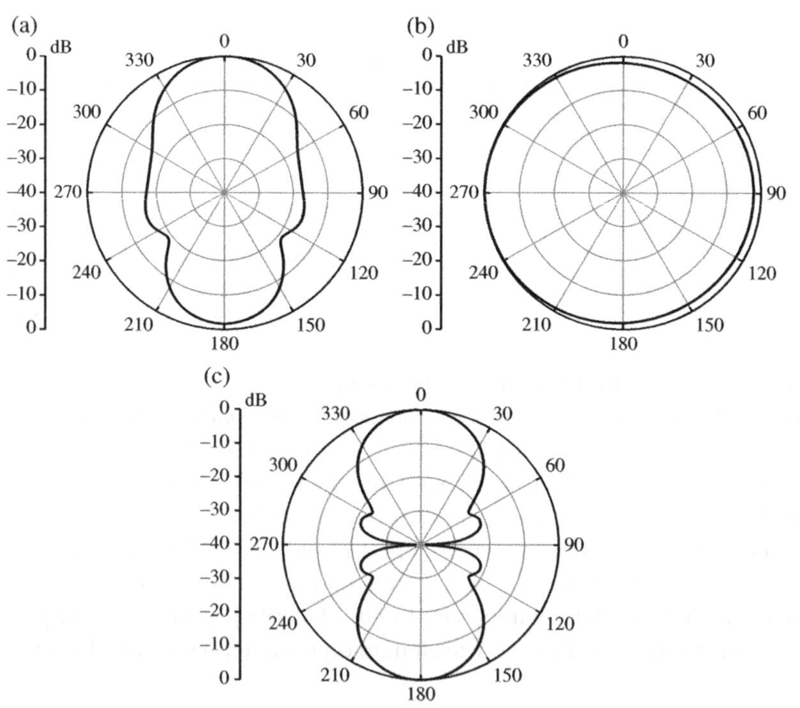

Figure 2.13 Calculated normalized patterns of loop-dipole composite antennas with $C = 0.5\lambda$ and $l = 1.0\lambda$ at 3 GHz. (a) On the *xoy*, (b) *xoz*, and (c) *xoz* planes.

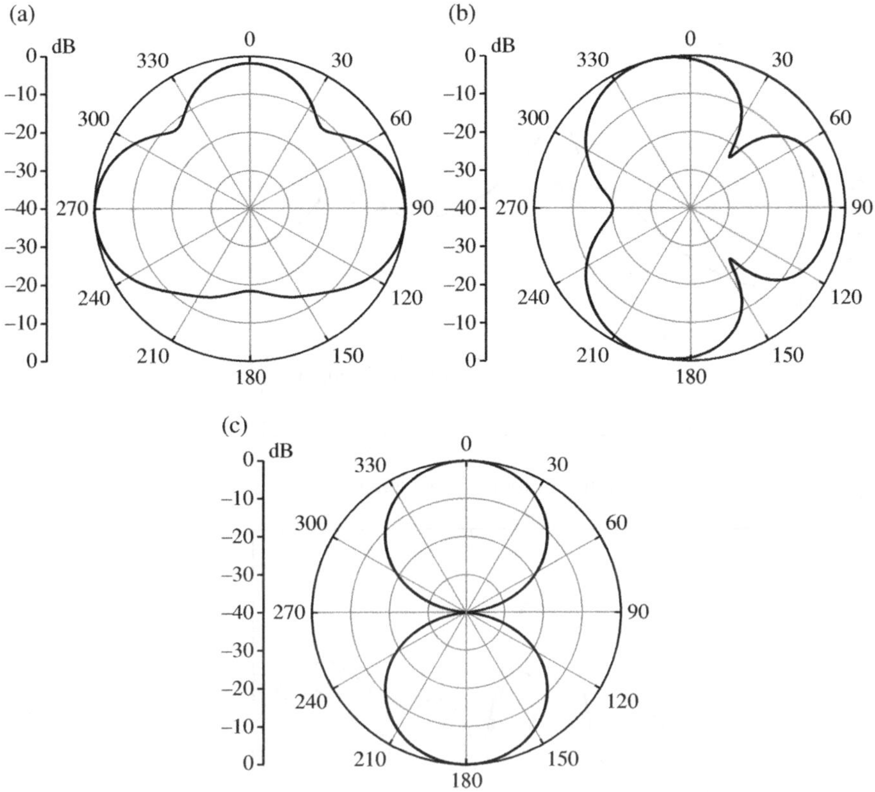

Figure 2.14 Calculated normalized patterns of loop-dipole composite antennas with $C = 2.0\lambda$ and $l = 1.0\lambda$ at 3 GHz. (a) On the *xoy*, (b) *xoz*, and (c) *xoz* planes.

by the radiation of the loop antenna. Figure 2.14 plots the calculated and normalized patterns on the three planes for the loop-dipole composite antenna of $C = 2.0\lambda$ and $l = 1.0\lambda$ at 3 GHz with a wire radius of 0.001λ [8]. Note that the patterns are not too different from those of the loop antenna with the circumference of two wavelengths because a negligible current flows to the one-wavelength dipole antenna for its huge impedance.

Figure 2.15a plots the calculated directivities for the case of the dipole length fixed at 0.5λ and the loop circumference varied. Note that the directivity of the loop-dipole composite antenna is almost the same as that of the loop antenna with the same circumference when the loop circumference is more than 0.8λ. Figure 2.15b plots the calculated and simulated directivities for the case of the loop circumference fixed at 0.5λ and the dipole length varied. Note that the directivity of

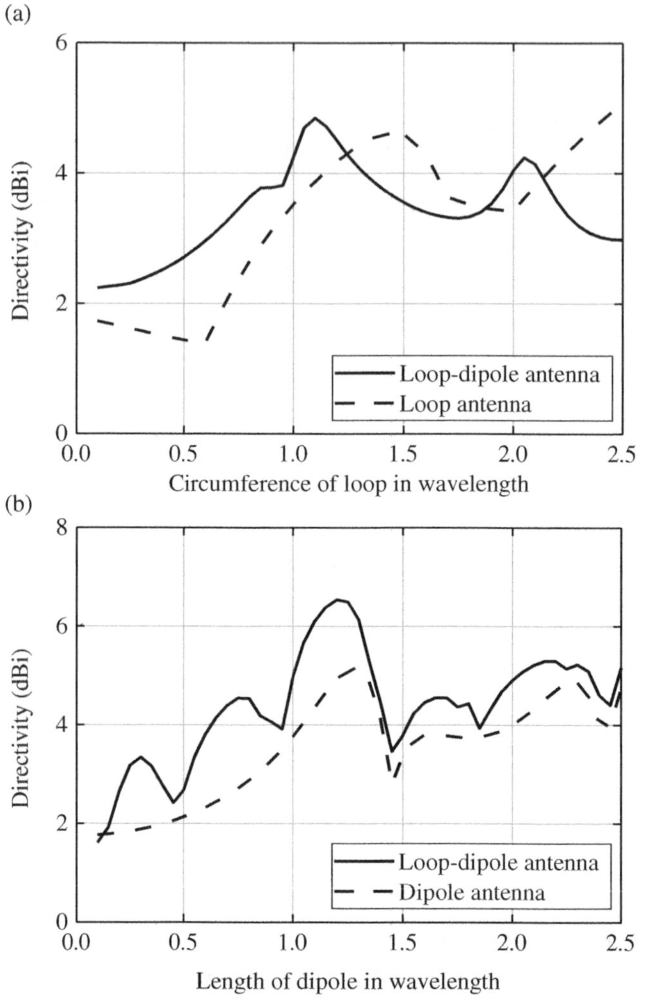

Figure 2.15 Calculated directivities of the loop-dipole composite antenna with different (a) circumferences of the loop and (b) lengths of the dipole.

the loop-dipole composite antenna is always higher than that of the dipole antenna alone [10].

The input impedance of the half-loop-monopole composite antenna is equal to one-half that of the loop-dipole antenna The half-loop-monopole antenna only radiates to the upper hemisphere. So, its directivity should be twice of the corresponding loop-dipole antenna [9].

2.6 Yagi-Uda and Half-Yagi-Uda Antennas

The Yagi-Uda antenna [10], as shown in Figure 2.16, consists of multiple parallel dipoles arranged along the x-axis. The dipole connected to a differential signal source V_d is the driver. To its left, the dipole acts as a reflector, while to its right, the dipoles function as directors. The reflector and directors are parasitic elements. They are collaborating with the driver through mutual coupling to guide the radiation toward endfire, that is, the x-axis [11–14]. The half-Yagi-Uda antenna, as shown in Figure 2.16, is modified from the Yagi-Uda antenna by adding a ground plane on the $z = 0$ surface and replacing the dipoles with monopoles, the differential signal source with a single-ended one V_s.

For the Yagi-Uda antenna, the current distribution over each element is assumed to be sinusoidal along the z-axis as

$$I_p(z) = I_p \frac{\sin\left[k_0\left(\frac{l_p}{2} - |z|\right)\right]}{\sin\left(\frac{k_0 l_p}{2}\right)} \tag{2.59}$$

where the current amplitude is given by

$$I_p = Z^{-1}V. \tag{2.60}$$

In (2.60), V is a vector of N elements that contains the feeding voltage at the center of each element and Z is the matrix of $N \times N$ dimension that contains the values of mutual impedance between the elements. Since the driver or second element is fed, there is a voltage V_2 at its center. The other elements are passive and have zero voltage at their centers. The mutual impedance between elements m and n in the matrix can be expressed as

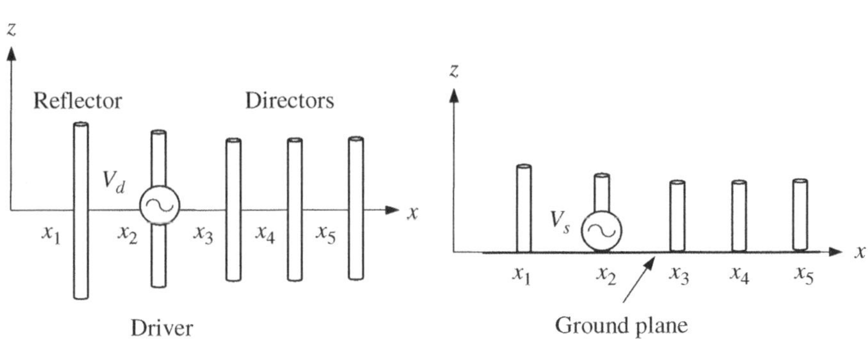

Figure 2.16 Yagi-Uda and half-Yagi-Uda antennas and coordinates.

$$z_{pq} = -\frac{j\eta_0}{4\pi \sin(k_0 l_p/2)\sin(k_0 l_q/2)} \int_{-l_q/2}^{l_q/2} \left[\frac{e^{-jk_0 R_+}}{R_+} + \frac{e^{-jk_0 R_-}}{R_-} - 2\cos(k_0 l_p/2)\frac{e^{-jk_0 R_0}}{R_0}\right] \sin\left[k_0\left(\frac{l_q}{2} - |z|\right)\right] dz \tag{2.61}$$

where

$$R_0 = \sqrt{d_{pq}^2 + z^2}, \tag{2.62}$$

$$R_+ = \sqrt{d_{pq}^2 + \left(z + \frac{l_p}{2}\right)^2}, \tag{2.63}$$

$$R_- = \sqrt{d_{pq}^2 + \left(z - \frac{l_p}{2}\right)^2} \tag{2.64}$$

with

$$d_{pq} = \sqrt{\left(x_p - x_q\right)^2 + \left(y_p - y_q\right)^2} \tag{2.65}$$

for $p \neq q$ and

$$d_{pq} = a_p \tag{2.66}$$

for $p = q$.

From the knowledge of the currents $I_p(z)$, the field expressions in the far-field region, where $k_0 r \gg 1$, can be obtained as

$$E_\theta \approx jk_0\eta_0 \frac{e^{-jk_0 r}}{2\pi r} F_z(\theta,\phi)\sin\theta \tag{2.67}$$

and

$$H_\phi \approx jk_0 \frac{e^{-jk_0 r}}{2\pi r} F_z(\theta,\phi)\sin\theta \tag{2.68}$$

where

$$F(\theta,\phi) = \sum_{p=1}^{N} \frac{2I_p}{k_0} e^{jk_0 x_p \sin\theta\cos\phi} \frac{\cos(k_0 h_p \cos\theta) - \cos(k_0 h_p)}{\sin(k_0 h_p)\sin^2\theta}. \tag{2.69}$$

It is seen from (2.69) that the field pattern factor is dependent on both azimuth and elevation angles of ϕ and θ. The power pattern is given by

$$P(\theta,\phi) = \left|\sum_{p=1}^{N} \frac{2I_p}{k_0} e^{jk_0 x_p \sin\theta\cos\phi} \frac{\cos(k_0 h_p \cos\theta) - \cos(k_0 h_p)}{\sin(kh_p)\sin\theta}\right|^2. \tag{2.70}$$

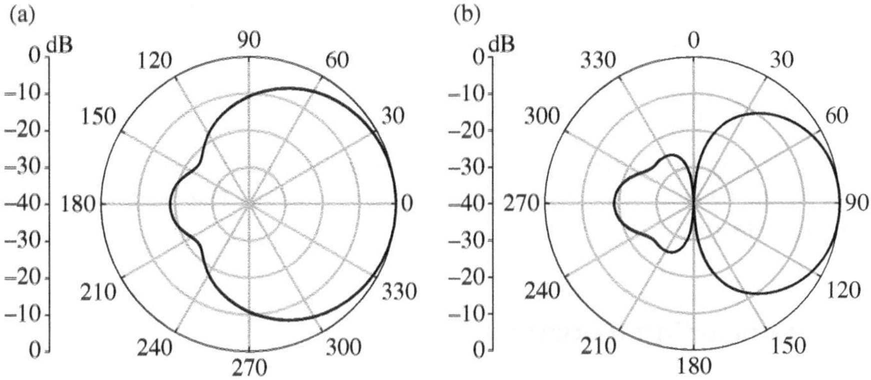

Figure 2.17 Normalized radiation patterns of the three-element Yagi-Uda antenna: (a) The *H* plane and (b) the *E* plane.

Here, we consider a three-element Yagi-Uda antenna with one reflector, one driver, and one director [14]. The element lengths, radii, and locations are in units of wavelength λ

$$l = \begin{bmatrix} l_1 \\ l_2 \\ l_3 \end{bmatrix} = \begin{bmatrix} 0.50 \\ 0.48 \\ 0.46 \end{bmatrix}, a = \begin{bmatrix} a_1 \\ a_2 \\ a_3 \end{bmatrix} = \begin{bmatrix} 0.003 \\ 0.003 \\ 0.003 \end{bmatrix}, d = \begin{bmatrix} x_1 \\ x_2 \\ x_3 \end{bmatrix} = \begin{bmatrix} -0.125 \\ 0 \\ 0.125 \end{bmatrix}. \tag{2.71}$$

Figure 2.17 plots the normalized *H*- and *E*-plane patterns for the three-element Yagi-Uda antenna. The maximum directivity is 8.18 dBi, which is 6.03 dB higher than that of a half-wave dipole antenna.

The impedance matrix is

$$Z = \begin{bmatrix} 73.07 + 41.37j & 60.47 - 0.97j & 36.25 - 25.53j \\ 60.47 - 0.97j & 64.93 + 11.75j & 53.72 - 2.71j \\ 36.25 - 25.53j & 53.72 - 2.71j & 57.65 - 17.01j \end{bmatrix}. \tag{2.72}$$

The input current and impedance of the driver are

$$I = \begin{bmatrix} I_1 \\ I_2 \\ I_3 \end{bmatrix} = \begin{bmatrix} -0.0290 + 0.0176j \\ 0.1062 - 0.0182j \\ -0.0801 - 0.0256j \end{bmatrix} \tag{2.73}$$

and

$$Z_{yu} = Z_2 = \frac{V_2}{I_2} = \frac{1}{I_2} = 9.15 + 1.57j. \tag{2.74}$$

The input impedance of the half-Yagi-Uda antenna is equal to one-half that of the Yagi-Uda antenna given by

$$Z_{hyu} = \frac{Z_{yu}}{2}. \tag{2.75}$$

The half-Yagi-Uda antenna only radiates to the upper hemisphere. So, its directivity should be twice the corresponding Yagi-Uda antenna. The maximum directivity is 11.18 dBi.

2.7 Concluding Remarks

This chapter summarized the important results of the most basic differential wire antennas such as the dipole, loop, loop-dipole, and Yagi-Uda antennas. It also considered the relationship between differential and single-ended wire antennas. It was shown that there exists an important ratio of 2 : 1 for the input impedance and 1 : 2 for the directivity between differential and single-ended wire antennas. For the single-ended wire antennas, we applied the single-ended voltage source to the feed gap through a single-wire transmission line. This was for the consistency in the analysis with the differential wire antennas. Actually, in practice, a coaxial transmission line is usually used to feed the single-ended antennas.

References

1 J. D. Kraus, *Antennas*. 2nd Edition, McGraw Hill, New York, 1988.
2 C. A. Balanis, *Antenna Theory: Analysis and Design*. 3rd Edition, Wiley, New Jersey, 2005.
3 J. E. Storer, "Impedance of thin-wire loop antennas," *AIEE Trans. Part I: Commun. Electron.*, vol. 75, no. 5, pp. 606–619, Nov. 1956.
4 E. Hallen, *Theoretical Investigations into the Transmitting and Receiving Qualities of Antennae*, Almqvist & Wiksells, Uppsala, Sweden, 1938.
5 B. R. Rao, "Far field patterns of large circular loop antennas: Theoretical and experimental results," *IEEE Trans. Antennas Propagat.*, vol. 16, no. 2, pp. 269–270, Mar. 1968.
6 G. A. Thiele, E. P. Ekelman, L. W. Henderson, "On the accuracy of transmission line model for folded dipole," *IEEE Trans. Antennas Propagat.*, vol. 28, no. 5, pp. 700–703, Sept. 1980.
7 D. H. Werner, "An exact integration procedure for vector potential of circular thin loop antennas," *IEEE Trans. Antennas Propagat.*, vol. 44, no. 2, pp. 157–165, Feb. 1996.

8 Y. L. Fang, Y. P. Zhang, "Theory and analysis of the Ω-shaped loop-dipole antenna," *IEEE Open J. Antennas Propagat.*, vol. 3, pp. 1161–1171, Oct. 2022.

9 G. P. Zhou, G. S. Smith, "An accurate theoretical model for the thin-wire circular half-loop antennas," *IEEE Trans. Antennas Propagat.*, vol. 39, no. 8, pp. 1157–1167, Aug. 1991.

10 H. Yagi, "Beam transmission of ultra-short waves," *Proc. IRE*, vol. 16, no. 6, pp. 715–740, Jun. 1928.

11 C. A. Chen, D. K. Cheng, "Optimum element lengths for Yagi-Uda arrays," *IEEE Trans. Antennas Propagat.*, vol. 23, no. 1, pp. 8–15, Jan. 1975.

12 D. K. Cheng, C. A. Cheng, "Optimum element spacings for Yagi-Uda arrays'," *IEEE Trans. Antennas Propagat.*, vol. 21, no. 5, pp. 615–623, Sept. 1973.

13 D. K. Cheng, "Gain optimization for Yagi-Uda arrays," *IEEE Antennas Propagat. Mag.*, vol. 33, no. 3, pp. 42–45, Jun. 1991.

14 J. Sophocles, *Electromagnetic Waves and Antennas*, www.ece.rutgers.edu/~orfanidi/ewa, 2016.

3

Differential Slot Antennas

3.1 Introduction

The slot antenna was patented by Blumlein in 1938 [1]. The papers about the slot antennas started to appear in open journals after the Second World War. In 1946, Booker published a paper on slot antennas and their relation to complementary dipole antennas [2]. By an extension of Babinet's principle from optics to electromagnetics, Booker postulated that the slot antenna should have the same field pattern as the dipole antenna, but with the electric and magnetic fields, interchanged and deduced the impedance relation between them. Also in 1946, Bailey published an approximate solution for the radiation field of a half-wavelength slot antenna and its radiation resistance seen from a shunt connection across its center [3]. In 1947, Mushiake obtained the same impedance relation for the general case of arbitrarily shaped and mutually complementary slot and planar antennas [4]. In 1948, Putman published measured input impedances of center-fed slot antennas near half-wavelength resonance [5]. In 1950, Begovich theoretically validated the impedance relation [6]. As a result, the impedance relation was well established, which has been adopted in antenna textbooks for the treatment of the input impedance of a slot antenna ever since [7, 8].

This chapter is devoted to the slot antenna. First, the slot antenna will be treated as complementary to a dipole antenna based on Booker's seminal paper [2]. An emphasis will be given to extending Booker's relation for differential complementary slot and dipole to single-ended counterparts. Then, the self-complementary antennas inspired by the slot antenna will be described in terms of Mushiake's work [4, 9]. Finally, the Yin-Yang antennas will be discussed according to the early and recent papers [10–18]. It will be seen that the slot antenna is a key component of the Yin-Yang antennas.

Differential Antennas: Theory and Practice, First Edition. Yueping Zhang.

3.2 Slot and Half-Slot Antennas

A slot antenna is cut in an infinitesimally thin, perfectly conducting, and infinitely large sheet. Figure 3.1 shows the configuration of the slot antenna on the *yoz* plane with its center at the origin of coordinates. It has a length of l and infinitely small width of w and is energized by a differential signal source V_d through a twin-wire transmission line connected between the midpoints of opposite edges. For the slot antenna, the nonzero electric and magnetic field components at any distance from it in cylindrical coordinates can be obtained from (2.2) to (2.4) for the dipole antenna with the equality given in Table 3.1 [6]. Note that the space dependence of electric and magnetic fields have interchanged with a sign reversal as suggested by Booker [2]. In Table 3.1, E_0 is the electric field strength across the center of the

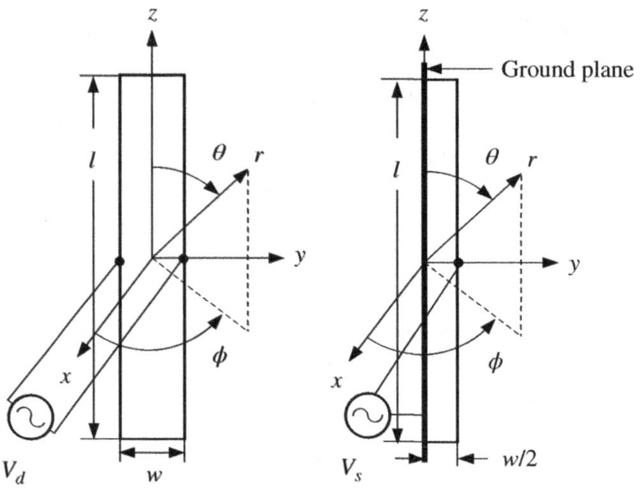

Figure 3.1 Slot and half-slot antennas and coordinates.

Table 3.1 Equality of field components for slot and dipole antennas.

Slot antenna	Dipole antenna
$-\dfrac{E_\phi}{E_0}$	$\dfrac{2H_\phi}{I_m}$
$\eta_0\dfrac{H_\rho}{E_0}$	$\dfrac{2E_\rho}{\eta_0 I_m}$
$\eta_0\dfrac{H_z}{E_0}$	$\dfrac{2E_z}{\eta_0 I_m}$

slot, which can be expressed as V_0/w where V_0 is the electric potential difference between the two driving points.

A half-slot antenna, as also shown in Figure 3.1, has a ground plane on the $y = 0$ surface. It has a length of l and a width of $w/2$ and is excited by a single-ended signal source V_s through a single-wire transmission line at the right edge. For the half-slot antenna, one can imagine that the half-slot antenna is realized from the slot antenna by inserting the ground plane, an infinite perfect electric conductor, on the $y = 0$ surface. Therefore, the image theory can be used here and the field expressions for the slot antenna can be directly applied to calculate the field distribution of the half-slot antenna to the right of the ground plane.

3.2.1 Input Impedances

The input impedance of the slot antenna Z_s can be calculated from the known input impedance of the complementary dipole antenna Z_d by Booker's relation [6]

$$Z_s Z_d = \frac{\eta_0^2}{4}. \tag{3.1}$$

Taking a half-wavelength slot antenna as an example, one can find its input impedance from the known input impedance of the half-wavelength dipole antenna as

$$Z_s = \frac{\eta_0^2}{4Z_d} = \frac{35476}{Z_d} = \frac{35476}{73 + j42.5} = 363 - j211\ \Omega. \tag{3.2}$$

There are, of course, slot complements to a folded dipole, loop, and Yagi-Uga antennas. The input impedance of the folded dipole antenna is four times that of the dipole antenna, while that of the folded slot antenna is one-quarter of that of the slot antenna. This is a useful way of reducing the rather high input impedance of a half-wavelength slot antenna [6].

The input impedance of the half-slot antenna Z_{hs} is one-half that of the related slot antenna Z_s as

$$Z_{hs} = \frac{Z_s}{2}. \tag{3.3}$$

Based on the impedance ratio in (3.3) and (2.12), an interesting extension of Booker's relation for differential complementary dipole and slot to single-ended counterparts was made by the author [15]. It simply substitutes Z_s with $2Z_{hs}$ and Z_d with $2Z_m$ in (3.1) to yield

$$Z_{hs} Z_m = \left(\frac{\eta_0}{4}\right)^2. \tag{3.4}$$

Note that the half-slot antenna in Figure 3.1 is not complementary to the monopole antenna in Figure 2.1. Neither do they follow Babinet's principle [15]. This is radically different from the slot and dipole antennas, which are complementary to each other and follow Babinet's principle [6]. Similarly, we also have

$$Z_{hs}Z_d = \frac{\eta_0^2}{8} \tag{3.5}$$

for the half-slot and dipole antennas and

$$Z_s Z_m = \frac{\eta_0^2}{8}. \tag{3.6}$$

for the monopole and slot antennas.

The relation (3.4) was simply deduced from the Booker's relation (3.3). In fact, it can be derived as follows [15]. The input impedance of the monopole antenna Z_m can be written as

$$Z_m = j\frac{\eta_0}{8\pi}\int_0^l \left[\frac{e^{-jk_0 z}}{z} + \frac{e^{-jk_0(l-z)}}{l-z}\right] \sin k_0 z \, dz \tag{3.7}$$

where l represents the doubled length of the monopole antenna [7]. The input impedance of the slot antenna was given by Begovich [6]. Keeping the phasor in [6] and transforming coordinates, we obtain the input impedance of the half-slot antenna as

$$Z_{hs} = \frac{\pi\eta_0}{2j} \frac{1}{\displaystyle\int_0^l \left[\frac{e^{-jk_0(l-z)}}{l-z} + \frac{e^{-jk_0 z}}{z}\right] \sin k_0 z \, dz}. \tag{3.8}$$

Multiplying both sides of (3.7) and (3.8), we obtain

$$Z_m Z_{hs} = j\frac{\eta_0}{8\pi}\frac{\pi\eta_0}{2j} \frac{\displaystyle\int_0^l \left[\frac{e^{-jk_0 z}}{z} + \frac{e^{-jk_0(l-z)}}{l-z}\right] \sin k_0 z \, dz}{\displaystyle\int_0^l \left[\frac{e^{-jk_0(l-z)}}{l-z} + \frac{e^{-jk_0 z}}{z}\right] \sin k_0 z \, dz} = \frac{\eta_0\eta_0}{16} = \left(\frac{\eta_0}{4}\right)^2. \tag{3.9}$$

Let z_{hs} and z_m represent Z_s and Z_m normalized to 30π, we get

$$z_{hs}z_m = 1 \tag{3.10}$$

or

$$z_{hs} = \frac{1}{z_m} = y_m, \tag{3.11}$$

which states in fact that the impedance for the half-slot antenna is proportional to the admittance for the noncomplementary monopole antenna, the constant of

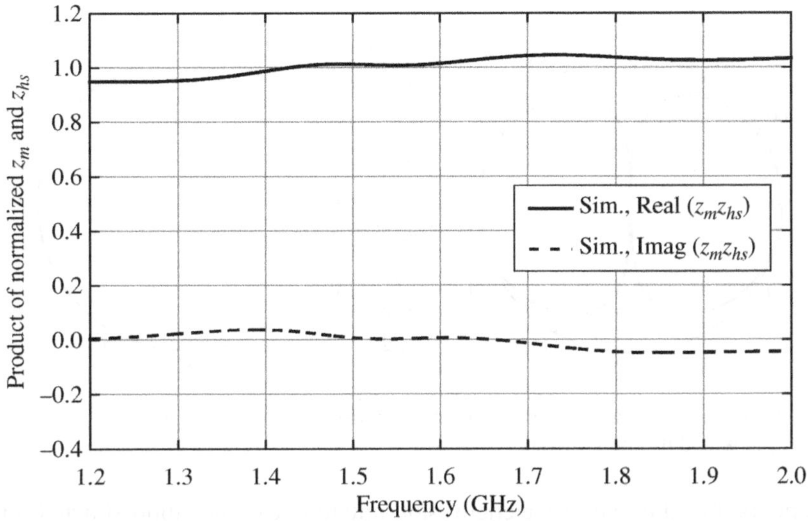

Figure 3.2 Product of the simulated z_m and z_{hs} for the monopole and half-slot antennas.

proportionality being independent of frequency, assuming of course that the surrounding medium is not dispersive. This important statement also holds for the slot antenna and the complementary dipole antenna with their impedances normalized to 60π.

The relation (3.4) has been numerically validated [16]. Figure 3.2 shows the product of the simulated z_m and z_{hs} for the monopole and half-slot antennas as a function of frequency from 1.2 to 2.0 GHz. It was assumed in the simulation that the half-slot and monopole antennas would be cut from semi-infinite, plane, perfectly conducting, thin sheets and installed on and above the perfectly conducting ground plane with a finite size of 50 cm × 50 cm, respectively. It is evident from the figure that the simulated product is quite close to the calculated product by the relation (3.4). The negligible errors can be attributed to the differences between the antennas for theory and simulation.

3.2.2 Radiation Characteristics

A half-wavelength ($l = \lambda/2$) slot antenna is the most common slot antenna. The electric and magnetic field components of the half-wavelength slot antenna can be obtained from (2.13) to (2.15) for the half-wavelength dipole antenna with the equality given in Table 3.1. For the far-field region where $k_0 r \gg 1$, the field

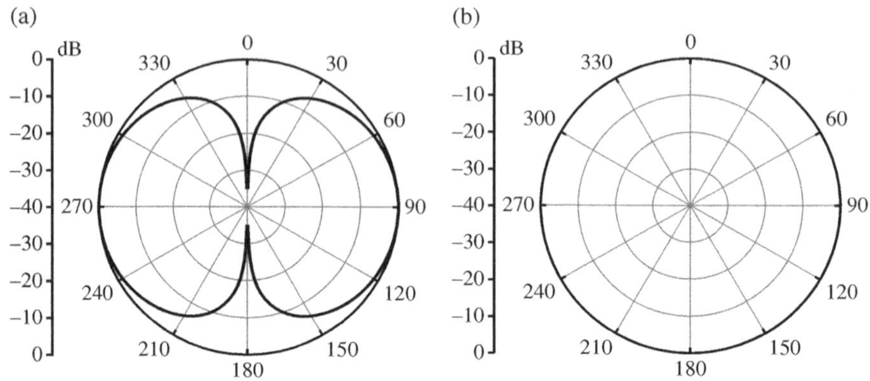

Figure 3.3 Normalized radiation patterns of the half-wavelength slot antenna. (a) The H plane and (b) the E plane.

components for the half-wavelength slot antenna can be approximated in spherical coordinates as

$$H_\theta = H_\rho \cos\theta - H_z \sin\theta \approx j\eta_0 \frac{E_0 e^{-jk_0 r}}{2\pi r}\left[\frac{\cos\left(\frac{\pi}{2}\cos\theta\right)}{\sin\theta}\right] \tag{3.12}$$

and

$$E_\phi \approx -j\frac{E_0 e^{-jk_0 r}}{2\pi r}\left[\frac{\cos\left(\frac{\pi}{2}\cos\theta\right)}{\sin\theta}\right]. \tag{3.13}$$

Note that the field pattern factor in (3.12) and (3.13) is the same as (2.18) for the half-wavelength dipole antenna. However, the polarization of radiation from the half-wavelength slot antenna is orthogonal to that from the half-wavelength dipole antenna. Figure 3.3 plots the normalized H- and E-plane radiation patterns. It is noted from the figure that the radiation is directional on the elevation plane. The directivity is 1.64 or 2.15 dBi and the half-power or 3-dB beamwidth is 78°.

The directivity of the half-slot antenna is twice of the corresponding slot antenna. For the half-wavelength half-slot antenna, the directivity is 3.28 or 5.15 dBi and the half-power or 3-dB beamwidth is 39°.

3.3 Self-Complementary Antennas

A self-complementary antenna is an antenna where the shape of the planar conducting sheet for a slot antenna is exactly identical to the shape of its complementary planar antenna. In other words, the self-complementary antenna is an

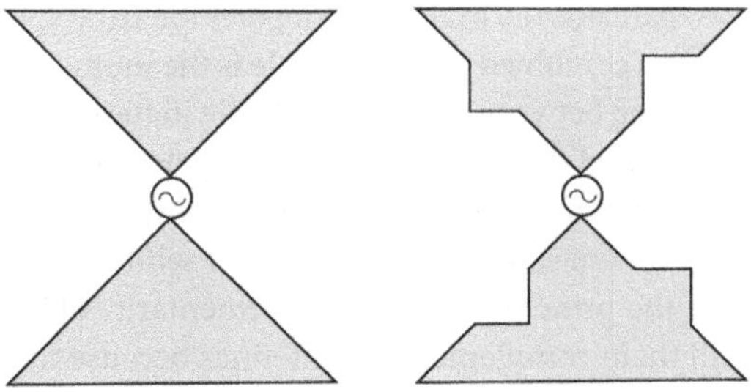

Figure 3.4 Two examples of self-complementary antennas.

antenna that is complementary to the original antenna itself. This is remarkable as there is an infinity of self-complementary antennas. Figure 3.4 shows two examples of self-complementary antennas. Note that the conducting sheets are infinitely extended. The input impedance of such an antenna can be expressed as

$$Z = Z_1 = Z_2 = \frac{\eta_0}{2}, \tag{3.14}$$

which in free space reduces to

$$Z = Z_1 = Z_2 = 60\pi \approx 188.4\,\Omega. \tag{3.15}$$

Equation (3.14) is known as Mushiake's relation, which shows that the input impedance of self-complementary antennas is always constant and independent of the source frequency [9]. However, the constant-impedance property of self-complementary antennas will be degraded because the conducting sheets cannot be infinitely extended in practical realization. The truncation of the conducting sheets will disturb the current distributions. The reflected currents from the truncated ends cause some harmful effects at the feed point. Hence, the reduction of the truncation effects is critical, and the truncation must be made for lower frequencies at a location where the current distribution density is sufficiently attenuated. The location can be determined from the full-wave simulations at larger dimensions corresponding to the wavelength.

3.4 Yin-Yang Antennas

The slot and dipole antennas in Booker's paper are mutually complementary antennas. Booker suggested that slot and dipole antennas might be combined to produce antennas with special polarization properties [2]. Clavin reported in 1954 such antennas for both linear and circular polarizations [19]. Itoh and Cheng presented in 1972 a crossed slot with a monopole to obtain a steerable cardioid pattern [10]. Mayes et al. published in 1972 the monopole slot for application as a broadband unidirectional antenna [11]. Clavin et al. described in 1974 an antenna consisting of a slot plus two parasitically excited monopoles for array applications [12]. A fundamental issue for any combined slot and dipole is the mutual coupling. Surtees studied the mutual coupling between a slot and a dipole in the early 1950s [13], while King and Owyang considered the slot with coupled dipoles in the late 1950s [14].

The antennas in [10–18] are certainly not self-complementary antennas because they do not satisfy the definition of self-complementary antennas, nor do they follow the principle of self-complementarity [19]. It is not rigorous either to simply call them complementary antennas because they do not follow the extended Babinet's principle. To highlight the importance of the complementary concept and mutual coupling associated with these antennas, we call them Yin-Yang antennas. This is because the principle of Yin and Yang includes the complementary concept and mutual coupling. In this section, we discuss Yin-Yang antennas [17, 18].

3.4.1 Basic Differential Structure

A Yin-Yang antenna is a combined slot-wire structure, which may be realized with single or multiple slots and wires. Figure 3.5 shows a basic Yin-Yang antenna. It consists of a slot plus two monopoles [12]. The slot has a length of l and a width of w.

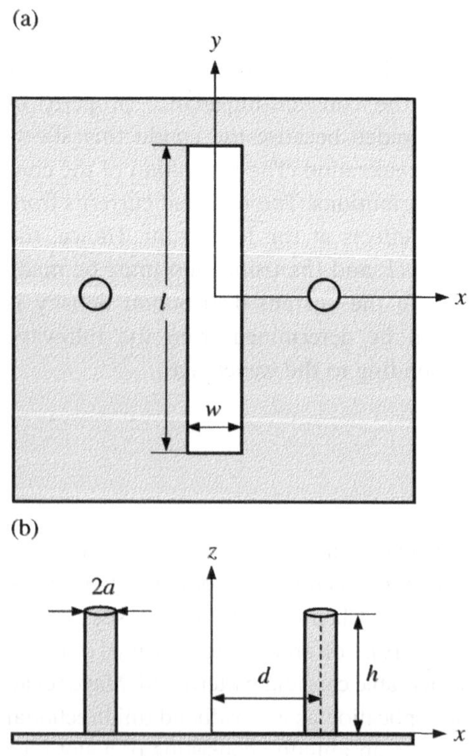

Figure 3.5 A basic Yin-Yang antenna. (a) Top view and (b) side view.

The two monopoles have the same height of h and are located symmetrically on either side of the slot with a distance of d. Assume that the slot is energized by a differential signal source V_d through a twin-wire transmission line connected between the midpoints of opposite edges, the two monopoles will be parasitically excited for the same amplitude but out of phase with one another. Alternatively, the two monopoles can be driven differentially, and the slot will be parasitically excited.

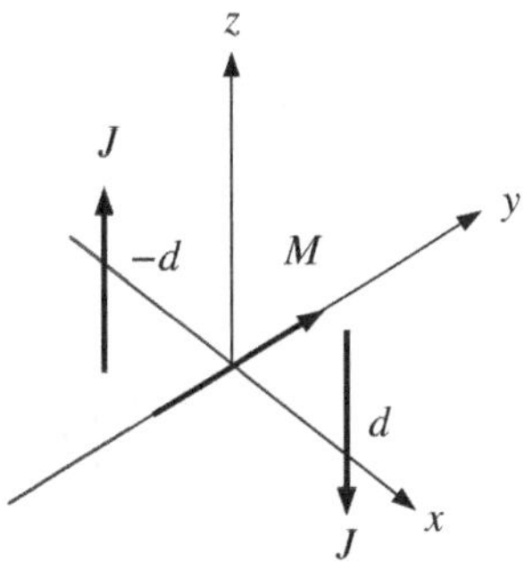

Figure 3.6 The array model of the basic Yin-Yang antenna.

3.4.2 The Array Model and the Distribution of Current

Based on the complementary concept, the image theory, and the equivalent principle, the basic Yin-Yang antenna can be treated as an array of the magnetic dipole located at the original position of the slot and the two electric dipoles located at the original positions of the two monopoles as shown in Figure 3.6. A good approximation for the current in a dipole can be made by assuming the current to be in the same form as that commonly assumed for a dipole of the same length. The magnetic and electric currents are thus given by

$$I_1^m(y) \approx I_m \sin k_0\left(\frac{l}{2} - |y|\right), \quad -\frac{l}{2} \le y \le \frac{l}{2}, \tag{3.16}$$

$$I_2^e(z) \approx I_e \sin k_0(h - |z|), \qquad x = -d, \quad -h \le z \le h, \tag{3.17}$$

$$I_3^e(z) \approx -I_e \sin k_0(h - |z|), \qquad x = d, \quad -h \le z \le h. \tag{3.18}$$

The above current expressions have been found to be valid for the slot with the width $w \ll \lambda$ and the monopole with the radius $a \ll \lambda$.

3.4.3 Electromagnetic Field

To obtain the electromagnetic field at any point for the basic Yin-Yang antenna, it is advantageous to consider the isolated magnetic dipole and the isolated electric dipole separately in the rectangular coordinate system.

For the magnetic dipole, the current in (3.16) will yield the field components

$$E_x = -j\frac{I_m \cos\phi}{4\pi\sqrt{x^2+z^2}}\left[e^{-jk_0R_1} + e^{-jk_0R_2} - 2\cos\left(\frac{k_0 l}{2}\right)e^{-jk_0R_0}\right], \tag{3.19}$$

$$E_y = 0, \tag{3.20}$$

$$E_z = j\frac{I_m \sin\phi}{4\pi\sqrt{x^2+z^2}}\left[e^{-jk_0R_1} + e^{-jk_0R_2} - 2\cos\left(\frac{k_0 l}{2}\right)e^{-jk_0R_0}\right], \tag{3.21}$$

$$H_x = j\frac{I_m \sin\phi}{4\pi\eta_0}\left[\left(\frac{y-\frac{l}{2}}{R_1}\right)e^{-jk_0R_1} + \left(\frac{y+\frac{l}{2}}{R_2}\right)e^{-jk_0R_2} - 2y\cos\left(\frac{k_0 l}{2}\right)\frac{e^{-jk_0R_0}}{R_0}\right], \tag{3.22}$$

$$H_z = j\frac{I_m \cos\phi}{4\pi\eta_0}\left[\left(\frac{y-\frac{l}{2}}{R_1}\right)e^{-jk_0R_1} + \left(\frac{y+\frac{l}{2}}{R_2}\right)e^{-jk_0R_2} - 2y\cos\left(\frac{k_0 l}{2}\right)\frac{e^{-jk_0R_0}}{R_0}\right], \tag{3.23}$$

$$H_y = -j\frac{I_m}{4\pi\eta_0}\left[\frac{e^{-jk_0R_1}}{R_1} + \frac{e^{-jk_0R_2}}{R_2} - 2\cos\left(\frac{k_0 l}{2}\right)\frac{e^{-jk_0R_0}}{R_0}\right] \tag{3.24}$$

where

$$R_0 = \sqrt{x^2 + y^2 + z^2}, \tag{3.25}$$

$$R_1 = \sqrt{x^2 + z^2 + \left(y - \frac{l}{2}\right)^2}, \tag{3.26}$$

$$R_2 = \sqrt{x^2 + z^2 + \left(y + \frac{l}{2}\right)^2}, \tag{3.27}$$

$$\phi = \tan^{-1}\frac{x}{z}. \tag{3.28}$$

For the electric dipole, the current in (3.17) will yield the field components

$$E_x = j\frac{\eta_0 I_e \cos\phi}{4\pi\sqrt{(x+d)^2 + y^2}}\left[\left(\frac{z-h}{R_4}\right)e^{-jk_0R_4} + \left(\frac{z+h}{R_5}\right)e^{-jk_0R_5} - 2z\cos(k_0 h)\frac{e^{-jk_0R_3}}{R_3}\right], \tag{3.29}$$

$$E_y = j\frac{\eta_0 I_e \sin\phi}{4\pi\sqrt{(x+d)^2 + y^2}}\left[\left(\frac{z-h}{R_4}\right)e^{-jk_0R_4} + \left(\frac{z+h}{R_5}\right)e^{-jk_0R_5} - 2z\cos(k_0 h)\frac{e^{-jk_0R_3}}{R_3}\right], \tag{3.30}$$

$$E_z = -j\frac{\eta_0 I_e}{4\pi}\left[\frac{e^{-jk_0R_4}}{R_4} + \frac{e^{-jk_0R_5}}{R_5} - 2\cos(k_0 h)\frac{e^{-jk_0R_3}}{R_3}\right], \tag{3.31}$$

$$H_x = -j\frac{I_e \sin\phi}{4\pi\sqrt{(x+d)^2 + y^2}}\left[e^{-jk_0R_4} + e^{-jk_0R_5} - 2\cos(k_0 h)e^{-jk_0R_3}\right], \tag{3.32}$$

$$H_y = j\frac{I_e\cos\phi}{4\pi\sqrt{(x+d)^2+y^2}}\left[e^{-jk_0R_4}+e^{-jk_0R_5}-2\cos(k_0h)e^{-jk_0R_3}\right], \tag{3.33}$$

$$H_z = 0 \tag{3.34}$$

where

$$R_3 = \sqrt{(x+d)^2+y^2+z^2}, \tag{3.35}$$

$$R_4 = \sqrt{(x+d)^2+y^2+(z-h)^2}, \tag{3.36}$$

$$R_5 = \sqrt{(x+d)^2+y^2+(z+h)^2}, \tag{3.37}$$

$$\phi = \tan^{-1}\frac{y}{x+d}. \tag{3.38}$$

For the electric dipole, the current in (3.18) will yield the field components

$$E_x = -j\frac{\eta_0 I_e\cos\phi}{4\pi\sqrt{(x-d)^2+y^2}}\left[\left(\frac{z-h}{R_7}\right)e^{-jk_0R_7}+\left(\frac{z+h}{R_8}\right)e^{-jk_0R_8}-2z\cos(k_0h)\frac{e^{-jk_0R_6}}{R_6}\right], \tag{3.39}$$

$$E_y = -j\frac{\eta_0 I_e\sin\phi}{4\pi\sqrt{(x-d)^2+y^2}}\left[\left(\frac{z-h}{R_7}\right)e^{-jk_0R_7}+\left(\frac{z+h}{R_8}\right)e^{-jk_0R_8}-2z\cos(k_0h)\frac{e^{-jk_0R_6}}{R_6}\right], \tag{3.40}$$

$$E_z = j\frac{\eta_0 I_e}{4\pi}\left[\frac{e^{-jk_0R_7}}{R_7}+\frac{e^{-jk_0R_8}}{R_8}-2\cos(k_0h)\frac{e^{-jk_0R_6}}{R_6}\right], \tag{3.41}$$

$$H_x = j\frac{I_e\sin\phi}{4\pi\sqrt{(x-d)^2+y^2}}\left[e^{-jk_0R_7}+e^{-jk_0R_8}-2\cos(k_0h)e^{-jk_0R_6}\right], \tag{3.42}$$

$$H_y = -j\frac{I_e\cos\phi}{4\pi\sqrt{(x-d)^2+y^2}}\left[e^{-jk_0R_7}+e^{-jk_0R_8}-2\cos(k_0h)e^{-jk_0R_6}\right], \tag{3.43}$$

$$H_z = 0 \tag{3.44}$$

where

$$R_6 = \sqrt{(x-d)^2+y^2+z^2}, \tag{3.45}$$

$$R_7 = \sqrt{(x-d)^2+y^2+(z-h)^2}, \tag{3.46}$$

$$R_8 = \sqrt{(x-d)^2+y^2+(z+h)^2}, \tag{3.47}$$

$$\phi = \tan^{-1}\frac{y}{x-d}. \tag{3.48}$$

The electromagnetic field at any point for the basic Yin-Yang antenna can thus be obtained by superposition of the separate fields derived above.

3.4.4 The Input Impedance

To determine the input impedance of the basic Yin-Yang antenna, a three-port network is considered with the following impedance matrix equation

$$\begin{bmatrix} V_1 \\ V_2 \\ V_3 \end{bmatrix} = \begin{bmatrix} Z_{11}\, Z_{12}\, Z_{13} \\ Z_{21}\, Z_{22}\, Z_{23} \\ Z_{31}\, Z_{32}\, Z_{33} \end{bmatrix} \begin{bmatrix} I_1 \\ I_2 \\ I_3 \end{bmatrix} \tag{3.49}$$

where V_1 is the driving electric potential difference at the center of the slot, V_2 and V_3 are the driving electric potential differences at the centers of the dipoles; I_1, I_2, and I_3 are the electric currents; Z_{11} is the self-impedance at the driving points of the slot, Z_{12} or Z_{21} is the mutual impedance between the driving points of the slot and the dipole at $x = -d$, Z_{13} or Z_{31} is the mutual impedance between the driving points of the slot and the dipole at $x = d$; Z_{22} and Z_{33} are the self-impedances at the driving points of the two dipoles, respectively, and Z_{23} and Z_{32} are the mutual impedances between the driving points of the two dipoles, respectively. Based on the symmetry and the reciprocity, one can find that $Z_{22} = Z_{33}$ and $Z_{23} = Z_{32}$. For the case of parasitic dipoles, it can be shown that $Z_{12} = -Z_{13} = Z_{21} = -Z_{31}$ because $I_2 = -I_3$. Thus, we can simplify (3.49) as

$$\begin{bmatrix} V_1 \\ 0 \\ 0 \end{bmatrix} = \begin{bmatrix} Z_{11} & Z_{21} & -Z_{21} \\ Z_{21} & Z_{22} & Z_{23} \\ -Z_{21} & Z_{23} & Z_{22} \end{bmatrix} \begin{bmatrix} I_1 \\ I_2 \\ -I_2 \end{bmatrix}. \tag{3.50}$$

The solution of (3.50) yields the input impedance at the driving points of the slot of the basic Yin-Yang antenna as

$$Z_{in} = Z_{11} - \frac{2Z_{21}^2}{Z_{22} - Z_{23}} \tag{3.51}$$

where $Z_{11} \approx Z_s$, $Z_{22} \approx Z_d/2$, Z_{21} and Z_{23} can be calculated as

$$Z_{21} = \frac{jZ_{11}}{2\pi \sin\left(\frac{k_0 l}{2}\right) \sin(k_0 h)} \int_0^h \left(e^{-jk_0\sqrt{d^2 + \left(\frac{l}{2}\right)^2 + z^2}} + e^{-jk_0\sqrt{d^2 + \left(\frac{l}{2}\right)^2 + z^2}} \right.$$

$$\left. - 2\cos\left(\frac{k_0 l}{2}\right) e^{-jk_0\sqrt{d^2 + z^2}} \right) \frac{d}{(d^2 + z^2)} \sin k_0(h - z)dz, \tag{3.52}$$

$$Z_{23} = -\frac{j\eta_0}{4\pi \sin(k_0 h)\sin(k_0 h)} \int_0^h \left(\frac{e^{-jk_0\sqrt{(2d)^2+(z-h)^2}}}{\sqrt{(2d)^2+(z-h)^2}} + \frac{e^{-jk_0\sqrt{(2d)^2+(z+h)^2}}}{\sqrt{(2d)^2+(z+h)^2}} \right.$$

$$\left. -2\cos(k_0 h)\frac{e^{-jk_0\sqrt{(2d)^2+z^2}}}{\sqrt{(2d)^2+z^2}} \right) \sin k_0(h-z)dz. \tag{3.53}$$

Let us consider the basic Yin-Yang antenna operating at 10 Hz with $l = 0.5\lambda$, $w = 0.002\lambda$, $a = 0.001\lambda$, and $h = 0.3\lambda$. Figures 3.7 and 3.8 plot the Z_{21} and Z_{23} as a function of d/λ. It is evident from the figures that the mutual impedances have damping oscillatory behaviors with separation distance. For the mutual impedances Z_{21} and Z_{23}, they are essentially due to the factors of $e^{-jk_0 d}$ and $e^{-j2k_0 d}$, respectively.

Figure 3.9 plots the input impedance Z_{in} as a function of d/λ. When the monopole is close to the slot antenna, inductive reactance is introduced into the slot antenna, and the capacitive reactance of the slot antenna approaches 0. As the distance increases, this effect becomes smaller and smaller. But the overall impact is also a sinusoidal change, which is caused by the mutual resistance of the two.

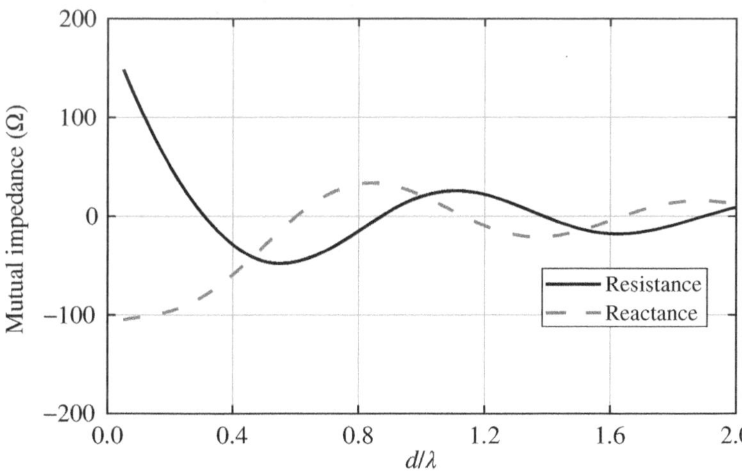

Figure 3.7 Mutual impedance between the half-wavelength slot and monopoles as a function of d/λ.

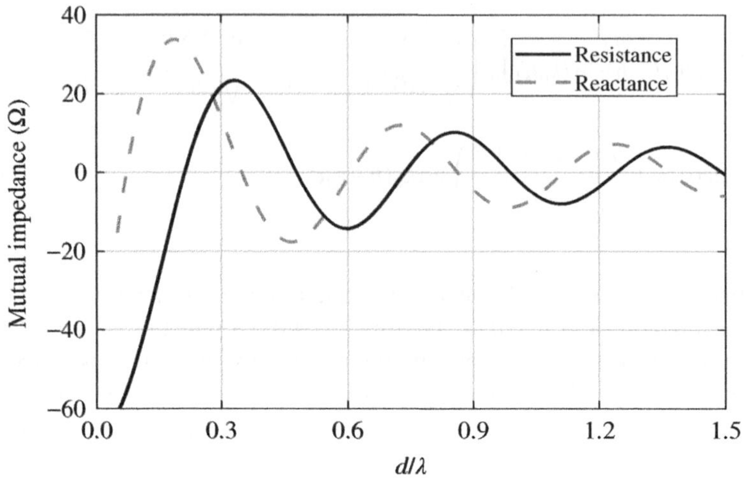

Figure 3.8 Mutual impedance between the two monopoles as a function of d/λ.

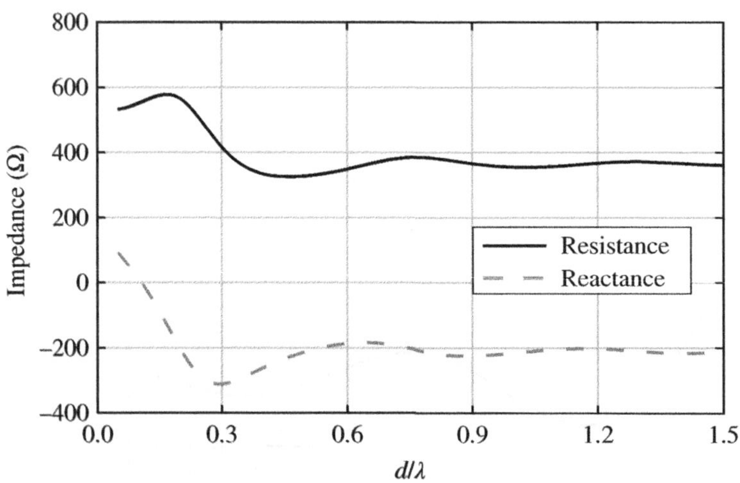

Figure 3.9 The input impedance of the basic Yin-Yang antenna as a function of d/λ.

3.4.5 Radiation Characteristics

The length of the slot is set to be 0.5λ and the amplitudes of the currents in (3.16)–(3.18) can be found as

$$I_m = 2V_1, \tag{3.54}$$

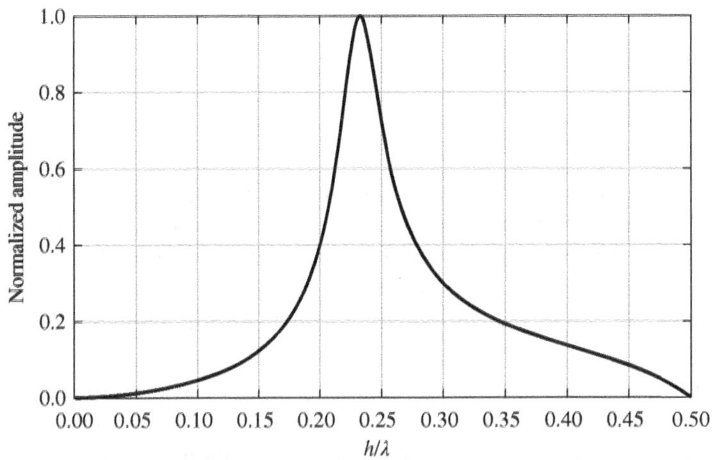

Figure 3.10 Normalized current amplitude I_e as a function of h/λ with d fixed at 0.25λ.

$$I_e \sin(k_0 h) = I_2 = -\frac{Z_{21}}{Z_{22} - Z_{23}} \frac{V_1}{Z_i}. \tag{3.55}$$

Figure 3.10 shows the normalized current amplitude I_e coupled from the slot as a function of h/λ with d fixed at 0.25λ. It is seen that the normalized current amplitude I_e first increases with h and then decreases with h. The normalized current amplitude reaches the maximum for $h = 0.23\lambda$.

The rectangular components of the field given in Section 3.4.3 are readily transformed into spherical components. For the far-field region where $k_0 r \gg 1$, the field components can be approximately expressed as

$$E_\phi = \frac{jV_1}{\pi r} e^{-jk_0 r} \left[\frac{\cos\left(\frac{\pi}{2} \cdot \sin\theta \cdot \sin\phi\right)}{1 - \sin^2\phi \sin^2\theta} \right] \cos\theta \sin\phi \tag{3.56}$$

and

$$E_\theta = -\frac{jV_1}{\pi r} e^{-jk_0 r} \left[\frac{\cos\left(\frac{\pi}{2} \cdot \sin\theta \cdot \sin\phi\right)}{1 - \sin^2\phi \sin^2\theta} \right] \cos\phi + \frac{120 I_2 e^{-jk_0 r}}{r \sin(k_0 h)} \left[\frac{\cos(k_0 h \cdot \cos\theta) - \cos(k_0 h)}{\sin\theta} \right] \sin(k_0 d \sin\theta \cos\phi). \tag{3.57}$$

In the E plane ($\phi = 0°$),

$$E_\phi = 0, \tag{3.58}$$

$$E_\theta = -\frac{jV_1}{\pi r} e^{-jk_0 r} + \frac{120 I_2 e^{-jk_0 r}}{r \sin(k_0 h)} \left[\frac{\cos\left(\frac{\pi}{2} \cdot \cos\theta\right)}{\sin\theta} \right] \sin(k_0 d \sin\theta). \tag{3.59}$$

In the H plane ($\phi = 90°$),

$$E_\phi = \frac{jV_1}{\pi r} e^{-jk_0 r} \left[\frac{\cos\left(\frac{\pi}{2} \cdot \sin\theta\right)}{\cos\theta} \right], \tag{3.60}$$

$$E_\theta = 0. \tag{3.61}$$

It is apparent that the radiation pattern of the basic Yin-Yang antenna can be thought of as that of the slot which has been modified by the presence of an array of two monopoles. The modification will only occur in the E plane but not in the H plane. This is because the radiation from the two monopoles has a null in the H plane of the slot but can cancel the slot radiation in the end-fire directions of its E plane.

To achieve identical E-plane and H-plane radiation patterns, the key is the phase difference between the far-fields of the slot and the monopoles. Figure 3.11 shows the phase difference between the far-fields of the slot and the monopoles as a function of h/λ, with $d = 0.25\lambda$, $a = 0.001\lambda$, and $w = 0.002\lambda$. It is seen that the far-fields of the slot and the monopoles tend to be in phase when $h < 0.2\lambda$, and tend to be out of phase when $0.25\lambda < h < 0.5\lambda$. Consequently, to maintain the far-fields of the slot and the monopoles out of phase, it is essential to control the height of the monopole within the range of 0.25λ to 0.5λ.

The distance d between the slot and monopole affects the phase and amplitude of the far-field of the monopoles. Figure 3.12 shows the phase difference between

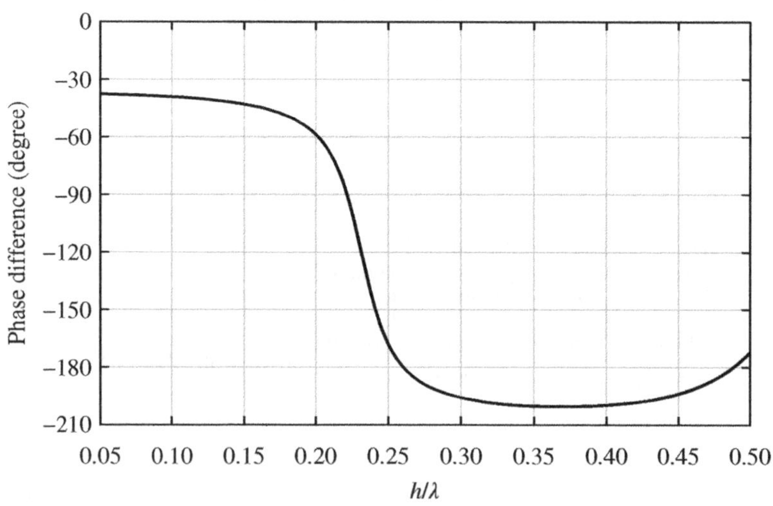

Figure 3.11 Phase difference between the far-fields of the slot and the monopole as a function of h/λ. ($d = 0.3\lambda$, $a = 0.001\lambda$, $w = 0.002\lambda$.)

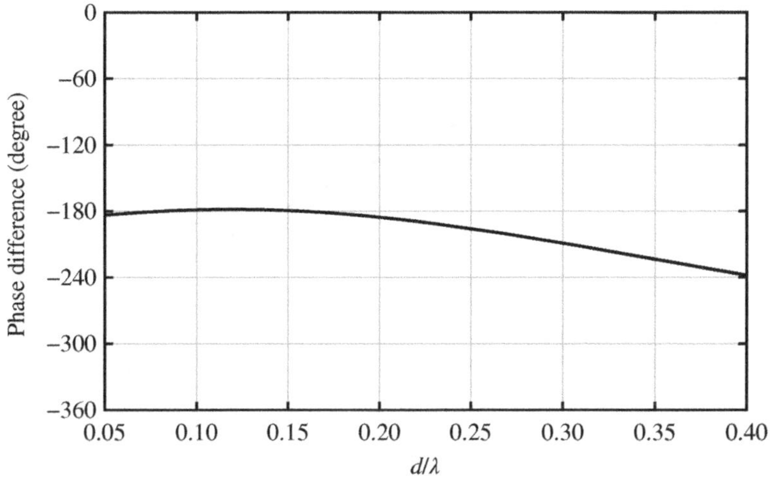

Figure 3.12 Phase difference between the far-fields of the slot and the monopoles as a function of d. ($a = 0.001\lambda$, $w = 0.002\lambda$.)

the far-fields of the slot and the monopoles as a function of d/λ, with $a = 0.001\lambda$ and $w = 0.002\lambda$. It is seen that the phase difference remains approximately $-180°$ for $d < 0.25\lambda$, but gradually decreases as d increases from 0.25λ. Figure 3.13 shows the normalized amplitude of the far-fields of the slot and the monopoles as a function of d/λ. It is seen that the normalized amplitude decreases rapidly as d increases. The distance d should be less than or equal to 0.25λ for effective pattern modulation of the Yin-Yang antenna.

The radius of the monopoles and the width of the slot alter the phase and amplitude of the far-fields of the slot and the monopoles. Figures 3.14 and 3.15 show the phase difference and the normalized amplitude as a function of a/λ, respectively. It is interesting to note that the phase difference remains approximately $-180°$ but the normalized amplitude increases from 0.7 to 1.0 for $w = 2a$ and as a increases from 0.001λ to 0.015λ. The radius a and width w should be larger for effective pattern modulation of the Yin-Yang antenna.

Figure 3.16 plots the normalized E- and H-plane patterns of the basic Yin-Yang antenna for $h = 0.3\lambda$ with $d = 0.15\lambda$, 0.3λ, and 0.5λ. As expected, the H-plane patterns do not change with d, while the E-plane patterns change with d. The change in the E plane only happens in the upper space but not in the lower space. This is because the monopoles only exist in the upper space. Figure 3.17 compares the normalized E- and H-plane patterns of the basic Yin-Yang antenna for $h = 0.3\lambda$ with $d = 0.15\lambda$. They are approximately equal in the upper space between $\pm 68°$.

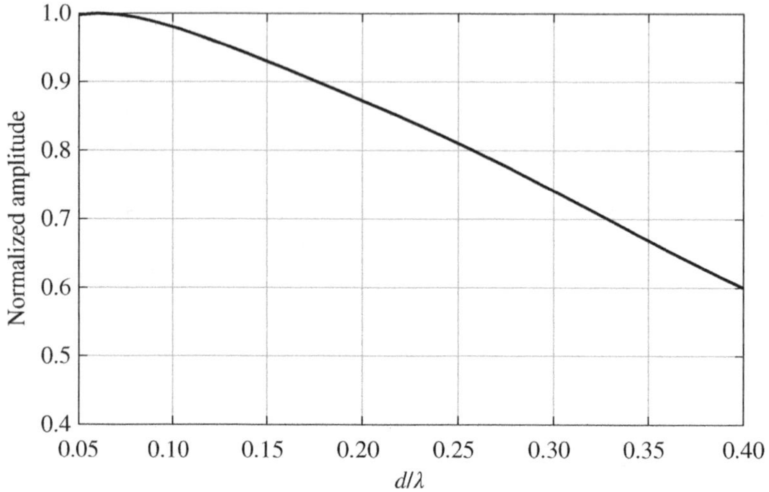

Figure 3.13 Normalized amplitude of the far-fields of the slot and the monopoles as a function of d. ($a = 0.001\lambda$, $w = 0.002\lambda$.)

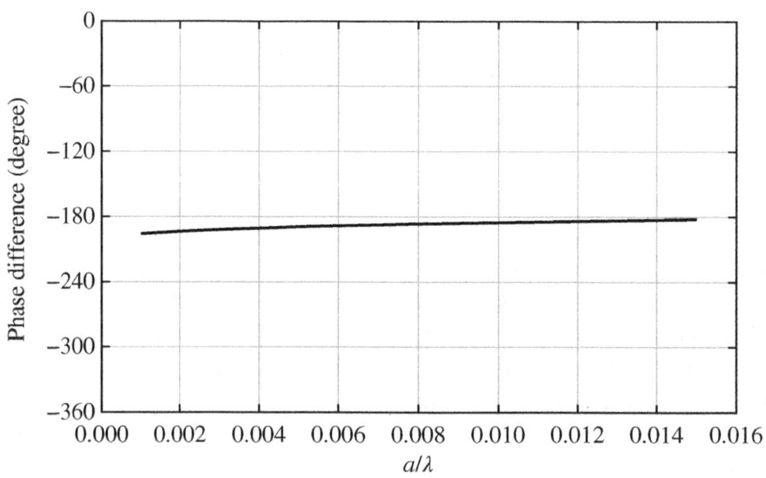

Figure 3.14 Phase difference between the far-fields of the slot and the monopoles as a function of a. ($h = 0.3\lambda$, $d = 0.25\lambda$.)

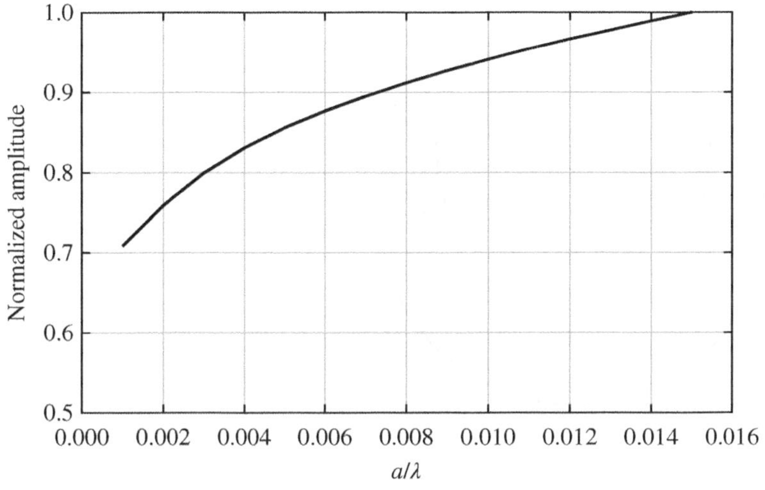

Figure 3.15 Normalized amplitude of the far-fields of the slot and the monopoles as a function of a. ($h = 0.3\lambda$, $d = 0.25\lambda$.)

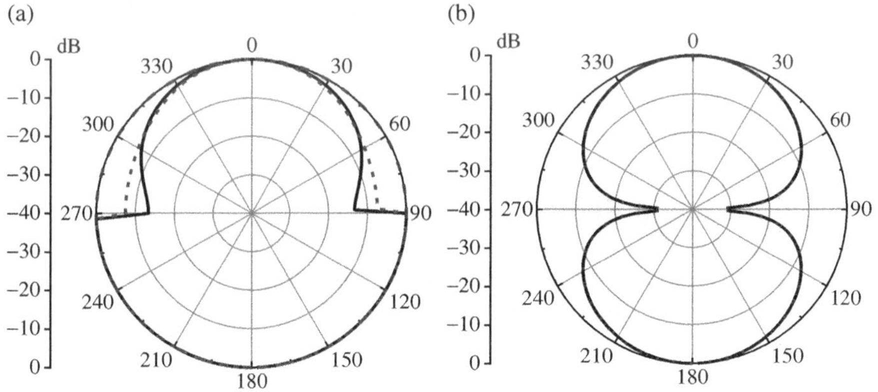

Figure 3.16 Normalized radiation patterns of the basic Yin-Yang antenna for $h = 0.3\lambda$, $d = 0.15\lambda$, 0.3λ, and 0.5λ. (a) The E plane and (b) the H plane.

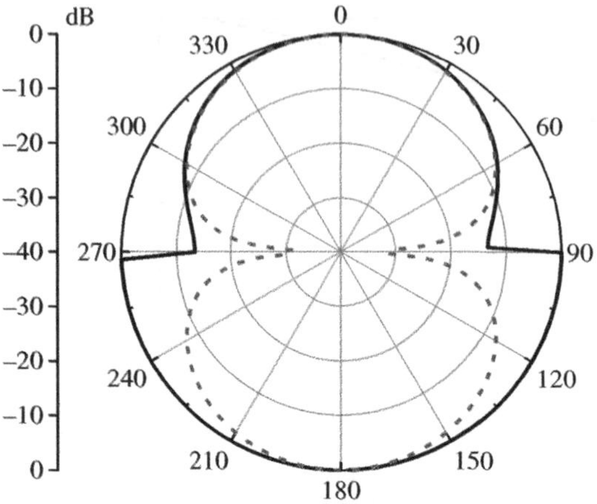

Figure 3.17 Normalized radiation patterns of the basic Yin-Yang antenna in the E and H planes for $h = 0.3\lambda$ with $d = 0.15\lambda$.

It should be noted that the normalized E- and H-plane patterns of the basic Yin-Yang antenna in the lower space have quite different beamwidth.

3.4.6 Basic Single-Ended Structure

The differential structure in Figure 3.5 can be easily modified into the single-ended structure by introducing a ground plane on the *yoz* plane and changing the differential signal source V_d to a single-ended signal source V_s. Alternatively, the monopole can be driven by the single-ended signal source V_s and the half slot will be parasitically excited.

3.5 Concluding Remarks

This chapter considered the slot antenna as complementary to a dipole antenna based on Booker's seminal paper. An interesting extension of Booker's relation for differential complementary slot and dipole to single-ended counterparts was highlighted. The self-complementary antennas were briefly described in terms of Mushiake's work. The Yin-Yang antennas in which the slot is a key component were discussed according to the early and recent papers. An attempt was made to treat basic Yin-Yang antenna as an array of the magnetic dipole located at the original position of the slot and the two electric dipoles located at the original positions of the two monopoles theoretically.

References

1 A. Blumlein, "Improvements in or relating to high frequency electrical conductors or radiators," British patent No. 515684 (1938).
2 H. G. Booker, "Slot aerials and their relations to complementary wire aerials," *JIEE (London)*, vol. 93, pt. IIIA, no. 4, pp. 620–626, 1946.
3 C. E. G. Bailey, "Slot feeders and slot aerials," *JIEE* (London), vol. 93, pt. IIIA, no. 4, pp. 615–619, 1946.
4 Y. Mushiake, "The input impedance of a slit antenna," *Joint Convention Record of Tohoku Sections of IEE and IECE of Japan*, pp. 25–26, Jun. 1948.
5 J. L. Putman, "Input impedance of centre-fed slot aerials near half-wave resonance," *JIEE* (London), vol. 95, pt. III, pp. 290–294, Jul. 1948.
6 N. A. Begovich, "Slot radiators," *Proc. IRE*, vol. 38, no. 7, pp. 803–806, Jul. 1950.
7 J. D. Kraus, *Antennas*. 2nd Edition, McGraw Hill, New York, 1988.
8 C. A. Balanis, *Antenna Theory: Analysis and Design*. 3rd Edition, Wiley, New Jersey, 2005.
9 Y. Mushiake, *Self-Complementary Antennas: Principle of Self-Complementarity for Constant Impedance*, Springer, New York, 1996.
10 K. Itoh, D. K. Cheng, "A novel slots-and-monopole antenna with a steerable cardioid pattern," *IEEE Trans. Aerosp. Electron. Syst.*, vol. 8, no. 2, pp. 130–134, Mar. 1972.
11 P. E. Mayes, W. T. Warren, F. M. Wiesenmeyer, "The monopole-slot: A small broadband unidirectional antenna," *IEEE Trans. Antennas Propagat.*, vol. 20, no. 4, pp. 1092–1099, Jul. 1972.
12 A. Clavin, D. A. Huebner, F. J. Kilburg, "An improved element for use in array antennas," *IEEE Trans. Antennas Propagat.*, vol. 22, no. 4, pp. 521–526, Jul. 1974.
13 W. J. Surtees, "Mutual coupling of a slot with a dipole antenna," *Proc. IRE.*, vol. 40, no. 2, pp. 208–211, Feb. 1952.
14 R. W. P. King, G. H. Owyang, "The slot antenna with coupled dipoles," *IRE Trans. Antennas Propagat.*, vol. 8, no. 2, pp. 136–143, Mar. 1960.
15 Y. P. Zhang, "Impedance relations for differential antennas and single-ended counterparts," *IEEE Trans. Antennas Propagat.*, vol. 70, no. 2, pp. 953–959, Feb. 2022.
16 J. Shu, Y. P. Zhang, "An impedance relation and its application in broadband antenna design," *IEEE Trans. Antennas Propagat.*, vol. 70, no. 11, pp. 10286–10293, Nov. 2022.
17 J. M. Ding, M. Tang, Y. P. Zhang, "Theory of input impedance of a slot antenna with two parasitically excited wires," *IEEE Antennas Wirel. Propagat. Lett.*, Feb. 2024. http://dx.doi.org/10.1109/LAWP.2024.3369700.

18 J. M. Ding, M. Tang, Y. P. Zhang, "Theory of radiation characteristics of a slot antenna with two parasitically excited wires." *IEEE Trans. Antennas Propagat.*, submitted.

19 A. Clavin, "A new antenna feed having equal E- and H-plane patterns," *IRE Trans. Antennas Propagat.*, vol. 2, no. 3, pp. 113–119, Jul. 1954.

4

Differential Microstrip Patch Antennas

4.1 Introduction

Microstrip consists of a conducting strip separated from a ground plane by a dielectric substrate. It was first published by Grieg and Engelmann in December 1952 as the planar version of a parallel-wire transmission line [1]. Since then, it has been used in the microwave industry to construct compact, rugged, and inexpensive transmission lines and components.

As an open structure, microstrip inevitably radiates, and so it can be designed as an antenna. Although it is credited in many references that Deschamps and Sichak published the first microstrip antenna paper in 1953 [2], Gutton and Baissinot filed the first microstrip antenna patent in 1955 [3], but carefully read the paper and the patent, one would find that in fact neither of them gave the microstrip antenna structure and radiation mechanism that we know today [4]. The microstrip antenna in the title of the paper actually refers to the microstrip-fed antenna [2]. The patented microstrip antenna actually explores discontinuity to strengthen microstrip line radiation [3]. In addition, Denlinger published the research results of radiation of rectangular and circular microstrip resonators in 1969 [5]. He aimed to reduce radiation in order to achieve a high-quality factor of the resonator. If he did not intend to suppress the radiation of the microstrip resonator, but strengthened it intentionally, then the crown of the inventor of the microstrip antenna would be worn on his head.

In fact, the microstrip antenna structure was first published by Byron in 1970 [6]. Figure 4.1 shows the printed circuit antenna proposed by Byron [7]. It was realized by printing a circular conductor patch on a dielectric substrate with a conductor ground plane and feeding it through two coaxial cables. Observing its structure, the current flow on the circular conductor patch, and the distributions of the stray electric field in the near region and the radiated electric field in the far zone, we can conclude that it is the circular microstrip patch antenna that we recognize

Differential Antennas: Theory and Practice, First Edition. Yueping Zhang.

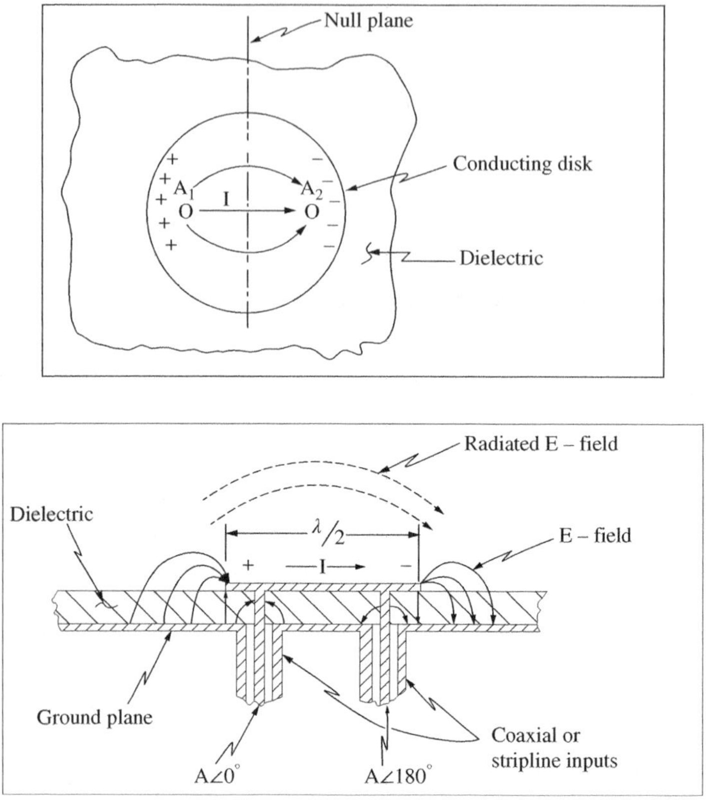

Figure 4.1 Illustration of the microstrip antenna structure proposed by Byron in 1970. *Source:* Cheston et al. [7]. (© 1970 IEEE).

today. It is interesting to note that the first microstrip patch antenna was also a differential microstrip patch antenna, which is the topic of this chapter.

The microstrip patch antenna concept aroused the attention of the antenna community, mainly due to the papers published, respectively by Munson and Howell from 1972 to 1974 [8–11]. Munson not only explained the radiation mechanism of rectangular microstrip patch antenna but also established a transmission line model for rectangular microstrip patch antenna. Munson also proposed and experimentally verified the concept of microstrip phased arrays [8, 9].

Howell designed single-feed linearly polarized rectangular, square, and circular microstrip patch antennas, and dual-feed circularly polarized square and circular microstrip patch antennas and published the experiment data. Howell also advocated the idea that the microstrip patch antenna can be treated as a resonant cavity with electric walls on the top and bottom, a magnetic wall on the periphery, and dielectric loading. Furthermore, Howell took the lead in exploring the radiation characteristics of the high-order modes of the microstrip patch antenna and the degree of mutual coupling between the two microstrip patch antennas [10, 11].

However, it was Munson's work that showed microstrip patch antennas a practical concept to solve many antenna system problems, gave birth to a new antenna industry, and thereby won him the reputation of the inventor of microstrip patch antennas [12, 13].

Byron, Howell, and Munson's papers on microstrip patch antennas blew a breath of fresh air into the antenna community. Soon we saw an avalanche of papers and books on microstrip patch antennas, which by the 1980s was in full flow and has been continued in the antenna field for over half a century. Compared with many papers on single-ended microstrip patch antennas, there were a few papers on differential microstrip patch antennas in the 1980s [14–18]. Nevertheless, they clearly show the advantage of differential microstrip patch antennas for suppressing higher-order modes, reducing cross-polarized radiation, removing pattern distortion, and producing an improved axial ratio (AR) for circular polarization. This chapter extends the cavity model theory to provide physical insight into the fundamentals of differential microstrip patch antennas. It also presents useful design guidelines for differential microstrip patch antennas. Furthermore, it compares differential with single-ended microstrip patch antennas to highlight unique properties of differential microstrip patch antennas. Section 4.2 describes the cavity model for analyzing differential microstrip patch antennas [19–21]. Sections 4.3–4.4 formulate the basic characteristics of differential rectangular and circular, and show how to design them, respectively [22–25]. Section 4.5 demonstrates how to design differential microstrip stacked patch antennas for enhanced impedance bandwidth [26]. Section 4.6 discusses differential microstrip patch arrays [15–18]. Section 4.7 gives some application examples [27–30]. Finally, Section 4.8 provides concluding remarks.

4.2 Cavity Model

The cavity model for microstrip patch antennas was developed and improved by Lo et al. [20, 21]. The model treats the region between the patch and the ground plane as a dielectric-loaded cavity bounded by the electric walls and a magnetic wall along the periphery of the patch. It has the advantage of being simple and providing physical insight. In this section, we describe the general formulation of the cavity model for the case of a microstrip patch antenna fed by dual coaxial probes. The formulation will be modified specifically in Sections 4.3–4.4 for differential rectangular and circular microstrip patch antennas.

4.2.1 Electrically Thin or Thick Substrate

For a dielectric substrate, the electrical thickness is defined as the ratio of the physical thickness to the wavelength in free space. If the ratio is less than 0.03, the dielectric substrate is electrically thin, while if the ratio is larger than 0.03, the dielectric substrate is electrically thick. The cavity model is applicable to analyze and design a microstrip patch antenna on an electrically thin substrate. While for a microstrip patch antenna on an electrically thick substrate, it requires a full-wave method such as the finite element method for analysis and design.

4.2.2 Fields in the Cavity

The geometry of the microstrip patch antenna and the coordinate system are shown in Figure 4.2. The patch on the surface of the grounded dielectric substrate of thickness h, dielectric constant ε_r, and loss tangent δ is fed by dual coaxial probes at (x_1, y_1) and (x_2, y_2). In the figure, C and A denote the periphery and area of the patch, respectively, and $\hat{n}$ is a unit vector, outer normal to the periphery. The cavity model is based on the following observations. (1) Electric field E has only the z

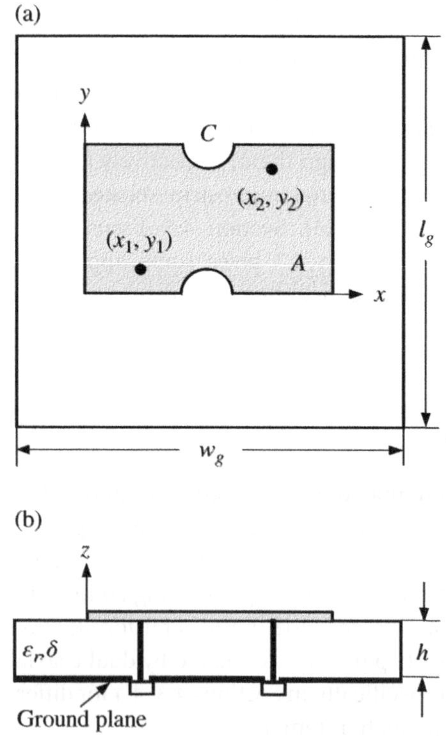

Figure 4.2 Geometry of the microstrip patch antenna and the coordinate system. (a) Top view and (b) side view.

component and magnetic field H has only the x and y components in the cavity. (2) Fields in the cavity do not vary with z. (3) The tangential component of H along the patch edge is negligible.

The electromagnetic fields excited with the z-directed current sources at (x_1, y_1) and (x_2, y_2) in the cavity must satisfy the following equations

$$(\nabla_t^2 + k^2)E_z = -j\omega\mu[J_z(x_1,y_1) + J_z(x_2,y_2)], \tag{4.1}$$

$$\boldsymbol{H} = \frac{j}{\omega\mu}\nabla_t \times (\hat{z}E_z) \tag{4.2}$$

and the boundary conditions, where ∇_t is the transverse component with respect to the z-axis of the Del operator, $\hat{z}$ is the unit vector in the z direction, and k is the wavenumber in the cavity. In equation (4.1), E_z can be expressed as

$$E_z = j\omega\mu \sum_{m=0}^{\infty}\sum_{n=0}^{\infty} \frac{1}{k^2 - k_{mn}^2} \frac{\langle [J_z(x_1,y_1) + J_z(x_2,y_2)]\psi_{mn}\rangle}{\langle \psi_{mn}\psi_{mn}\rangle}\psi_{mn} \tag{4.3}$$

where k_{mn} and ψ_{mn} are the eigenvalues and eigenfunctions of homogeneous equation (4.4), respectively

$$(\nabla_t^2 + k^2)E_z = 0, \tag{4.4}$$

$$\langle [J_z(x_1,y_1) + J_z(x_2,y_2)]\psi_{mn}\rangle = \iint [J_z(x_1,y_1) + J_z(x_2,y_2)]\psi_{mn}^* ds, \tag{4.5}$$

$$\langle \psi_{mn}\psi_{mn}\rangle = \iiint \psi_{mn}\psi_{mn}^* dv \tag{4.6}$$

where $\psi^*{}_{mn}$ is the complex conjugate of ψ_{mn}. The surface integration in (4.5) is over the equivalent current ribbons at the dual feeds and the volume integration in (4.6) is over the entire cavity. Forcing either $J_z(x_1, y_1)$ or $J_z(x_2, y_2)$ be zero in (4.1), (4.3), and (4.5), they degenerate as the expressions for the fields in the cavity for a single-ended microstrip patch antenna fed with a single coaxial probe.

It is noted from (4.3) that E_z would be infinity when the wavenumber k were real and equal to the eigenvalues k_{mn}. In the realistic case, k is complex due to dielectric loss,

$$k = k_0\sqrt{\varepsilon_r(1-j\delta)}, \tag{4.7}$$

E_z will not be infinite but quite large. Richard et al. found that k should be replaced with an effective wavenumber k_e for better agreement with experiment,

$$k_e = k_0\sqrt{\varepsilon_r(1-j\delta_e)} \tag{4.8}$$

where δ_e is the effective loss tangent that considers all losses. Now how to compute δ_e becomes the question, which will be answered in Section 4.2.5. This phenomenon of increased E_z is known as resonance. Hence, the fields in the cavity can also be thought as the superposition of the resonant modes ψ_{mn} of (4.4).

4.2.3 Radiation Fields

To calculate the radiation fields of the microstrip patch antenna, one can first obtain Huygen's magnetic current source on the magnetic wall of the cavity as

$$\boldsymbol{M} = 2\hat{n} \times E_z \tag{4.9}$$

where the factor of 2 accounts for the presence of the ground plane. The electric vector potential of $\boldsymbol{M}$ in the spherical coordinate system is

$$\boldsymbol{F}(\mathbf{r}) = \frac{\varepsilon_0}{4\pi}\int_C \frac{\boldsymbol{M}(\boldsymbol{r}')}{|\,\boldsymbol{r}-\boldsymbol{r}'\,|} e^{-jk_0|\boldsymbol{r}-\boldsymbol{r}'|} dl(\boldsymbol{r}') \tag{4.10}$$

and the radiation fields at $\boldsymbol{r}$ are

$$E_\theta = \eta_0 H_\phi = -j\omega\eta_0 F_\phi = j\omega\eta_0 (F_x \sin\phi - F_y \cos\phi), \tag{4.11}$$

$$E_\phi = -\eta_0 H_\theta = j\omega\eta_0 F_\theta = j\omega\eta_0 (F_x \cos\phi + F_y \sin\phi)\cos\theta \tag{4.12}$$

where the effect of the dielectric substrate on the radiation fields is neglected, due to the electrically thin substrate and the magnetic current source radiates in free space with impedance $\eta_0 = 120\pi\,\Omega$.

4.2.4 Polarization

A microstrip patch antenna can be designed to radiate single linearly or circularly polarized wave as well as dual linearly or circularly polarized waves. If the microstrip patch antenna is designed for dual polarizations, it needs dual ports, which are often intended to operate independently. Therefore, it is clear that the dual ports should be isolated. The isolation between the dual ports is defined as the ratio between the power incident upon one port and the power delivered to another port when it is terminated by a matched load. Good isolation will promise uncorrelated transmission of signals on both ports. An isolation of 30 dB can be realized between the dual ports of a microstrip patch antenna. If the microstrip patch antenna is designed for circular polarization, the AR is used to characterize it as a good or poor circularly polarized antenna. The AR is defined as the ratio between the minor and major axes of the polarization ellipse. The closer the AR to 0 dB, the better is the circularly polarized antenna. The 3-dB AR bandwidth is usually narrower than the 10-dB impedance bandwidth of the microstrip patch antenna.

When an antenna radiates a wave of desirable co-polarization, it will also radiate wave of undesirable cross-polarization. A purely polarized antenna will have low cross-polarized radiation. A measure of how the purely polarized antenna radiates is the cross-polarization level, which is quantified in a specified plane as the ratio of the maximum magnitude of E_{cpol} to the maximum magnitude of E_{xpol} as

$$XPL = 20\log\left|\frac{E_{cpol}}{E_{xpol}}\right| \tag{4.13}$$

where E_{cpol} and E_{xpol} are calculated according to the third definition by Ludwig [31] as

$$E_{cpol} = E_\theta \cos\phi - E_\phi \sin\phi, \tag{4.14}$$

$$E_{xpol} = E_\theta \sin\phi + E_\phi \cos\phi. \tag{4.15}$$

The major factors to cause radiation of cross-polarization by a microstrip patch antenna are excitation of higher-order modes and spurious emission of coaxial probes.

4.2.5 Directivity and Radiation Efficiency

Directivity is a fundamental antenna parameter. It is related to the electrical size of the antenna. The smaller the electrical size, the lower is the antenna directivity. The smallest directivity would be 1 or 0 dBi. The directivity of an antenna is defined as

$$D = \frac{\frac{1}{2}\frac{|E_\theta|^2}{120\pi}}{\frac{P_r}{4\pi r^2}} = \frac{|E|^2 r^2}{60P_r} \tag{4.16}$$

where E is the electric field strength at r in the direction of the maximum radiation and P_r is the radiated power and can be calculated as

$$\begin{aligned} P_r &= \frac{1}{2}\,\mathrm{Re}\int_0^{\pi/2}\int_0^{2\pi}\left(E_\theta H_\phi^* - E_\phi H_\theta^*\right) r^2 \sin\theta\, d\phi d\theta \\ &= \frac{1}{2\eta_0}\int_0^{\pi/2}\int_0^{2\pi}\left(|E_\theta|^2 + |E_\phi|^2\right) r^2 \sin\theta\, d\phi d\theta. \end{aligned} \tag{4.17}$$

The directivity of a microstrip patch antenna is typically several dBi.

Radiation efficiency is another fundamental antenna parameter, which measures how efficiently an antenna can convert the power fed to the antenna to the power radiated to free space. The radiation efficiency of the microstrip patch antenna is defined as

$$e = \frac{P_r}{P_r + P_c + P_d + P_{sw}} \tag{4.18}$$

where P_c and P_d are dissipated power due to conductor and dielectric losses and P_{sw} is power carried by the surface wave. P_{sw} is found to be less significant for microstrip patch antennas on electrically thin substrates. For a single-ended microstrip patch antenna with a single coaxial probe feed, P_{sw} can be estimated with the analytical formula given in [32] as

$$P_{sw} = \frac{3.4\frac{h}{\lambda}\sqrt{\varepsilon_r - 1}}{1 - 3.4\frac{h}{\lambda}\sqrt{\varepsilon_r - 1}} P_r = \alpha_s P_r \tag{4.19}$$

where the denominator of course should be larger than 0. For a differential microstrip patch antenna with dual feeds of coaxial probes, P_{sw} is smaller and can be estimated with the method given in [33]. P_c and P_d are respectively written as

$$P_c = \omega\mu \frac{\delta_s}{2h} \iint_s \left(|H_x|^2 + |H_y|^2\right) ds, \tag{4.20}$$

$$P_d = \frac{\omega\delta\varepsilon_r\varepsilon_0}{2} \iiint_v \left(|E_z|^2\right) dv \tag{4.21}$$

where δ_s is the skin depth of the patch and ground plane. The radiation efficiency of a microstrip patch antenna can be as high as 90%.

As suggested previously, all losses P_r, P_{sw}, P_c, and P_d should be lumped into a single effective dielectric loss with effective loss tangent δ_e. Lee et al. showed that δ_e can be found by an iterative procedure

$$\delta_{i+1} = \delta_i \frac{[P_r(\delta_i) + P_{sw}(\delta_i) + P_c(\delta_i) + P_d(\delta_i)]}{P_d(\delta_i)} \tag{4.22}$$

where the initial value is $\delta_0 = \delta$ [34]. The iterative process continues until the test $(\delta_{i+1} - \delta_i)/\delta_i$ is smaller than a percentage, which is usually chosen to be 2%.

4.2.6 Input Impedances

The microstrip patch antenna has dual feeds, which can be considered two single-ended input ports. The two single-ended input ports can also be configured into one differential or one common-mode input port. Hence, there are different types of input impedances for the microstrip patch antenna with dual feeds. They can be derived from the following impedance matrix equation

$$\begin{bmatrix} V_1 \\ V_2 \end{bmatrix} = \begin{bmatrix} Z_{11} & Z_{12} \\ Z_{21} & Z_{22} \end{bmatrix} \begin{bmatrix} I_1 \\ I_2 \end{bmatrix} \tag{4.23}$$

where V_1 and V_2 are the voltages across ports 1 and 2, I_1 and I_2 are the currents into ports 1 and 2, Z_{11} and Z_{22} are the self-impedances at ports 1 and 2, respectively, and

Z_{12} and Z_{21} are the mutual impedances between ports 1 and 2. Z_{11} and Z_{12} can be calculated, respectively, as

$$Z_{11} = -\frac{E_{av}(x_1, y_1)h}{I_1}, \tag{4.24}$$

$$Z_{12} = -\frac{E_{av}(x_1, y_1)h}{I_2} \tag{4.25}$$

where E_{av} is the average value of the electric field at feed 1. Similarly, Z_{22} and Z_{21} can be calculated. The differential and common-mode input impedances are given by

$$Z_{dif} = Z_{11} - Z_{12} - Z_{21} + Z_{22}, \tag{4.26}$$

$$Z_{com} = \frac{1}{2}(Z_{11} + Z_{12} + Z_{21} + Z_{22}). \tag{4.27}$$

The single-ended input impedances are given by

$$Z_{sig1} = Z_{11} - \frac{Z_{12}Z_{21}}{Z_L + Z_{22}}, \tag{4.28}$$

$$Z_{sig2} = Z_{22} - \frac{Z_{21}Z_{12}}{Z_L + Z_{11}} \tag{4.29}$$

where Z_L represents the load impedance that can be changed for tuning, matching, and modifying radiation patterns. If Z_L is infinitely small, that is, the other feed is short-circuited, equations (4.28) and (4.29) simplify to

$$Z_{sig1} = Z_{11} - \frac{Z_{12}Z_{21}}{Z_{22}}, \tag{4.30}$$

$$Z_{sig2} = Z_{22} - \frac{Z_{21}Z_{12}}{Z_{11}}, \tag{4.31}$$

while if Z_L is infinitely large, that is, the other feed is open-circuited, equations (4.28) and (4.29) simplify to

$$Z_{sig1} = Z_{11}, \tag{4.32}$$

$$Z_{sig2} = Z_{22}. \tag{4.33}$$

If the microstrip patch antenna is symmetric, (4.26), (4.29), (4.32), and (4.33) reduce to

$$Z_{dif} = 2(Z_{11} - Z_{12}) = 2(Z_{22} - Z_{21}), \tag{4.34}$$

$$Z_{com} = Z_{11} + Z_{12} = Z_{22} + Z_{21}, \tag{4.35}$$

$$Z_{sig1} = Z_{sig2} = Z_{sig} = Z_{11}. \tag{4.36}$$

It is interesting to examine the relationships between Z_{dif} and Z_{com} with Z_{sig} of the microstrip patch antenna. We show that there exist simple relationships between them [35]. Quantitative relations can be deduced as follows

$$\frac{Z_{dif}}{Z_{sig}} = \begin{cases} 2, & for \quad Z_{12} = 0 \\ 4, & for \quad Z_{12} = -Z_{11} \\ 0, & for \quad Z_{12} = Z_{11} \end{cases} \tag{4.37}$$

and

$$\frac{Z_{com}}{Z_{sig}} = \begin{cases} 1, & for \quad Z_{12} = 0 \\ 2, & for \quad Z_{12} = Z_{11} \\ 0, & for \quad Z_{12} = -Z_{11} \end{cases}, \tag{4.38}$$

which reveal that one impedance can be determined from the other whose solution is already known. Also, it should be mentioned that the patch size in the above derivation remains the same for both differential and single-ended operations. Hence, the size ratio is 1. Actually, the size ratio can be larger than 1, which will become clear when we deal with, for example, rectangular microstrip patch antennas.

Now, it is worthwhile recalling the important ratio of 2 : 1, which was commonly believed to be true between a differential structure and its single-ended counterpart. It is indeed true for differential and single-ended wire antennas, slot antennas, and BJT amplifiers as discussed in Chapters 1, 2, and 3. However, (4.37) clearly shows that it is untrue between the differential and single-ended microstrip patch antennas.

4.2.7 Reflection Coefficients

As discussed earlier, the microstrip patch antenna with dual feeds has single-ended, differential-mode, and common-mode input impedances, which leads to different types of reflection coefficients. For the single-ended input impedance, the reflection coefficient is the standard S_{11} as given by

$$S_{11} = \frac{Z_{sig} - Z_0}{Z_{sig} + Z_0} \tag{4.39}$$

where Z_0 is the characteristic impedance, which typically is 50 Ω. For the differential input impedance, the corresponding reflection coefficient is

$$S_{d11} = \frac{Z_{dif} - Z_{d0}}{Z_{dif} + Z_{d0}} \tag{4.40}$$

where Z_{d0} typically is $2Z_0$. For the common-mode input impedance, the resultant reflection coefficient is

$$S_{c11} = \frac{Z_{com} - Z_{c0}}{Z_{com} + Z_{c0}} \tag{4.41}$$

where Z_{c0} typically is $0.5Z_0$.

4.2.8 Resonance and Electrical Separation

The microstrip patch antenna is a resonant antenna that supports a great number of resonant modes. The resonant frequencies of resonant modes are given by

$$f_{mn} = \frac{k_{mn}}{2\pi\sqrt{\mu\epsilon}} \tag{4.42}$$

where the eigenvalues k_{mn} are calculated with the patch edge extended slightly to account for the fringing field. Generally, the microstrip patch antenna is designed to operate near the resonant frequency of the dominant mode. It is found that the resonance of a differential microstrip patch antenna depends on the electrical separation of the dual-probe feeds. The electrical separation is defined as the ratio of the distance between the dual-probe feeds to the wavelength in free space, which is similar to what we define as the electrical thickness. Only if the electrical separation is large enough, the input reactance can be zero, and the resonance will occur. Thus, the electrical separation is an important parameter in the design of a differential microstrip patch antenna [36].

4.2.9 Quality Factor and VSWR Bandwidth

The quality factor or Q factor is a dimensionless parameter that is defined as the ratio of the peak energy stored in the resonator in a cycle of oscillation to the energy lost per radian of the cycle. For the microstrip patch antenna, Q factor consists of four components, which are given by

$$\frac{1}{Q} = \frac{1}{Q_r} + \frac{1}{Q_d} + \frac{1}{Q_c} + \frac{1}{Q_{sw}} \tag{4.43}$$

where Q_r, Q_d, Q_c, and Q_{sw} are quality factors due to the radiation, dielectric, conductor, and surface wave losses, respectively. Q_r will be given when we discuss the microstrip patch antenna with a specific regular shape. Q_{sw} is very large for a single-ended microstrip patch antenna on an electrically thin substrate, which will become even larger for the differential counterpart. Hence, Q_{sw} is far less significant for the total Q. Q_d and Q_c are given by

$$Q_d = \frac{1}{\delta}, \tag{4.44}$$

$$Q_c = \pi h\sqrt{\frac{120\sigma}{\lambda}} \tag{4.45}$$

where σ is the conductivity of the patch and ground plane.

Q factor is a fundamental parameter because it is related to both radiation and impedance characteristics. The radiation efficiency of the microstrip patch antenna defined in equation (4.18) can also be calculated as

$$e = \frac{Q}{Q_r}. \tag{4.46}$$

The impedance bandwidth of the microstrip patch antenna is defined according to the value of the voltage standing wave ratio (VSWR) that can be tolerated. If the value of VSWR is less than S, the usable bandwidth of the antenna is related to Q factor by

$$BW = \frac{1}{Q}\frac{S-1}{\sqrt{S}}. \tag{4.47}$$

S is usually chosen to be 2 for determining the VSWR bandwidth.

4.3 Rectangular Patch

The rectangular microstrip patch antenna is the most widely used microstrip patch antenna. Figure 4.3 shows the geometry and coordinate system of the rectangular microstrip patch antenna with dual feeds of coaxial probes. It is seen that the

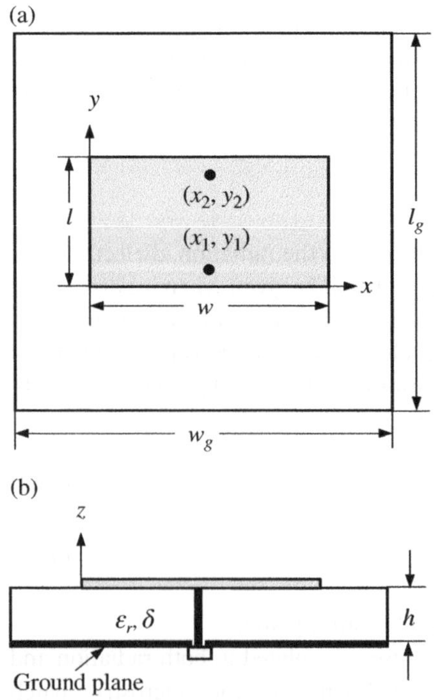

Figure 4.3 Geometry of the rectangular microstrip patch antenna and the coordinate system. (a) Top view and (b) side view.

rectangular microstrip patch antenna has dimensions w along the x-axis and l along the y-axis. Note that l is the resonant dimension, or the direction of current flow and the width w is chosen to be larger than l to get a wider bandwidth. The dual feeds of coaxial probes are located at points (x_1, y_1) and (x_2, y_2), respectively. The feeds along the center-line at $x_1 = x_2 = w/2$ and $y_2 = l - y_1$ are the most common, which help minimize higher-order modes and cross-polarized radiation.

4.3.1 Resonant Modes

For the rectangular microstrip patch antenna, it can be modelled as a rectangular cavity. The eigenfunctions and eigenvalues of the cavity can be written, respectively, as

$$\psi_{mn}(x, y) = \cos\left(\frac{m\pi x}{w_e}\right)\cos\left(\frac{n\pi y}{l_e}\right), \tag{4.48}$$

$$k_{mn}^2 = \left(\frac{m\pi}{w_e}\right)^2 + \left(\frac{n\pi}{l_e}\right)^2 \tag{4.49}$$

where m and n are the mode indices in the x and y directions, respectively. In (4.48) and (4.49), w_e and l_e are the effective values of w and l, taking into account the fringing fields at the edges of the rectangular microstrip patch [37]. l_e may be approximated as follows.

$$l_e = l + \left(\frac{w_{eq} - w}{2}\right)\frac{\varepsilon_e(w) + 0.3}{\varepsilon_e(w) - 0.258} \tag{4.50}$$

where $\varepsilon_e(w)$ is the effective dielectric constant and w_{eq} is the equivalent width calculated from the planar waveguide model. $\varepsilon_e(w)$ can be calculated as

$$\varepsilon_e(w) = \frac{\varepsilon_r + 1}{2} + \frac{\varepsilon_r - 1}{2}\left(1 + 10\frac{h}{w}\right)^{-1/2} \tag{4.51}$$

and w_{eq} as

$$w_{eq} = \frac{120\pi h}{Z(w)\sqrt{\varepsilon_e(w)}} \tag{4.52}$$

where $Z(w)$, the impedance of a microstrip line of width w and thickness h, is given by

$$Z(w) = \frac{60\pi}{\sqrt{\varepsilon_r}}\left\{\frac{w}{2h} + 0.441 + 0082\left[\frac{\varepsilon_r - 1}{\varepsilon_r^2}\right] + \frac{\varepsilon_r + 1}{2\pi\varepsilon_r}\left[1.415 + \ln\left(\frac{w}{2h} + 0.94\right)\right]\right\}^{-1}. \tag{4.53}$$

Similarly, w_e can be calculated by replacing l, w_{eq}, w, $Z(w)$, and $\varepsilon_e(w)$ in equations (4.50), (4.51), (4.52), and (4.53) with w, l_{eq}, l, $Z(l)$, and $\varepsilon_e(l)$, respectively.

The dual feeds of coaxial probes can be treated as z-directed current ribbons centered at the feed points, respectively, as

$$J_z(x_1,y_1) = \begin{cases} \frac{I}{d_e}, & x_1 - \frac{d_e}{2} < x < x_1 + \frac{d_e}{2},\ y = y_1 \\ 0, & elsewhere \end{cases}, \tag{4.54}$$

$$J_z(x_2,y_2) = \begin{cases} -\frac{I(1-\alpha)\cos\beta}{d_e}, & x_2 - \frac{d_e}{2} < x < x_2 + \frac{d_e}{2},\ y = y_2 \\ 0, & elsewhere \end{cases} \tag{4.55}$$

where I is the current to the feed, and d_e is the effective probe diameter of the feed. Usually, d_e is several times larger than d and can be chosen to obtain good agreement between the theoretical and experimental impedances [20]. In (4.55), α and β represent the amplitude and phase imbalances at feed 2 with respect to those at feed 1. Ideally, $\alpha = 0$ and $\beta = 0°$. Practically, α can be up to 0.1 and β to 10° [23]. Substituting (4.48), (4.54), and (4.55) into (4.5) yields

$$\begin{aligned}
&\langle [J_z(x_1,y_1) + J_z(x_2,y_2)]\psi_{mn}(x,y)\rangle \\
&= \frac{I}{d_e}\left[\cos\left(\frac{n\pi y_1}{l_e}\right)\int_{x_1-\frac{d_e}{2}}^{x_1+\frac{d_e}{2}}\cos\left(\frac{m\pi x}{w_e}\right)dx - (1-\alpha)\cos\beta\cos\left(\frac{n\pi y_2}{l_e}\right)\int_{x_2-\frac{d_e}{2}}^{x_2+\frac{d_e}{2}}\cos\left(\frac{m\pi x}{w_e}\right)dx\right] \\
&= I\left[\cos\left(\frac{m\pi x_1}{w_e}\right)\cos\left(\frac{n\pi y_1}{l_e}\right) - (1-\alpha)\cos\beta\cos\left(\frac{m\pi x_2}{w_e}\right)\cos\left(\frac{n\pi y_2}{l_e}\right)\right]j_0\left(\frac{m\pi d_e}{2w_e}\right)
\end{aligned} \tag{4.56}$$

where $j_0(x)$ is the Sinc function given by

$$j_0(x) = \frac{\sin x}{x}. \tag{4.57}$$

For the rectangular microstrip patch antenna with dual feeds of coaxial probes located at the most common positions and under an ideal differential excitation, (4.56) simplifies to

$$\begin{aligned}
&\langle [J_z(x_1,y_1) + J_z(x_2,y_2)]\psi_{mn}(x,y)\rangle \\
&= I\cos\left(\frac{m\pi x_1}{w_e}\right)\cos\left(\frac{n\pi y_1}{l_e}\right)[1 - \cos(n\pi)]j_0\left(\frac{m\pi d_e}{2w_e}\right).
\end{aligned} \tag{4.58}$$

Similarly, substituting (4.48) into (4.6) yields

$$\langle \psi_{mn}(x,y)\psi_{mn}(x,y)\rangle = \int_0^{w_e}\cos^2\left(\frac{m\pi x}{w_e}\right)dx\int_0^{l_e}\cos^2\left(\frac{m\pi y}{l_e}\right)dy = \frac{w_e l_e}{\epsilon_{0m}\epsilon_{0n}} \tag{4.59}$$

where $\epsilon_{0m} = 1$ or 2 when m is or is not 0, respectively, so is ϵ_{0n}.

The vertical electric field E_z given in (4.3) can be specifically written for the rectangular microstrip patch cavity as

$$E_z = j\omega\mu \sum_{m=0}^{\infty}\sum_{n=0}^{\infty} \frac{1}{k_e^2 - k_{mn}^2} \frac{\langle [J_z(x_1, y_1) + J_z(x_2, y_2)]\psi_{mn}(x, y)\rangle}{\langle \psi_{mn}\psi_{mn}\rangle}\psi_{mn}(x, y)$$
$$= \sum_{m=0}^{\infty}\sum_{n=0}^{\infty} B_{mn}[1 - \cos(n\pi)]\cos\left(\frac{m\pi x}{w_e}\right)\cos\left(\frac{n\pi y}{l_e}\right) \tag{4.60}$$

where

$$B_{mn} = j\omega\mu I \frac{\sqrt{\frac{\epsilon_{0m}\epsilon_{0n}}{w_e l_e}}}{k_e^2 - k_{mn}^2}\cos\left(\frac{m\pi x_1}{w_e}\right)\cos\left(\frac{n\pi y_1}{l_e}\right) j_0\left(\frac{m\pi d_e}{2w_e}\right). \tag{4.61}$$

According to (4.2), the vertical electric field E_z will be accompanied by the horizontal magnetic field components H_x and H_y. They can be expressed, respectively, as

$$H_x = -\frac{j}{\omega\mu}\sum_{m=0}^{\infty}\sum_{n=0}^{\infty} B_{mn}[1 - \cos(n\pi)]\left(\frac{n\pi}{l_e}\right)\cos\left(\frac{m\pi x}{w_e}\right)\cos\left(\frac{n\pi y}{l_e}\right), \tag{4.62}$$

$$H_y = \frac{j}{\omega\mu}\sum_{m=0}^{\infty}\sum_{n=0}^{\infty} B_{mn}[1 - \cos(n\pi)]\left(\frac{m\pi}{w_e}\right)\cos\left(\frac{m\pi x}{w_e}\right)\cos\left(\frac{n\pi y}{l_e}\right). \tag{4.63}$$

As the magnetic field is transverse to the z-axis and the resonant modes are termed TM_{mn} modes. It should be noted from (4.60) that the cancellation mechanism exists in the differential rectangular microstrip patch antenna, which can be explored to suppress some TM_{mn} modes to reduce cross-polarized radiation. For instance, the TM_{01} mode can be excited because $[1 - \cos(n\pi)] = 2$, while the TM_{02} mode cannot be excited because $[1 - \cos(n\pi)] = 0$. Likewise, the TM_{10} mode can be excited, while the TM_{20} mode cannot be excited. The first two TM_{01} and TM_{10} modes are particularly important because either is generally used as the operating mode. For example, if the differential rectangular microstrip patch antenna uses the TM_{01} mode, it is excited at the resonant frequency of the TM_{01} mode. The amplitude of the TM_{01} mode is the largest, and the TM_{01} mode is the dominant mode.

4.3.2 Resonant Frequencies

The resonant frequencies of the rectangular microstrip patch cavity are given by

$$f_{mn} = \frac{c}{2\pi\sqrt{\varepsilon_r}}\sqrt{\left(\frac{m\pi}{w_e}\right)^2 + \left(\frac{n\pi}{l_e}\right)^2}. \tag{4.64}$$

It was found that the calculation of the resonant frequency based on the above formula for the fundamental mode is accurate to within 3% of measured values.

4.3.3 Radiation Characteristics

The radiation of the differential rectangular microstrip patch antenna can be considered from the magnetic current flowing around the cavity walls at $x = 0$, w and $y = 0$, l. The magnetic current is obtained by substituting (4.56) into (4.9) as

$$\boldsymbol{M} = \hat{n} \times E_z = \hat{n} \times \sum_{m=0}^{\infty} \sum_{n=0}^{\infty} B_{mn}[1 - \cos(n\pi)] \cos\left(\frac{m\pi x}{w_e}\right) \cos\left(\frac{n\pi y}{l_e}\right). \tag{4.65}$$

At $x = 0$,

$$\boldsymbol{M} = -\hat{y} \sum_{m=0}^{\infty} \sum_{n=0}^{\infty} B_{mn}[1 - \cos(n\pi)] \cos\left(\frac{n\pi y}{l_e}\right). \tag{4.66}$$

At $x = w_e$,

$$\boldsymbol{M} = \hat{y} \sum_{m=0}^{\infty} \sum_{n=0}^{\infty} B_{mn}[1 - \cos(n\pi)](-1)^m \cos\left(\frac{n\pi y}{l_e}\right). \tag{4.67}$$

At $y = 0$,

$$\boldsymbol{M} = -\hat{x} \sum_{m=0}^{\infty} \sum_{n=0}^{\infty} B_{mn}[1 - \cos(n\pi)] \cos\left(\frac{m\pi x}{w_e}\right). \tag{4.68}$$

At $y = l_e$,

$$\boldsymbol{M} = \hat{x} \sum_{m=0}^{\infty} \sum_{n=0}^{\infty} B_{mn}[1 - \cos(n\pi)](-1)^n \cos\left(\frac{m\pi x}{w_e}\right). \tag{4.69}$$

For the TM_{01} mode, the magnetic currents along the edges at $y = 0$ and l are constant and in phase, while those along the edges at $x = 0$ and w vary cosinusoidally and are out of phase. For this reason, the edges at $y = 0$ and l are known as radiating edges because the magnetic currents along the edges contribute to radiation predominantly, while the edges at $x = 0$ and w are known as nonradiating edges. Similarly, for the TM_{10} mode, the edges at $x = 0$ and w are radiating edges, while the edges at $y = 0$ and l are nonradiating edges.

4.3.3.1 Radiated Fields

The electric vector potential of the magnetic current can be calculated by evaluating the resultant integral of (4.10) as

$$\boldsymbol{F} = \frac{\varepsilon_0 h e^{-jk_0 r}}{2\pi r}\int \boldsymbol{M} e^{jk_0(x\sin\theta\cos\phi + y\sin\theta\sin\phi)}dxdy \tag{4.70}$$

where a factor of 2 has been included to take into account the image effect of the ground plane. Substituting (4.66), (4.67), (4.68), and (4.69) into (4.70), we have

$$\begin{aligned}\boldsymbol{F} = &\frac{\varepsilon_0 h e^{-jk_0 r}}{2\pi r}\sum_{m=0}^{\infty}\sum_{n=0}^{\infty} B_{mn}[1-\cos(n\pi)]\Big\{\hat{x}\left[1-(-1)^n e^{jk_0 l_e\sin\theta\sin\phi}\right]\\ &\int_0^{w_e}\cos\left(\frac{m\pi x}{w_e}\right)e^{jk_0 x\sin\theta\cos\phi}dx + \hat{y}\left[1-(-1)^m e^{jk_0 w_e\sin\theta\cos\phi}\right]\\ &\int_0^{l_e}\cos\left(\frac{n\pi y}{l_e}\right)e^{jk_0 y\sin\theta\sin\phi}dy\Big\}.\end{aligned} \tag{4.71}$$

Finding the integrations in (4.71), we obtain

$$\begin{aligned}\boldsymbol{F} = &\frac{2\varepsilon_0 h e^{-jk_0 r}}{\pi r}\sum_{m=0}^{\infty}\sum_{n=0}^{\infty} B_{mn}[1-\cos(n\pi)]e^{j\left(\frac{u+m\pi}{2}+\frac{v+n\pi}{2}\right)}\\ &\sin\left(\frac{u+m\pi}{2}\right)\sin\left(\frac{v+n\pi}{2}\right)\left\{-\hat{x}\left[\frac{w_e u}{u^2-(m\pi)^2}\right]+\hat{y}\left[\frac{l_e v}{v^2-(n\pi)^2}\right]\right\}\end{aligned} \tag{4.72}$$

where

$$u = k_0 w_e\sin\theta\cos\phi, \tag{4.73}$$

$$v = k_0 l_e\sin\theta\sin\phi. \tag{4.74}$$

Thus, the radiation fields in the far zone at r are evaluated from (4.11) and (4.12) as

$$\begin{aligned}E_\theta &= \eta_0 H_\phi = j\omega\eta_0\left(F_x\sin\phi - F_y\cos\phi\right)\\ &= \frac{4k_0 e^{-jk_0 r}}{\lambda r}\sum_{m=0}^{\infty}\sum_{n=0}^{\infty} V_{mn}e^{j\left(\frac{u+m\pi}{2}+\frac{v+n\pi}{2}\right)}\sin\left(\frac{u+m\pi}{2}\right)\sin\left(\frac{v+n\pi}{2}\right)\\ &\left\{\left[\frac{w_e^2}{u^2-(m\pi)^2}\right]+\left[\frac{l_e^2}{v^2-(n\pi)^2}\right]\right\}\sin\theta\sin\phi\cos\phi,\end{aligned} \tag{4.75}$$

$$\begin{aligned}E_\phi &= -\eta_0 H_\theta = j\omega\eta_0\left(F_x\cos\phi + F_y\sin\phi\right)\cos\theta\\ &= \frac{4k_0 e^{-jk_0 r}}{\lambda r}\sum_{m=0}^{\infty}\sum_{n=0}^{\infty} V_{mn}e^{j\left(\frac{u+m\pi}{2}+\frac{v+n\pi}{2}\right)}\sin\left(\frac{u+m\pi}{2}\right)\sin\left(\frac{v+n\pi}{2}\right)\\ &\left\{\left[\frac{w_e^2\cos^2\phi}{u^2-(m\pi)^2}\right]+\left[\frac{l_e^2\sin^2\phi}{v^2-(n\pi)^2}\right]\right\}\sin\theta\cos\theta\end{aligned} \tag{4.76}$$

where V_{nm} is the voltage difference between the dual feeds at (x_1, y_1) and (x_2, y_2), which is given by

$$V_{mn} = hB_{mn}[1 - \cos(n\pi)]. \tag{4.77}$$

Equations (4.75), (4.76), and (4.77) reveal that the radiation of the TM_{mn} mode reaches the peak in the broadside direction when n is an odd number and the radiation of the TM_{mn} mode is suppressed in all directions when n is an even number due to the cancellation mechanism introduced by the differential excitation.

For the dominant TM_{01} mode, (4.75) and (4.76) simplify, respectively, to

$$E_\theta = j\frac{4k_0 e^{-jk_0 r}}{\lambda r} V_{01} e^{j\left(\frac{u+v}{2}\right)} \sin\left(\frac{u}{2}\right) \cos\left(\frac{v}{2}\right) \left\{ \left[\frac{w_e^2}{u^2}\right] + \left[\frac{l_e^2}{v^2 - \pi^2}\right] \right\} \sin\theta \sin\phi \cos\phi, \tag{4.78}$$

$$E_\phi = j\frac{4k_0 e^{-jk_0 r}}{\lambda r} V_{01} e^{j\left(\frac{u+v}{2}\right)} \sin\left(\frac{u}{2}\right) \cos\left(\frac{v}{2}\right) \left\{ \left[\frac{w_e^2 \cos^2\phi}{u^2}\right] + \left[\frac{l_e^2 \sin^2\phi}{v^2 - \pi^2}\right] \right\} \sin\theta \cos\theta. \tag{4.79}$$

For the E plane ($\phi = 90°$), equations (4.78) and (4.79) reduce, respectively, to

$$E_\theta = j\frac{2we^{-jk_0 r}}{\lambda r} V_{01} e^{j\left(\frac{k_0 l_e}{2}\sin\theta\right)} \cos\left(\frac{k_0 l_e}{2}\sin\theta\right), \tag{4.80}$$

$$E_\phi = 0. \tag{4.81}$$

For the H plane ($\phi = 0°$), equations (4.78) and (4.79) reduce, respectively, to

$$E_\theta = 0, \tag{4.82}$$

$$E_\phi = j\frac{2we^{-jk_0 r}}{\lambda r} V_{01} e^{j\left(\frac{k_0 w_e}{2}\sin\theta\right)} \frac{\sin\left(\frac{k_0 w_e}{2}\sin\theta\right)}{\frac{k_0 w_e}{2}\sin\theta} \cos\theta. \tag{4.83}$$

Figure 4.4 shows the radiation patterns in the E and H planes of a differential rectangular microstrip patch antenna with w = 60 mm and l = 40 mm on an RT/duroid 5880 substrate ($\varepsilon_r = 2.2$, $\delta = 0.0009$) at the resonant frequency 2.42 GHz of the dominant TM_{01} mode. The substrate size is $w_g \times l_g \times h = 125 \times 125 \times 0.787\ \text{mm}^3$. The diameters of the probes and openings are 0.5 and 2 mm, respectively. The two probes are separated by 15 mm. Note that the radiation of the dominant TM_{01} mode indeed reaches the peak in the broadside direction. The agreement between the calculated and simulated radiation patterns is acceptable in the forward region. The difference in the side and back lobe regions is because the radiation patterns are calculated for an infinite ground plane but they are simulated on a finite ground plane.

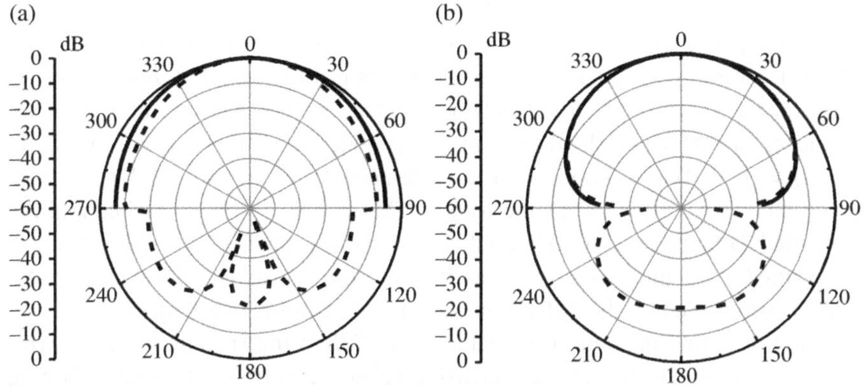

Figure 4.4 Normalized radiation patterns for the differential rectangular microstrip patch antenna. (a) The *E* plane and (b) the *H* plane.

4.3.3.2 Cross Polarization

A higher-order mode orthogonal to the operating dominant mode is a major source of cross-polarized radiation. The differential rectangular microstrip patch antenna operates in the dominant TM_{01} mode and suppresses the orthogonal and significant higher-order TM_{20} mode. Hence, it will have low cross-polarized radiation [22].

It is well known that the *XPL* is minimum in the diagonal (or *D*) planes $\phi = \pm 45°$ of a rectangular microstrip patch antenna. Hence, it has been recommended to use the *D* planes in any *XPL* discussions for microstrip patch antennas [38]. Figure 4.5

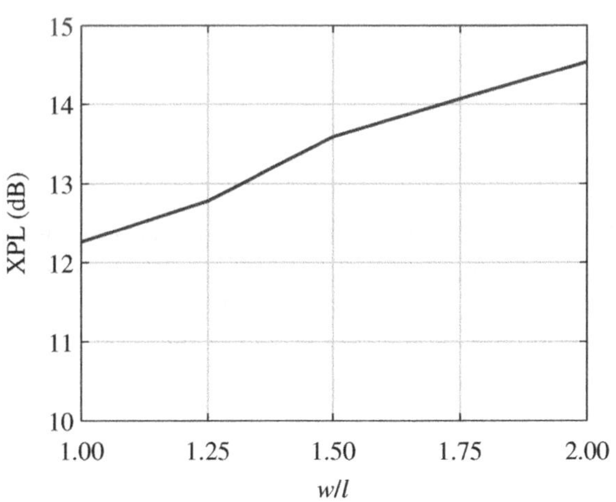

Figure 4.5 XPL in the *D* plane as a function of *w*/*l* for the differential rectangular microstrip patch antenna.

shows the calculated *XPL* in the *D* plane $\phi = 45°$ for a differential rectangular microstrip patch with $1 \leq w/l \leq 2$ as an independent variable where l is 25 mm. The differential rectangular microstrip patch is printed on RT/duroid 5880 substrate ($\varepsilon_r = 2.2$, $\delta = 0.0009$, $h = 0.787$ mm). It is designed to operate at the resonant frequency of 3.8 GHz of the dominant TM_{01} mode. The dual feeds are located at $(x_1, y_1) = (w/2, 0)$ and $(x_2, y_2) = (w/2, l)$, respectively. As shown, *XPL* gets larger as w/l increases. For $w/l = 1.0$, the wanted TM_{01} mode is strongly excited and the unwanted TM_{10} mode is weakly excited, leading to a smaller XPL. For $w/l = 2.0$, the wanted TM_{01} mode is strongly excited and the unwanted higher-order TM_{20} mode is suppressed by the differential excitation, resulting in a larger XPL.

In addition, the calculations reveal that the *XPL* in the *D* plane increases with decreasing substrate thickness, increasing substrate permittivity, and enlarging the separation of dual feeds for differential rectangular microstrip patch antennas.

4.3.3.3 Radiation Conductance and Directivity

The radiated power can be calculated by substituting equations (4.78) and (4.79) into (4.17), which, however, requires a numerical integration. Thouroude et al. made some approximations [39], which enables an analytical integration to calculate the radiated power as

$$P_r = \frac{1}{2\eta_0}\int_0^{\pi/2}\int_0^{2\pi}\left(|E_\theta|^2 + |E_\phi|^2\right)r^2 \sin\theta\, d\phi d\theta \approx \frac{A\pi^4}{46\,080}|V_{01}|^2 \left[(1-B)\left(1-\frac{A}{15}+\frac{A^2}{420}\right)+\frac{B^2}{5}\left(1-\frac{A}{7}+\frac{A^2}{189}\right)\right] \tag{4.84}$$

where

$$A = \left(\frac{\pi w}{\lambda}\right)^2, \tag{4.85}$$

$$B = \left(\frac{2l}{\lambda}\right)^2. \tag{4.86}$$

It was found that equation (4.84) predicts the value of radiated power with an error less than 2.5% for $l/\lambda = 0.3$ with a limit of $w/\lambda = 0.6$ and 4% for $l/\lambda = 0.15$ with a limit of $w/\lambda = 0.3$. It should be noted that the radiation fields of the dominant mode were used to replace the summation of all modes in the above calculation of the radiated power. This approximation is acceptable because the dominant mode contributes most significantly to the radiated power. The radiation conductance is then

$$G_r = \frac{2P_r}{|V_{01}|^2} \approx \frac{A\pi^4}{23\,040}\left[(1-B)\left(1-\frac{A}{15}+\frac{A^2}{420}\right)+\frac{B^2}{5}\left(1-\frac{A}{7}+\frac{A^2}{189}\right)\right]. \tag{4.87}$$

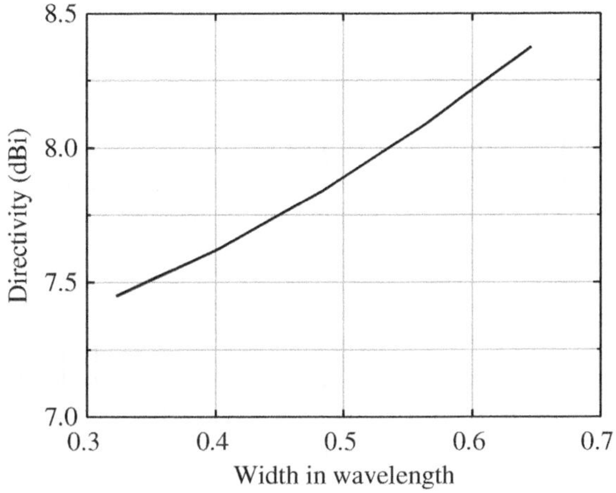

Figure 4.6 Directivity of the differential rectangular microstrip patch antenna operating in the dominant TM_{01} mode as a function of the electrical width.

Based on the equations from (4.80) to (4.87) and using the definition of (4.14), the directivity of the differential rectangular microstrip patch antenna can be expressed as

$$D = \frac{|E|^2 r^2}{60P_r} = \frac{\left(\frac{2w}{\lambda r}\right)^2 r^2}{30|V_{01}|^2 G_r} = \frac{2}{15G_r}\left(\frac{w}{\lambda}\right)^2 \tag{4.88}$$

Figure 4.6 shows the calculated directivity of the differential rectangular microstrip patch antenna operating in the dominant TM_{01} mode as a function of the electrical width. It is seen that the directivity increases with the electrical width. It approaches 7.4 dBi for the electrical width of 0.325 and reaches about 8.4 dBi for the case of the electrical width of 0.65.

4.3.3.4 Radiation Efficiency and Effective Loss Tangent

The dissipated power due to the dielectric loss can be determined by integrating the electric field inside the cavity over the volume as

$$P_d = \frac{\omega\delta\varepsilon_r\varepsilon_0}{2}\iiint_v \left(|E_z|^2\right)dv \approx \frac{\omega\delta\varepsilon_r\varepsilon_0}{2h}|V_{mn}|^2\int_0^{w_e}\int_0^{l_e}\cos^2\left(\frac{m\pi x}{w_e}\right) \cos^2\left(\frac{n\pi y}{l_e}\right)dxdy = \frac{\omega\delta\varepsilon_r\varepsilon_0 w_e l_e}{2\epsilon_{0m}\epsilon_{0n}h}|V_{mn}|^2. \tag{4.89}$$

The conductance associated with the dielectric loss is then

$$G_d = \frac{2P_d}{|V_{nm}|^2} = \frac{\omega\delta\varepsilon_r\varepsilon_0 w_e l_e}{\epsilon_{0m}\epsilon_{0n}h}, \tag{4.90}$$

which reduces to

$$G_d = \frac{\omega\delta\varepsilon_r\varepsilon_0 w_e l_e}{2h} \tag{4.91}$$

for the dominant TM_{01} mode.

The dissipated power due to the conductor loss can be calculated by integrating over the patch and ground plane areas as

$$\begin{aligned} P_c &= \omega\mu\frac{\delta_s}{2h}\iint_s \left(|H_x|^2 + |H_y|^2\right) ds = \frac{\delta_s}{2}\frac{1}{\omega\mu h^2}|V_{mn}|^2 \\ &\left\{\int_0^{w_e}\int_0^{l_e}\left(\frac{n\pi}{l_e}\right)^2\cos^2\left(\frac{m\pi x}{w_e}\right)\cos^2\left(\frac{n\pi y}{l_e}\right)dxdy + \int_0^{w_e}\int_0^{l_e}\left(\frac{m\pi}{w_e}\right)^2\cos^2\left(\frac{m\pi x}{w_e}\right)\cos^2\left(\frac{n\pi y}{l_e}\right)dxdy\right\} \\ &= \frac{\delta_s}{2}\frac{1}{\omega\mu h^2}\frac{w_e l_e}{\epsilon_{0m}\epsilon_{0n}}|V_{mn}|^2\left[\left(\frac{m\pi}{w_e}\right)^2 + \left(\frac{n\pi}{l_e}\right)^2\right]. \end{aligned} \tag{4.92}$$

The conductance associated with the conductor loss is then

$$G_c = \frac{2P_c}{|V_{nm}|^2} = \delta_s\frac{1}{\omega\mu h^2}\frac{w_e l_e}{\epsilon_{0m}\epsilon_{0n}}\left[\left(\frac{m\pi}{w_e}\right)^2 + \left(\frac{n\pi}{l_e}\right)^2\right], \tag{4.93}$$

which reduces to

$$G_c = \delta_s\frac{\pi^2}{\omega\mu h^2}\frac{w_e}{2l_e} \tag{4.94}$$

for the dominant TM_{01} mode.

The total conductance for the differential rectangular microstrip patch antenna operating in the dominant TM_{01} mode is the summation of three kinds of conductance given in equations (4.87), (4.91), and (4.94).

Based on equations (4.87), (4.91) and (4.94) as well as using the definition of (4.18), the radiation efficiency of the differential rectangular microstrip patch antenna can be calculated. Figure 4.7 shows the calculated radiation efficiencies of the differential rectangular microstrip patch antennas operating in the dominant TM_{01} mode as a function of the substrate electrical thickness. Note that the radiation efficiency of the differential rectangular microstrip patch antenna on RT/duroid 5880 substrate is higher than that on LTCC Ferro A6M substrate because the former has smaller conductor loss, substrate loss, and dielectric constant than those of the latter. It is more important to note that there exists the optimum substrate electrical thickness of about 0.02 for which the radiation efficiency reaches the maximum.

The electric energy stored in the region between the rectangular patch and the ground plane can be calculated as

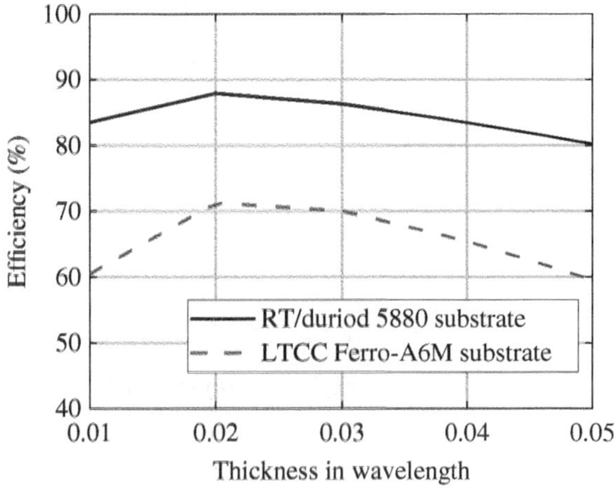

Figure 4.7 Radiation efficiency of the differential rectangular microstrip patch antenna operating in the dominant TM_{01} mode as a function of the electrical thickness.

$$
\begin{aligned}
U_e &= \frac{\varepsilon_r \varepsilon_0}{4} \iiint_v (|E_z|^2) dv \approx \frac{\varepsilon_r \varepsilon_0}{4h} |V_{mn}|^2 \int_0^{w_e} \int_0^{l_e} \cos^2\left(\frac{m\pi x}{w_e}\right) \cos^2\left(\frac{n\pi y}{l_e}\right) dxdy \\
&= \frac{\varepsilon_r \varepsilon_0}{4h} \frac{w_e l_e}{\epsilon_{0m} \epsilon_{0n}} |V_{mn}|^2.
\end{aligned}
\tag{4.95}
$$

At resonance, the electric stored energy W_e is equal to the magnetic stored energy W_m, which leads to the total stored energy as

$$
U = U_e + U_m = 2U_e = \frac{\varepsilon_r \varepsilon_0}{2h} \frac{w_e l_e}{\epsilon_{0m} \epsilon_{0n}} |V_{mn}|^2. \tag{4.96}
$$

Thus, the effective loss tangent can be expressed as

$$
\begin{aligned}
\delta_e = \frac{P_r + P_c + P_d}{\omega U} &= \delta + \frac{1}{\pi h} \sqrt{\frac{\lambda_0}{120\sigma}} \\
&+ \frac{A\pi^4 h}{11520 \omega \varepsilon_r \varepsilon_0 w l} \left[(1 - B)\left(1 - \frac{A}{15} + \frac{A^2}{420}\right) + \frac{B^2}{5}\left(1 - \frac{A}{7} + \frac{A^2}{189}\right) \right].
\end{aligned}
\tag{4.97}
$$

It was shown in Section 4.2.5 that the calculation of δ_e requires an iterative process, which typically starts by setting $\delta_e = \delta$ [34].

4.3.4 Impedance Characteristics

The impedance matrix elements for the rectangular microstrip patch antenna with dual feeds of coaxial probes are given by

$$Z_{11} = -j\omega\mu h \sum_{m=0}^{\infty}\sum_{n=0}^{\infty} \frac{j_0^2\left(\frac{m\pi d_e}{2w_e}\right)}{\left[k_0^2\varepsilon_r(1-j\delta_e) - k_{mn}^2\right]} \cos^2\left(\frac{m\pi x_1}{w_e}\right)\cos^2\left(\frac{n\pi y_1}{l_e}\right)$$
$$= \sum_{m=0}^{\infty}\sum_{n=0}^{\infty} Z_{11mn}, \tag{4.98}$$

$$Z_{12} = -j\omega\mu h \sum_{m=0}^{\infty}\sum_{n=0}^{\infty} \frac{j_0^2\left(\frac{m\pi d_e}{2w_e}\right)}{\left[k_0^2\varepsilon_r(1-j\delta_e) - k_{mn}^2\right]} \cos\left(\frac{m\pi x_1}{w_e}\right)\cos\left(\frac{n\pi y_1}{l_e}\right)$$
$$\cos\left(\frac{m\pi x_2}{w_e}\right)\cos\left(\frac{n\pi y_2}{l_e}\right) = \sum_{m=0}^{\infty}\sum_{n=0}^{\infty} Z_{12mn}, \tag{4.99}$$

$$Z_{22} = -j\omega\mu h \sum_{m=0}^{\infty}\sum_{n=0}^{\infty} \frac{j_0^2\left(\frac{m\pi d_e}{2w_e}\right)}{\left[k_0^2\varepsilon_r(1-j\delta_e) - k_{mn}^2\right]} \cos^2\left(\frac{m\pi x_2}{w_e}\right)\cos^2\left(\frac{n\pi y_2}{l_e}\right)$$
$$= \sum_{m=0}^{\infty}\sum_{n=0}^{\infty} Z_{22mn} \tag{4.100}$$

for $0 < x_1 < w$ and $0 < y_1 < l/2$ [20]. Equations (4.98), (4.99), and (4.100) indicate that the self-impedances at the dual feeds and the mutual impedances between the dual feeds are the superposition of mode impedances.

4.3.4.1 Differential Input Impedance

Substituting (4.98), (4.99), and (4.100) into (4.22), we can obtain the general expression for the differential input impedance. If the condition that was used to simplify (4.48) into (4.50) is used again, the differential input impedance can be simplified as

$$Z_d = -2j\omega\mu h \sum_{m=0}^{\infty}\sum_{n=0}^{\infty} \frac{j_0^2\left(\frac{m\pi d_e}{2w_e}\right)[1-\cos(n\pi)]}{\left[k_0^2\varepsilon_r(1-j\delta_e) - k_{mn}^2\right]} \cos^2\left(\frac{m\pi x_1}{w_e}\right)\cos^2\left(\frac{n\pi y_1}{l_e}\right). \tag{4.101}$$

If the mode index n is even in equation (4.101), the corresponding mode impedance is zero, implying that the differential rectangular microstrip patch antenna will excite the dominant mode TM_{01} but will not excite the static TM_{00} mode and higher-order TM_{02} mode.

Consider a patch printed on a substrate with $\varepsilon_r = 2.2$ and $\delta = 0.0009$. The patch dimensions are $w = 18.522$ mm and $l = 18$ mm. The substrate size is $w_g \times l_g \times h = 25 \times 25 \times 0.635\ \text{mm}^3$. It is found δ_e to be 0.2 at 5.2 GHz. Let the feeds at $(x_1, y_1) = (0.0054w, 0.0056l)$ and $(x_2, y_2) = (0.0054w, 0.9944l)$ to excite the TM_{10}

mode. Figure 4.8 shows the calculated and simulated differential input impedance. It is evident from the figure that the microstrip patch antenna resonates at 5.3 GHz and excellent agreement is obtained between the calculation and simulation. Also let the feeds at $(x_1, y_1) = (0.0054w, 0.0056l)$ and $(x_2, y_2) = (0.9946w, 0.9944l)$ to excite both TM_{10} and TM_{01} modes to realize circular polarization. Figure 4.9 shows the

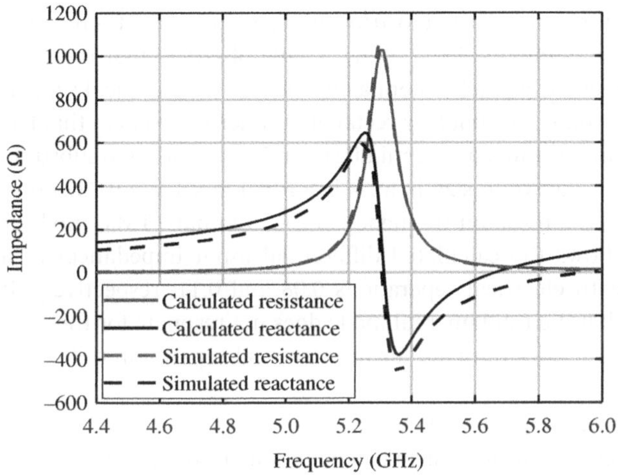

Figure 4.8 Differential input impedance of the rectangular microstrip patch antenna with dual feeds of coaxial probes for the TM_{10} mode.

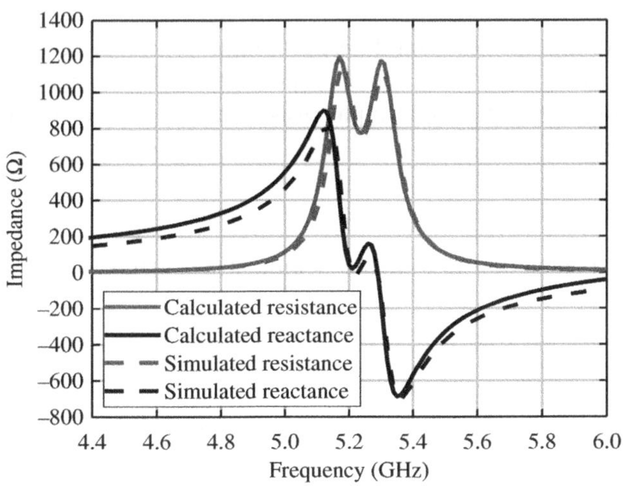

Figure 4.9 Differential input impedance of the rectangular microstrip patch antenna with dual feeds of coaxial probes for the TM_{10} and TM_{01} modes.

calculated and simulated differential input impedance. It is interesting to note from the figure that the microstrip patch antenna resonates at 5.17 GHz of the TM_{10} mode and 5.3 GHz of the TM_{01} mode and excellent agreement is obtained between the theory and simulation. Left-hand circular polarization (LHCP) occurs at 5.25 GHz.

4.3.4.2 Fundamental Resonance and Electrical Separation

Generally, the differential rectangular microstrip patch antenna operates near the resonant frequency of the dominant TM_{01} mode. Zhang observed that the resonance of a differential microstrip patch antenna depends on the electrical separation of dual-probe feeds [36]. For example, a rectangular microstrip patch that has $w = 63$ mm and $l = 40$ mm is printed on the middle of an RT/duroid 5880 substrate ($\varepsilon_r = 2.2$, $\delta = 0.0009$). The substrate size is $w_g \times l_g \times h = 125 \times 125 \times 0.787\ \text{mm}^3$. This patch size will yield the resonant frequency of the dominant TM_{01} mode at 2.42 GHz. Figure 4.10 shows the calculated differential input impedances as a function of frequency with electrical separations 0.08 and 0.11, respectively. It is clear from the figure that the fundamental mode does not resonate for the electrical separation of 0.08 but does resonate for the electrical separation of 0.11.

4.3.4.3 Resonant Differential Input Resistance

When the differential rectangular microstrip patch antenna resonates at the dominant TM_{01} mode, the resonant differential input resistance can be obtained from equation (4.97) as

$$R_{d01} = \frac{4\omega\mu h k_0^2 \varepsilon_r \delta_e}{\left(k_0^2 - k_{01}^2\right)^2 + \left(k_0^2 \varepsilon_r \delta_e\right)^2} \cos^2\left(\frac{\pi y_1}{l_e}\right). \tag{4.102}$$

Equation (4.102) shows that the resonant differential input resistance reaches the maximum when the dual feeds are at the opposite edges of the rectangular microstrip patch. The resonant differential input resistance gets smaller when the dual feeds are closer.

4.3.5 Quality Factor

The quality factor due to radiation loss in the dominant TM_{01} mode can be found as

$$Q_r = \frac{\omega U}{P_r} = \frac{2\omega U_e}{P_r} = \frac{\omega \varepsilon_r \varepsilon_0 w l}{h G_r}. \tag{4.103}$$

The calculation from equation (4.103) reveals that the value of Q_r decreases with increasing h/λ or reducing ε_r. The total quality factor for the rectangular microstrip patch antenna operating in the dominant TM_{01} mode is given by

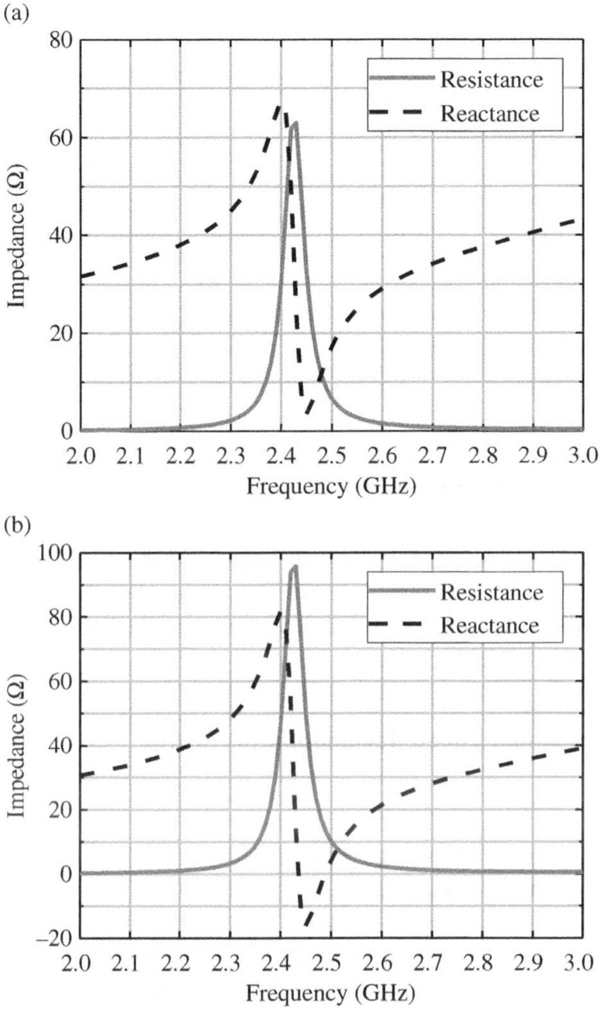

Figure 4.10 Simulated input impedance as a function of frequency for the differential rectangular microstrip antenna with different electrical separations of (a) 0.08 and (b) 0.11.

$$Q = \left(\frac{hG_r}{\omega\varepsilon_0\varepsilon_r wl} + \frac{1}{\pi h}\sqrt{\frac{\lambda}{120\sigma}} + \delta \right)^{-1}. \tag{4.104}$$

The differential input impedance bandwidths for various rectangular microstrip patch antennas on substrates of different thicknesses and dielectric constants can be more conveniently calculated using the quality factor and VSWR. The

Table 4.1 Input impedance bandwidths of differential rectangular microstrip patch antennas.

	BW% for VSWR ≤ 2				
h/λ	$\varepsilon_r = 1.00$	$\varepsilon_r = 2.21$	$\varepsilon_r = 2.55$	$\varepsilon_r = 4.70$	$\varepsilon_r = 10.2$
0.005	1.16	0.81	0.85	0.76	0.5
0.010	1.87	1.12	1.13	0.88	0.56
0.015	2.70	1.58	1.52	1.12	0.66
0.020	3.55	2.06	1.96	1.38	0.79
0.025	4.43	2.57	2.41	1.67	0.95
0.030	5.32	3.08	2.88	1.98	1.11
0.035	6.23	3.61	3.36	2.30	1.29
0.040	7.14	4.15	3.84	2.62	1.47

calculated results are listed in Table 4.1. Note that differential input impedance bandwidth increases with increasing substrate electrical thickness but decreases with increasing substrate dielectric constant. The differential input impedance bandwidth is typically less than a few percent.

4.3.6 Differential Versus Single-Ended Rectangular Microstrip Patch Antennas

There have been much fewer studies about differential as compared with single-ended rectangular microstrip patch antennas. An early paper on differential rectangular microstrip patch antennas was about the suppression of unwanted probe radiation for the patch fabricated on a relatively thick and low-permittivity substrate [40]. This section compares differential with single-ended rectangular microstrip patch antennas in terms of fundamental differences and performance characteristics.

4.3.6.1 Excitation and Suppression of Resonant Modes

There are numerous resonant modes in a rectangular microstrip patch cavity. The fundamental difference between differential and single-ended rectangular microstrip patch antennas lies in the excitation and suppression of resonant modes. Table 4.2 lists the first six resonant modes. Note that all these modes can be generated in a single-ended rectangular microstrip patch antenna but only four of them can be excited in a differential rectangular microstrip patch antenna. If $w > l$, the TM_{01} mode is the dominant mode. Both differential and single-ended

Table 4.2 First six resonant modes.

Resonant modes	TM_{01}	TM_{02}	TM_{03}	TM_{10}	TM_{20}	TM_{30}
Differential	Excited	Suppressed	Excited	Excited	Suppressed	Excited
Single-ended	Excited	Excited	Excited	Excited	Excited	Excited

rectangular microstrip patch antennas are commonly designed to operate in this mode. If $w/2 > l$, the TM_{02} mode is the higher-order mode closest to the dominant TM_{01} mode, which will distort the radiation pattern in the E plane, and increase the cross-polarized radiation. Fortunately, this undesirable higher-order mode is automatically suppressed in the differential rectangular microstrip patch antenna. The other higher-order modes are far away from the dominant TM_{01} mode. They will not cause serious problems and thus can be ignored.

4.3.6.2 Input Impedance Ratio

For the rectangular microstrip patch antenna with dual feeds of coaxial probes, the differential input impedance was given by equation (4.101) and the single-ended input impedance at feed 1 with feed 2 being opened is actually Z_{11} given by equation (4.98). It is meaningful to examine the relationship between the differential and single-ended input impedances. It can be found from (4.98) and (4.99) that

$$\frac{Z_{11,mn}}{Z_{12,mn}} = \frac{\psi_{mn}(x_1, y_1)}{\psi_{mn}(x_2, y_2)} = \frac{\cos\left(\frac{m\pi x_1}{w}\right)\cos\left(\frac{n\pi y_1}{l}\right)}{\cos\left(\frac{m\pi x_2}{w}\right)\cos\left(\frac{n\pi y_2}{l}\right)}. \tag{4.105}$$

Since $x_1 = x_2$ and $y_2 = l - y_1$, thus

$$\frac{Z_{11,mn}}{Z_{12,mn}} = \frac{\psi_{mn}(x_1, y_1)}{\psi_{mn}(x_2, y_2)} = \frac{\cos\left(\frac{n\pi y_1}{l}\right)}{\cos\left(\frac{n\pi(l-y_1)}{l}\right)} = \begin{cases} 1 & for \quad n = 2, 4, 6... \\ -1 & for \quad n = 1, 3, 5... \end{cases}. \tag{4.106}$$

As a result, the differential input impedance Z_d is

$$Z_d = 2\sum_{m,n=0}^{\infty}(Z_{11,mn} - Z_{12,mn}) = 4\sum_{m=0,n=2i+1}^{\infty} Z_{11,mn} \tag{4.107}$$

where $i = 0, 1, 2, 3, ...$ It is important to note that Z_d has no term of n being an even number. As for the single-ended input impedance Z_s, it is given by

$$Z_s = \sum_{m=0}^{\infty}\sum_{n=0}^{\infty} Z_{11mn}. \tag{4.108}$$

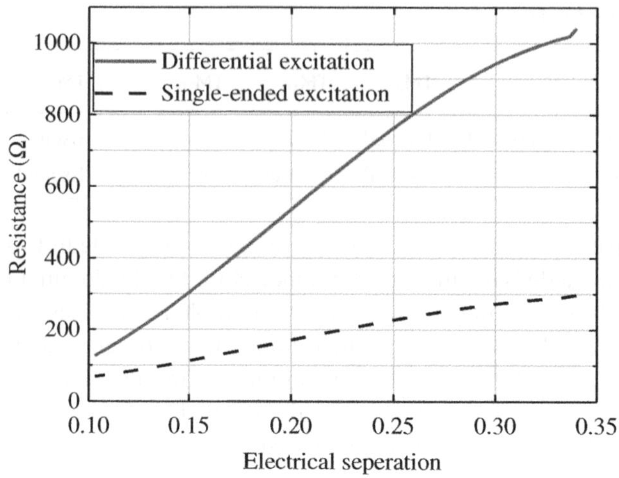

Figure 4.11 Variation of input resistances at the resonant frequency of the dominant TM_{01} mode with the electrical separation.

Note that Z_s has terms regardless of n being an odd or even number. Thus, the ratio of Z_d to Z_s is

$$\frac{Z_d}{Z_s} = 4\left(\sum_{m=0}^{\infty}\sum_{n=2i+1}^{\infty} Z_{11mn}\right) \Bigg/ \left(\sum_{m=0}^{\infty}\sum_{n=0}^{\infty} Z_{11mn}\right) \approx 4. \tag{4.109}$$

Figure 4.11 shows the calculated differential and single-ended input resistances at the resonant frequency of the dominant TM_{01} mode as a function of electrical separation. Note that both input resistances increase with the electrical separation or the probes moving to the edge of the patch. It is interesting to note that the differential input resistance is indeed about four times the single-ended input resistance. This means that if the dual-feed antenna is designed to well match to a 50-Ω signal source for the single-ended excitation, it will not well match to a 100-Ω but a 200-Ω signal source for the differential excitation.

4.3.6.3 Reduction of Cross-Polarized Radiation

The cross-polarized radiation of a single-ended rectangular microstrip patch antenna has been investigated [40–45]. It was found that the cross-polarized radiation can be attributed to the spurious radiation of the feeding probe and the excitation of unwanted modes that have polarizations orthogonal to the wanted mode [44]. The cross-polarized radiation is especially remarkable in the H plane due to the fact that the feeding probe is asymmetrically positioned in the H plane. It is near one radiating edge but far from the other, which causes different time delays for the field to build up at the radiating edges. The field intensity differences at both edges for a given point of time generate the higher cross-polarized radiation in the *H* plane and destroy the deep null in the cross-polarized radiation in the *H* plane predicted by the cavity model [38].

The cross-polarized radiation of a differential rectangular microstrip patch antenna has also been investigated [22, 23, 38, 40, 45]. It was found that the differential rectangular microstrip patch antenna produces much lower cross-polarized radiation than its single-ended counterpart, which can be explained as follows. First, the spurious radiation of the dual feeding probes cancels out each other, which reduces the cross-polarized radiation, especially for the patch on a thick substrate. For instance, Petosa et al. showed that an L-band patch on a thick substrate reduced the peak cross-polarized radiation by 18 dB in the *H* plane and removed the pattern distortion in the *E* plane with the differential probe feeding [40]. Second, the positions of the dual feeding probes make the structure symmetrical in the *H* plane, which recovers the deep null in the cross-polarized radiation in the *H* plane predicted by the cavity model [38]. Third, the differential feeding suppresses some higher-order modes orthogonal to the operating dominant mode. This is particularly effective for the rectangular patch with the aspect ratio $w/l = 2$. Assuming that such a rectangular patch antenna operates in the dominant TM_{01} mode, the higher-order TM_{20} mode will unavoidably be strongly excited under the single-ended operation, and the cross-polarized radiation will be larger than the co-polarized radiation, making it operate improperly [44]. However, the higher-order TM_{20} mode will be automatically suppressed by the differential feeding. The cross-polarized radiation is definitely smaller than the co-polarized radiation in the differential rectangular microstrip patch antenna, making it operate properly [45].

4.3.6.4 Performance Improvement

For a rectangular microstrip patch antenna operating in the dominant TM_{01} mode, the patch length l is controlled for the resonant frequency and the patch width w is determined for wider bandwidth. The wider the width w, the larger are the bandwidth and radiation efficiency. However, an excessive width w degrades the original performance because the effect of higher-order modes becomes significant.

Let us denote the widths of the differential and single-ended rectangular microstrip patch antennas by w_d and w_s, respectively. First, we examine the rectangular patches with $w_d > w_s$. For this case, the typical choice is $w_d = 2.0l$ and $w_s = 1.5l$. Thus, it can be found that the impedance bandwidth ratio of BW_d to BW_s is

$$\frac{BW_d}{BW_s} = \frac{Q_s}{Q_{dif}} \approx \frac{w_d}{w_s} > 1 \tag{4.110}$$

and the radiation efficiency ratio of e_d to e_s is

$$\frac{e_d}{e_s} = \frac{Q_{dif}}{Q_s}\frac{Q_{rs}}{Q_{rdif}} \geq 1. \tag{4.111}$$

Hence, the differential rectangular microstrip patch antenna outperforms the single-ended rectangular microstrip patch antenna in all aspects. Then, we evaluate the rectangular patches with $w_d = w_s$. For this case, two usual choices are $w_d = w_s = 2.0l$ and $w_d = w_s = 1.5l$. For $w_d = w_s = 2.0l$, the differential rectangular microstrip patch antenna outperforms the single-ended counterpart in all aspects because the differential one has an optimum width but the single-ended one has an excessive width. For $w_d = w_s = 1.5l$, both differential and single-ended rectangular microstrip patch antennas have comparable performance. This is because the width for the single-ended rectangular microstrip patch antenna is optimal, while the width for the differential rectangular microstrip patch antenna is also suited. Finally, we consider the rectangular patches with $w_d < w_s$. For this case, the choice is $w_s = 1.5l$ and w_d varies from $1.2l$ to $1.4l$. As expected, the single-ended rectangular microstrip patch antenna outperforms the differential one.

4.3.7 Design Procedure

The design procedure for single-ended rectangular microstrip patch antennas has been extended for differential rectangular microstrip patch antennas [23]. It determines the patch size from the given substrate to meet the requirement of resonant frequency but totally fails to take the specification of bandwidth into account. As for the required bandwidth, the designer has to rely on a trial-and-error technique, for example, to try different substrates.

In the design of a differential microstrip patch antenna, electrical separation plays an important role. The resonance of the differential microstrip patch antenna occurs when the imaginary part of the differential input impedance goes to zero, which requires the electrical separation to be larger than an empirical value of 0.1 for rectangular patches [36]. Thus, the design formulas for a single-ended rectangular microstrip patch antenna [46] can be used for a differential rectangular microstrip patch antenna [23]. However, the electrical separation must be observed and satisfied. Usually, the electrical separation can be satisfied with an increased patch width w or a small rectangular slot cut between the dual feeds [36]. For the single-ended rectangular microstrip patch antenna, w is usually to be $1.5l$, while for the differential rectangular microstrip patch antenna, w can be equal to or larger than $2.0l$ [23]. However, if the rectangular patch antenna is designed for circular or dual-linear polarizations, w must be chosen to be equal to l.

Design procedure is demonstrated with an example to meet the specifications at 2 GHz for 2% impedance bandwidth and 3-dBi gain using FR4 substrate $\varepsilon_r = 4.4$,

Figure 4.12 Photo of the fabricated both differential and single-ended rectangular microstrip patch antennas.

$\delta = 0.025$, and $h = 1.6$ mm. It starts from the determination of the patch dimensions and feed locations using the formulas in Sections 4.3.1, 4.3.2, and 4.3.4.3 for their initial values and then optimizes them with a full-wave electromagnetic simulator. Finally, it ends with their optimal values for the patch dimensions as $w = 74.4$ mm, $l = 35$ mm, and feed locations at $(x_1, y_1) = (w/2, 9.654\text{ mm})$ and $(x_2, y_2) = (w/2, 25.346\text{ mm})$. The electrical separation is satisfied as $15.692/150 > 0.1$.

Figure 4.12 is the photo of the fabricated antennas. The right one is the differential rectangular microstrip patch antenna, while the left one is the single-ended one. Both the simulated and measured results confirm that the differential rectangular microstrip patch antenna resonates at 2 GHz. The simulated and measured fractional impedance bandwidths are 3.9% and 4.1% for differential as well as 1.3% and 1.9% for single-ended operations, respectively. The simulated and measured gains are 4.2 and 3.7 dBi for differential as well as both 1.2 dBi for single-ended operations, respectively. The simulated and measured radiation efficiencies are 42% and 49% for differential and 17.8% and 18.7% for single-ended operations, respectively [47–49]. Clearly, the differential rectangular microstrip patch antenna outperforms the single-ended counterpart. The differential rectangular microstrip patch antenna meets the design specifications, while the single-ended rectangular microstrip patch antenna fails.

4.4 Circular Patch

The circular microstrip patch antenna is the earliest microstrip patch antenna, which offers similar performance as the rectangular microstrip patch antenna does. Figure 4.13 shows the geometry and coordinate system of a circular

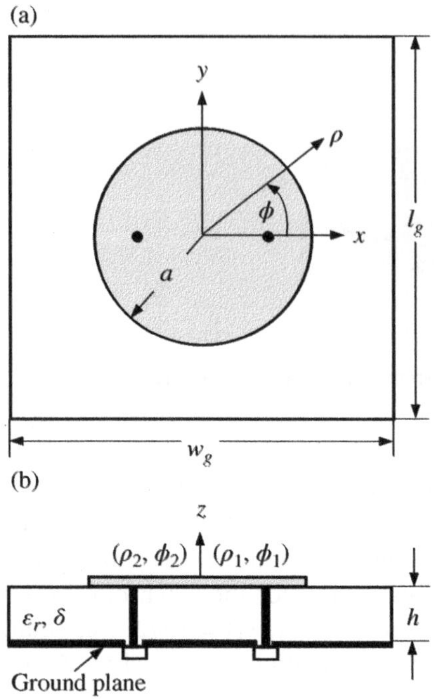

Figure 4.13 Geometry of the circular microstrip patch antenna and the coordinate system. (a) Top view and (b) side view.

microstrip patch antenna with dual feeds of coaxial probes. It is seen that the circular microstrip patch antenna has a radius of a. The dual feeds are located at points (ρ_1, ϕ_1) and (ρ_2, ϕ_2), respectively. For differential operation, the probe positions should satisfy such requirements as

$$\rho_1 = \rho_2 \, and \, \phi_2 - \phi_1 = \pi. \tag{4.112}$$

4.4.1 Resonant Modes

For the circular microstrip patch antenna, it can be modelled as a cylindrical cavity. The eigenfunctions and eigenvalues of the cavity can be written respectively as

$$\psi_{nm}(\rho, \phi) = J_n(k_{nm}\rho)\cos(n\phi), \tag{4.113}$$

$$k_{nm} = \frac{X_{nm}}{a_e} \tag{4.114}$$

where n is the mode index in the ϕ direction and X_{nm} is the mth zero of

$$J_n'(k_{nm}a_e) = 0. \tag{4.115}$$

In equation (4.115), the prime denotes derivative with respect to the argument and a_e is the effective radius, taking into account the fringing fields at the edge of the circular microstrip patch [50]. The effective radius a_e may be approximated from the physical radius a as

$$a_e = a\left[1 + \frac{2h}{\pi a \varepsilon_r} \ln\left(\frac{\pi a}{2h} + 1.7726\right)\right]^{1/2}. \tag{4.116}$$

The dual feeds of coaxial probes can be treated as z-directed current ribbons centered at the feed points, respectively, as

$$J_z(\rho_1, \phi_1) = J(\phi_1)\frac{I\delta(\rho - \rho_1)}{\rho_1 \phi_w}, \tag{4.117}$$

$$J(\phi_1) = \begin{cases} 1, & -\frac{\phi_w}{2} < \phi < \frac{\phi_w}{2}, \\ 0, & elsewhere \end{cases} \tag{4.118}$$

$$J_z(\rho_2, \phi_2) = J(\phi_2)\frac{[(1-\alpha)\cos\beta]\delta(\rho - \rho_2)}{\rho_2 \phi_w}, \tag{4.119}$$

$$J(\phi_2) = \begin{cases} -1, & \pi - \frac{\phi_w}{2} < \phi < \pi + \frac{\phi_w}{2} \\ 0, & elsewhere \end{cases} \tag{4.120}$$

where $\delta(\rho)$ is the Dirac delta function, ϕ_w is the effective width of the feed that needs to be chosen to obtain good agreement between the theoretical and experimental impedances, α and β represent the amplitude and phase imbalances at feed 2 with respect to those at feed 1. Ideally, $\alpha = 0$ and $\beta = 0°$. Practically, α can be up to 0.1 and β to 10° [23]. Substituting equations (4.113), (4.117), (4.118), (4.119), and (4.120) into (4.5) yields

$$\begin{aligned}
&\langle [J_z(\rho_1, \phi_1) + J_z(\rho_2, \phi_2)]\psi_{nm}(\rho, \phi)\rangle \\
&= I\left[\begin{aligned} &\int_0^{2\pi} J(\phi_1)\cos(n\phi)d\phi \int_0^{a_e} \frac{\delta(\rho-\rho_1)}{\rho_1\phi_w} J_n(k_{nm}\rho)\rho d\rho \\ &+ (1-\alpha)\cos\beta \int_0^{2\pi} J(\phi_2)\cos(n\phi)d\phi \int_0^{a_e} \frac{\delta(\rho-\rho_2)}{\rho_2\phi_w} J_n(k_{nm}\rho)\rho d\rho \end{aligned}\right] \\
&= I\left[J_n(k_{nm}\rho_1)j_0\left(\frac{n\phi_w}{2}\right) - (1-\alpha)\cos\beta\cos(n\pi)J_n(k_{nm}\rho_2)j_0\left(\frac{n\phi_w}{2}\right)\right] \\
&= I[J_n(k_{nm}\rho_1) - (1-\alpha)\cos\beta\cos(n\pi)J_n(k_{nm}\rho_2)]j_0\left(\frac{n\phi_w}{2}\right).
\end{aligned} \tag{4.121}$$

For the circular microstrip patch antenna with dual feeds of coaxial probes located at the positions defined by equation (4.112) and under an ideal differential excitation, equation (4.121) simplifies to

$$\langle [J_z(\rho_1, \phi_1) + J_z(\rho_2, \phi_2)]\psi_{nm}(\rho, \phi)\rangle = I[1 - \cos(n\pi)]J_n(k_{nm}\rho_1)j_0\left(\frac{n\phi_w}{2}\right). \tag{4.122}$$

Similarly, substituting equation (4.113) into (4.6) produces

$$\langle \psi_{nm}(\rho, \phi)\psi_{nm}(\rho, \phi)\rangle = \frac{\pi a_e^2}{\epsilon_{0n}} J_n^2(k_{nm}a_e)\left[1 - \frac{n^2}{(k_{nm}a_e)^2}\right]. \tag{4.123}$$

where $\epsilon_{0n} = 1$ for $n = 0$, $\epsilon_{0n} = 2$ for $n \neq 0$

The vertical electric field E_z given in equation (4.3) can be specifically written for the differential circular microstrip patch antenna as

$$\begin{aligned} E_z &= j\omega\mu \sum_{m=0}^{\infty}\sum_{n=0}^{\infty} \frac{1}{k_e^2 - k_{mn}^2} \frac{\langle [J_z(\rho_1, \phi_1) + J_z(\rho_2, \phi_2)]\psi_{nm}(\rho, \phi)\rangle}{\langle \psi_{nm}(\rho, \phi)\psi_{nm}(\rho, \phi)\rangle} \psi_{nm}(\rho, \phi) \\ &= \sum_{m=1}^{\infty}\sum_{n=0}^{\infty} B_{nm}[1 - \cos(n\pi)]J_n(k_{nm}\rho)\cos(n\phi) \end{aligned} \tag{4.124}$$

where

$$B_{nm} = j\omega\mu I \frac{k_{nm}^2}{k_e^2 - k_{nm}^2} \frac{\epsilon_{0n} J_n(k_{nm}\rho_1)j_0\left(\frac{n\phi_w}{2}\right)}{\pi J_n^2(k_{nm}a_e)\left[(k_{nm}a_e)^2 - n^2\right]}. \tag{4.125}$$

According to equation (4.2), the vertical electric field E_z will be accompanied by horizontal magnetic field components H_ρ and H_ϕ. They can be expressed, respectively, as

$$H_\rho = \frac{j}{\omega\mu}\frac{1}{\rho}\frac{\partial E_z}{\partial \phi} = -\frac{j}{\omega\mu}\frac{1}{\rho}\sum_{m=1}^{\infty}\sum_{n=0}^{\infty} nB_{nm}[1 - \cos(n\pi)]J_n(k_{nm}\rho)\sin(n\phi), \tag{4.126}$$

$$H_\phi = -\frac{j}{\omega\mu}\frac{\partial E_z}{\partial \rho} = -\frac{j}{\omega\mu}\sum_{m=1}^{\infty}\sum_{n=0}^{\infty} B_{nm}[1 - \cos(n\pi)]k_{nm}J_n'(k_{nm}\rho)\cos(n\phi). \tag{4.127}$$

As the magnetic field is transverse to the z-axis, the resonant modes are termed TM_{nm} modes. It should be noted from equations (4.124), (4.126), and (4.127) that the cancellation mechanism exists in the differential circular microstrip patch antenna, which can automatically suppress some TM_{nm} modes to reduce the cross-polarized radiation. For instance, the TM_{11} mode can be excited because $[1 - \cos(n\pi)] = 2$, while the TM_{21} mode cannot be excited because $[1 - \cos(n\pi)] = 0$. Thus, the differential circular microstrip patch antenna generally operates in the TM_{11} mode. As the TM_{11} mode requires the smallest patch radius for resonance, it is known as the dominant mode.

4.4.2 Resonant Frequencies

The resonant frequencies of resonant modes of a circular microstrip patch cavity can be calculated as

$$f_{nm} = \frac{cX_{nm}}{2\pi a_e \sqrt{\varepsilon_r}}. \tag{4.128}$$

The first seven values of X_{nm} are given in Table 4.3. It was found that the calculation of the resonant frequency based on the above formula for the TM_{11} mode for the differential circular patch antenna is accurate to within 2.5% of measured values [50].

4.4.3 Radiation Characteristics

The radiation of the differential circular microstrip patch antenna can be considered from the magnetic current flowing around the radiating edge at $\rho = a_e$. The magnetic current is obtained by substituting (4.123) into (4.9) as

$$\boldsymbol{M} = -2\hat{n} \times E_z = \hat{\phi}'2 \sum_{m=1}^{\infty} \sum_{n=0}^{\infty} B_{nm}[1 - \cos(n\pi)] J_n(k_{nm}a_e) \cos(n\phi') \tag{4.129}$$

where primed coordinates denote a point in the source region. It is seen that the magnetic current follows the $\cos(n\phi')$ distribution along the ϕ' direction. The magnetic current has $2n$ zeros around the patch perimeter for the TM_{nm} modes that can be excited. It will change the direction across a zero.

Table 4.3 First seven zeros used with equation (4.115).

(n, m)	**(0, 1)**	**(1, 1)**	**(2, 1)**	**(0, 2)**	**(3, 1)**	**(4, 1)**	**(1, 2)**
X_{nm}	0	1.841	3.054	3.832	4.201	5.317	5.331

4.4.3.1 Radiated Fields

The electric vector potential of the magnetic current can be calculated by evaluating the resultant integral of (4.10) as

$$\boldsymbol{F}=\frac{\varepsilon_0 h e^{-jk_0 r}}{2\pi r}\times\int_0^{2\pi}\hat{\phi}'\sum_{m=1}^{\infty}\sum_{n=0}^{\infty}B_{nm}[1-\cos(n\pi)]J_n(k_{nm}a_e)\cos(n\phi')e^{jk_0 a_e\sin\theta\cos(\phi'-\phi)}a_e d\phi' \tag{4.130}$$

where the unit vector in the cylindrical coordinate system can be expressed in the spherical coordinate system as

$$\hat{\phi}' = -\hat{r}\sin\theta\sin(\phi'-\phi)-\hat{\theta}\cos\theta\sin(\phi'-\phi)+\hat{\phi}\cos(\phi'-\phi). \tag{4.131}$$

Substituting (4.131) into (4.130), we have

$$F_\phi=\frac{\varepsilon_0 a_e h e^{-jk_0 r}}{2\pi r}\sum_{m=1}^{\infty}\sum_{n=0}^{\infty}B_{nm}[1-\cos(n\pi)]J_n(k_{nm}a_e)\times\int_0^{2\pi}\cos(n\phi')\cos(\phi'-\phi)e^{jk_0 a_e\sin\theta\cos(\phi'-\phi)}d\phi', \tag{4.132}$$

$$F_\theta=-\frac{\varepsilon_0 a_e h e^{-jk_0 r}}{2\pi r}\sum_{m=1}^{\infty}\sum_{n=0}^{\infty}B_{nm}[1-\cos(n\pi)]J_n(k_{nm}a_e)\times\cos\theta\int_0^{2\pi}\cos(n\phi')\sin(\phi'-\phi)e^{jk_0 a_e\sin\theta\cos(\phi'-\phi)}d\phi'. \tag{4.133}$$

Applying the integration formula

$$\int_0^{2\pi}e^{jk_0 a_e\sin\theta\cos(\phi'-\phi)}\begin{bmatrix}\cos(n\phi')\\ \sin(n\phi')\end{bmatrix}d\phi'=2\pi j^n J_n(k_0 a_e)\begin{bmatrix}\cos(n\phi)\\ \sin(n\phi)\end{bmatrix}, \tag{4.134}$$

we can find

$$\int_0^{2\pi}e^{jk_0 a_e\sin\theta\cos(\phi'-\phi)}\begin{bmatrix}\cos(\phi'-\phi)\\ \sin(\phi'-\phi)\end{bmatrix}\cos(n\phi')d\phi' = \pi j^{n+1}[J_{n+1}(k_0 a_e\sin\theta)\mp J_{n-1}(k_0 a_e\sin\theta)]\begin{bmatrix}\cos(n\phi)\\ \sin(n\phi)\end{bmatrix}. \tag{4.135}$$

So, equations (4.132) and (4.133) can be rewritten as

$$F_\phi = \frac{\varepsilon_0 a_e h e^{-jk_0 r}}{2r} \sum_{m=1}^{\infty} \sum_{n=0}^{\infty} B_{nm}[1 - \cos(n\pi)] J_n(k_{nm} a_e) \times j^{n+1}[J_{n+1}(k_0 a_e \sin\theta) - J_{n-1}(k_0 a_e \sin\theta)] \cos(n\phi), \tag{4.136}$$

$$F_\theta = -\frac{\varepsilon_0 a_e h e^{-jk_0 r}}{2r} \sum_{m=1}^{\infty} \sum_{n=0}^{\infty} B_{nm}[1 - \cos(n\pi)] J_n(k_{nm} a_e) \times j^{n+1} \cos\theta [J_{n+1}(k_0 a_e \sin\theta) + J_{n-1}(k_0 a_e \sin\theta)] \sin(n\phi). \tag{4.137}$$

Thus, the radiated fields in the far zone at r are evaluated from equations (4.11) and (4.12) as

$$E_\theta = -j\omega\eta_0 F_\phi = -j\frac{k_0 a_e h e^{-jk_0 r}}{2r} \sum_{m=1}^{\infty} \sum_{n=0}^{\infty} B_{nm}[1 - \cos(n\pi)] J_n(k_{nm} a_e) \times j^{n+1}[J_{n+1}(k_0 a_e \sin\theta) - J_{n-1}(k_0 a_e \sin\theta)] \cos(n\phi), \tag{4.138}$$

$$E_\phi = j\omega\eta_0 F_\theta = -j\frac{k_0 a_e h e^{-jk_0 r}}{2r} \sum_{m=1}^{\infty} \sum_{n=0}^{\infty} B_{nm}[1 - \cos(n\pi)] J_n(k_{nm} a_e) \times j^{n+1} \cos\theta [J_{n+1}(k_0 a_e \sin\theta) + J_{n-1}(k_0 a_e \sin\theta)] \sin(n\phi). \tag{4.139}$$

They can be written succinctly with a clearer physical meaning as

$$E_\theta = \frac{k_0 a_e e^{-jk_0 r}}{2r} \sum_{m=1}^{\infty} \sum_{n=0}^{\infty} V_{nm} j^n [J_{n+1}(k_0 a_e \sin\theta) - J_{n-1}(k_0 a_e \sin\theta)] \cos(n\phi), \tag{4.140}$$

$$E_\phi = \frac{k_0 a_e e^{-jk_0 r}}{2r} \sum_{m=1}^{\infty} \sum_{n=0}^{\infty} V_{nm} j^n \cos\theta [J_{n+1}(k_0 a_e \sin\theta) + J_{n-1}(k_0 a_e \sin\theta)] \sin(n\phi) \tag{4.141}$$

where V_{nm} is the voltage difference between the dual feeds at $(a, 0^\circ)$ and $(a, 180^\circ)$, which is given by

$$V_{nm} = hB_{nm}[1 - \cos(n\pi)] J_n(k_{nm} a_e). \tag{4.142}$$

Equations (4.140), (4.141), and (4.142) reveal that the radiation of all TM_{1m} modes reaches the peak in the broadside direction and the radiation of all TM_{2m} modes is suppressed in all directions due to the cancellation mechanism introduced by the differential excitation. Figure 4.14 shows the co-polarized

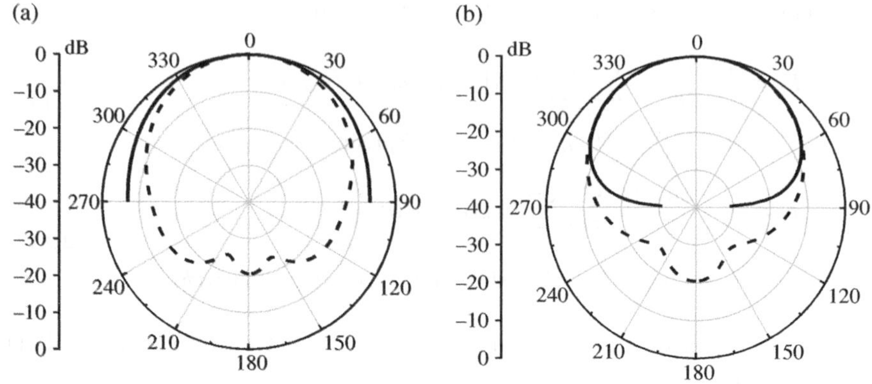

Figure 4.14 Normalized radiation patterns for the differential circular microstrip patch antenna. (a) The *E* plane and (b) the *H* plane.

radiation patterns in the *E* and *H* planes of a differential circular microstrip patch antenna with radius $a = 24.5$ mm on RT/duroid 5880 substrate ($\varepsilon_r = 2.2$, $\delta = 0.0009$) at resonant frequency 2.42 GHz of the dominant TM_{11} mode. The differential circular microstrip patch antenna has a size of $w_g \times l_g \times h = 125$ mm $\times$ 125 mm $\times$ 0.787 mm. The diameters of the probes and openings are 0.5 and 2 mm, respectively. The two probes are separated by 12.5 mm. Note that the radiation of the dominant TM_{11} mode indeed reaches the peak in the broadside direction. The agreement between the calculated and simulated radiation patterns is excellent in the forward region. The difference in the side and back lobe regions is because the radiation patterns are calculated for an infinite ground plane but they are simulated on a finite ground plane.

4.4.3.2 Cross Polarization

A higher-order mode orthogonal to the operating dominant mode is a major source of the cross-polarized radiation. The differential circular microstrip patch antenna operates in the dominant TM_{11} mode and suppresses the orthogonal and significant higher-order TM_{21} mode. In the *D* plane, the calculations reveal that the *XPL* increases with decreasing substrate thickness, increasing substrate permittivity, and enlarging the separation of dual feeds for differential circular microstrip patch antennas.

4.4.3.3 Radiation Conductance and Directivity

The radiated power can be calculated by substituting equations (4.140) and (4.141) into (4.17) as

$$\begin{aligned} P_r &= \frac{1}{2\eta_0}\int_0^{\pi/2}\int_0^{2\pi}\left(|E_\theta|^2 + |E_\phi|^2\right) r^2 \sin\theta \, d\phi d\theta \\ &\approx \frac{(k_0 a_e)^2}{960}|V_{nm}|^2 \int_0^{\pi/2} \{[J_{n+1}(k_0 a_e \sin\theta) - J_{n-1}(k_0 a_e \sin\theta)]^2 \\ &\quad + (\cos\theta)^2 [J_{n+1}(k_0 a_e \sin\theta) + J_{n-1}(k_0 a_e \sin\theta)]^2\} \sin\theta \, d\theta. \end{aligned} \tag{4.143}$$

It should be noted that the radiated fields of the resonant mode were used to replace the summation of all modes in the above calculation of the radiated power. This approximation is acceptable because the resonant mode contributes most significantly to the radiated power. Equation (4.143) can be written as

$$P_r \approx \frac{(k_0 a_e)^2}{960}|V_{nm}|^2 I_1 \tag{4.144}$$

where

$$\begin{aligned} I_1 = \int_0^{\pi/2} &\{[J_{n+1}(k_0 a_e \sin\theta) - J_{n-1}(k_0 a_e \sin\theta)]^2 + (\cos\theta)^2 [J_{n+1}(k_0 a_e \sin\theta) \\ &+ J_{n-1}(k_0 a_e \sin\theta)]^2\} \sin\theta \, d\theta. \end{aligned} \tag{4.145}$$

The radiation conductance is then

$$G_r = \frac{2P_r}{|V_{nm}|^2} \approx \frac{(k_0 a_e)^2}{480} I_1. \tag{4.146}$$

Based on the equations from (4.140) to (4.146) and using the definition of (4.14), the directivity of the differential circular microstrip patch antenna can be expressed as

$$D \approx \frac{(k_0 a_e)^2}{120 G_r} = \frac{4}{I_1}. \tag{4.147}$$

Figure 4.15 shows the calculated directivity of the differential circular microstrip patch antenna operating in the dominant TM_{11} mode as a function of the electrical radius. It is seen that the directivity fluctuates with the electrical radius between 4.8 and 12 dBi. It approaches 4.8 dBi for a very small electrical radius and reaches a maximum of 12 dBi in the case of the electrical radius of 0.9.

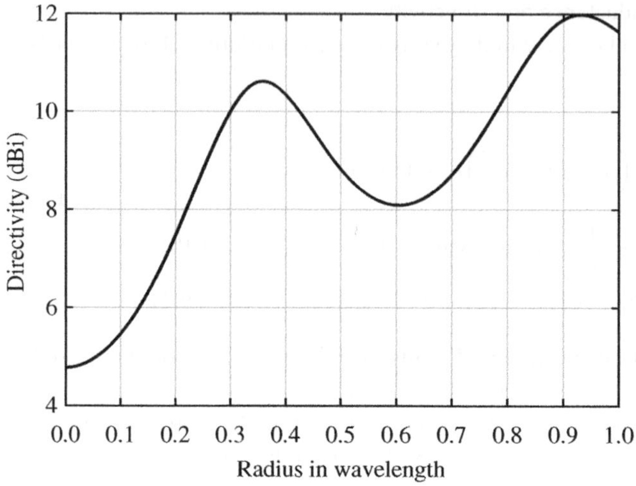

Figure 4.15 Directivity of the differential circular microstrip patch antenna operating in the dominant TM_{11} mode as a function of the electrical radius.

4.4.3.4 Radiation Efficiency and Effective Loss Tangent

The dissipated power due to the dielectric loss can be determined by integrating the electric field inside the cavity over the volume as

$$\begin{aligned} P_d &= \frac{\omega\delta\varepsilon_r\varepsilon_0}{2}\iiint_v (|E_z|^2)dv \approx \frac{\omega\delta\varepsilon_r\varepsilon_0}{2h}\frac{|V_{nm}|^2}{J_n^2(k_{nm}a_e)}\int_0^{a_e}\int_0^{2\pi} J_n^2(k_{nm}\rho)\cos^2(n\phi)\rho d\rho d\phi \\ &= \frac{\omega\delta\varepsilon_r\varepsilon_0}{2h}\frac{|V_{nm}|^2}{J_n^2(k_{nm}a_e)}\pi\int_0^{a_e} J_n^2(k_{nm}\rho)\rho d\rho \\ &= \frac{\epsilon_{0n}\omega\delta\varepsilon_r\varepsilon_0|V_{nm}|^2\pi a_e^2}{4h}\left[1-\frac{n^2}{(k_{nm}a_e)^2}\right]. \end{aligned} \tag{4.148}$$

The conductance associated with the dielectric loss is then

$$G_d = \frac{2P_d}{|V_{nm}|^2} = \frac{\epsilon_{0n}\omega\delta\varepsilon_r\varepsilon_0\pi a_e^2}{2h}\left[1-\frac{n^2}{(k_{nm}a_e)^2}\right]. \tag{4.149}$$

The dissipated power due to the conductor loss can be calculated by integrating over the patch and ground plane areas as

$$P_c = \omega\mu\frac{\delta_s}{2h}\iint_s \left(|H_\rho|^2 + |H_\phi|^2\right) ds = \frac{\mu}{\omega}\frac{\delta_s}{2}\left(\frac{1}{\mu h}\right)^2 \left[\frac{|V_{nm}|}{J_n(k_{nm}a_e)}\right]^2$$
$$\left\{\int_0^{a_e}\int_0^{2\pi} n^2\frac{J_n^2(k_{nm}\rho)}{\rho}\sin^2(n\phi)d\rho d\phi + \int_0^{a_e}\int_0^{2\pi} k_{nm}^2\left[J_n'(k_{nm}\rho)\right]^2\cos^2(n\phi)\rho d\rho d\phi\right\}$$
$$= \frac{\mu}{\omega}\frac{\delta_s}{2}\left(\frac{1}{\mu h}\right)^2\left[\frac{|V_{nm}|}{J_n(k_{nm}a_e)}\right]^2 \frac{\pi}{2}\epsilon_{0n}\left[(k_{nm}a_e)^2 - n^2\right]J_n^2(k_{nm}a_e)$$
$$= \frac{\mu}{\omega}\frac{\delta_s}{2}\left(\frac{1}{\mu h}\right)^2 |V_{nm}|^2\frac{\pi}{2}\epsilon_{0n}\left[(k_{nm}a_e)^2 - n^2\right]. \tag{4.150}$$

The conductance associated with the conductor loss is then

$$G_c = \frac{2P_c}{|V_{nm}|^2} = \frac{\mu}{\omega}\pi\frac{\delta_s}{2}\left(\frac{1}{\mu h}\right)^2 \epsilon_{0n}\left[(k_{nm}a_e)^2 - n^2\right]. \tag{4.151}$$

The total conductance is

$$G = G_r + G_d + G_c. \tag{4.152}$$

Based on equations (4.143), (4.148), and (4.150) as well as using the definition of (4.15), the radiation efficiency of the differential circular microstrip patch antenna can be calculated. Figure 4.16 shows the calculated radiation efficiency of the differential circular microstrip patch antenna operating in the dominant TM_{11} mode

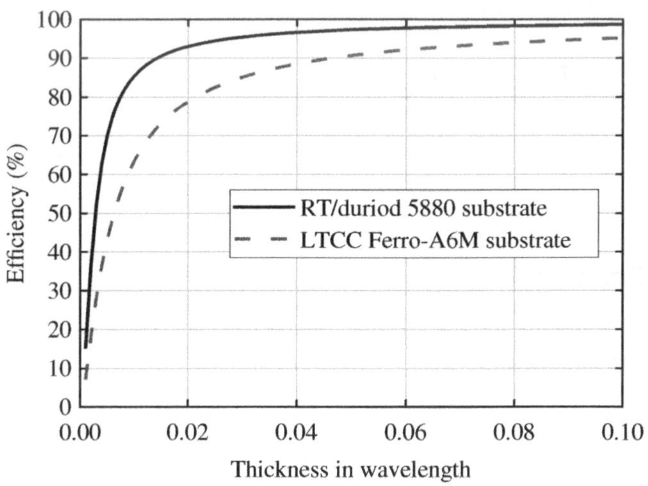

Figure 4.16 Radiation efficiency of the differential circular microstrip patch antenna operating in the dominant TM_{11} mode as a function of the electrical thickness.

as a function of the substrate electrical thickness. Note that the radiation efficiency of the differential circular microstrip patch antenna on RT/duroid 5880 substrate is higher than that on LTCC Ferro A6M substrate because the former has smaller conductor loss, substrate loss, and dielectric constant than those of the latter.

The electric energy stored in the region between the circular patch and the ground plane can be calculated as

$$\begin{aligned} U_e &= \frac{\varepsilon_r\varepsilon_0}{4}\iiint_v (|E_z|^2)dv \approx \frac{\varepsilon_r\varepsilon_0}{4h}\frac{|V_{nm}|^2}{J_n^2(k_{nm}a_e)}\int_0^{a_e}\int_0^{2\pi} J_n^2(k_{nm}\rho)\cos^2(n\phi)\rho d\rho d\phi \\ &= \frac{\omega\delta\varepsilon_r\varepsilon_0}{2h}\frac{|V_{nm}|^2}{J_n^2(k_{nm}a_e)}\pi\int_0^{a_e} J_n^2(k_{nm}\rho)\rho d\rho \\ &= \frac{\epsilon_{0n}\varepsilon_r\varepsilon_0|V_{nm}|^2\pi a_e^2}{4h}\left[1-\frac{n^2}{(k_{nm}a_e)^2}\right]. \end{aligned} \tag{4.153}$$

At resonance, the electric stored energy W_e is equal to the magnetic stored energy W_m, which leads to the total stored energy as

$$U = U_e + U_m = 2U_e = \frac{\epsilon_{0n}\varepsilon_r\varepsilon_0|V_{nm}|^2\pi a_e^2}{2h}\left[1-\frac{n^2}{(k_{nm}a_e)^2}\right]. \tag{4.154}$$

Thus, the effective loss tangent can be expressed as

$$\delta_e = \frac{P_r + P_c + P_d + P_{sw}}{\omega U} = \delta + \frac{1}{\pi h}\sqrt{\frac{\lambda_0}{120\sigma}} + \frac{h(k_0a_e)^2 I_1}{480\pi\left[(k_{nm}a_e)^2 - n^2\right]}. \tag{4.155}$$

4.4.4 Impedance Characteristics

The impedance matrix elements for the circular microstrip patch antenna with dual feeds of coaxial probes are given by

$$\begin{aligned} Z_{11} &= -j\omega\mu h\sum_{m=1}^{\infty}\sum_{n=0}^{\infty}\frac{\epsilon_{0n}}{\pi}\left\{\left[j_0\left(\frac{n\phi_w}{2}\right)\right]^2\frac{J_n^2(k_{nm}\rho_1)}{J_n^2(k_{nm}a_e)}\frac{k_{nm}^2\cos^2(n\phi_1)}{[k_0^2\varepsilon_r(1-j\delta_e)-k_{nm}^2](k_{nm}^2a_e^2-n^2)}\right\} \\ &= \sum_{m=1}^{\infty}\sum_{n=0}^{\infty} Z_{11nm}, \end{aligned} \tag{4.156}$$

$$Z_{12}=Z_{21}=-j\omega\mu h\sum_{m=1}^{\infty}\sum_{n=0}^{\infty}\frac{\epsilon_{0n}}{\pi}$$

$$\left\{\left[j_0\left(\frac{n\phi_w}{2}\right)\right]^2\frac{J_n(k_{nm}\rho_1)J_n(k_{nm}\rho_2)}{J_n^2(k_{nm}a_e)}\times\frac{k_{nm}^2\cos(n\phi_1)\cos(n\phi_2)}{\left[k_0^2\varepsilon_r(1-j\delta_e)-k_{nm}^2\right]\left(k_{nm}^2a_e^2-n^2\right)}\right\}$$

$$=\sum_{m=1}^{\infty}\sum_{n=0}^{\infty}Z_{12nm}=\sum_{m=1}^{\infty}\sum_{n=0}^{\infty}Z_{21nm}, \tag{4.157}$$

$$Z_{22}=-j\omega\mu h\sum_{m=1}^{\infty}\sum_{n=0}^{\infty}\frac{\epsilon_{0n}}{\pi}\left\{\left[j_0\left(\frac{n\phi_w}{2}\right)\right]^2\frac{J_n^2(k_{nm}\rho_2)}{J_n^2(k_{nm}a_e)}\frac{k_{nm}^2\cos^2(n\phi_2)}{\left[k_0^2\varepsilon_r(1-j\delta_e)-k_{nm}^2\right]\left(k_{nm}^2a_e^2-n^2\right)}\right\}$$

$$=\sum_{m=1}^{\infty}\sum_{n=0}^{\infty}Z_{22nm}. \tag{4.158}$$

Equations (4.156), (4.157), and (4.158) show that the self-impedances at the dual feeds and the mutual impedances between the dual feeds are the superposition of self and mutual mode impedances.

4.4.4.1 Differential Input Impedance

Substituting equations (4.156), (4.157), and (4.158) into (4.26), we can obtain the general expression for the differential input impedance. If the condition given in equation (4.112) is taken into account, the differential input impedance can be simplified as

$$Z_d = -2j\omega\mu h\sum_{m=1}^{\infty}\sum_{n=0}^{\infty}\frac{\epsilon_{0n}}{\pi}$$

$$\left\{\left[j_0\left(\frac{n\phi_w}{2}\right)\right]^2\frac{J_n^2(k_{nm}\rho_1)}{J_n^2(k_{nm}a_e)}\frac{k_{nm}^2\cos^2(n\phi_1)[1-\cos(n\pi)]}{\left[k_0^2\varepsilon_r(1-j\delta_e)-k_{nm}^2\right]\left(k_{nm}^2a_e^2-n^2\right)}\right\}. \tag{4.159}$$

If the mode index n is even in equation (4.159), the corresponding mode impedance is zero, implying that the differential circular microstrip patch antenna will excite the dominant mode TM_{11} but will not excite the static TM_{01} mode and higher-order TM_{21} mode. Figure 4.17 shows the calculated and simulated differential input impedance of the circular microstrip patch antenna with dual feeds of coaxial probes. The differential circular microstrip patch antenna is the same as that in Section 4.4.3.1. Generally, the calculated resonant frequency is slightly higher, the calculated resonant resistance is slightly lower, and the simulated input impedance is more inductive.

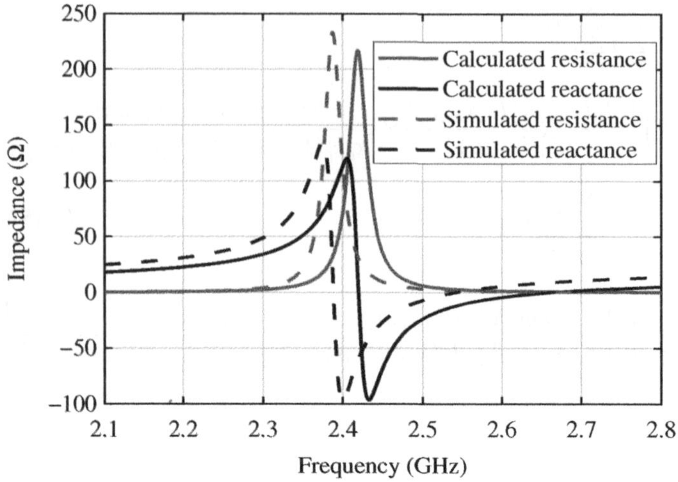

Figure 4.17 Differential input impedance of the circular microstrip patch antenna with dual feeds of coaxial probes.

4.4.4.2 Fundamental Resonance and Electrical Separation

Generally, the differential circular microstrip patch antenna operates near the resonant frequency of the dominant TM_{11} mode. Zhang observed that the fundamental resonance of a differential microstrip patch antenna depends on the electrical separation of dual-probe feeds [36]. Figure 4.18 shows the relationship between the resonance of the dominant TM_{11} mode with the electrical separation. Note that the resonance occurs when the electrical separation is larger than a specific value of 0.035. Otherwise, the differential input impedance is inductive and there is no resonance. It can also be determined that the specific value is smaller for the electrically thicker substrate of higher permittivity than that for an electrically thinner substrate of low permittivity.

4.4.4.3 Resonant Differential Input Resistance

When the differential circular microstrip patch antenna resonates in the dominant TM_{11} mode, the resonant differential input resistance can be obtained from equation (4.159) as

$$R_{d11} = 4\frac{2\omega\mu h}{\pi}\left[j_0\left(\frac{\phi_w}{2}\right)\right]^2 \frac{J_1^2(k_{11}\rho_1)}{J_1^2(k_{11}a_e)} \frac{k_{11}^2 k_0^2 \varepsilon_r \delta_{eff}}{\left(k_{11}^2 a_e^2 - 1\right)}, \tag{4.160}$$

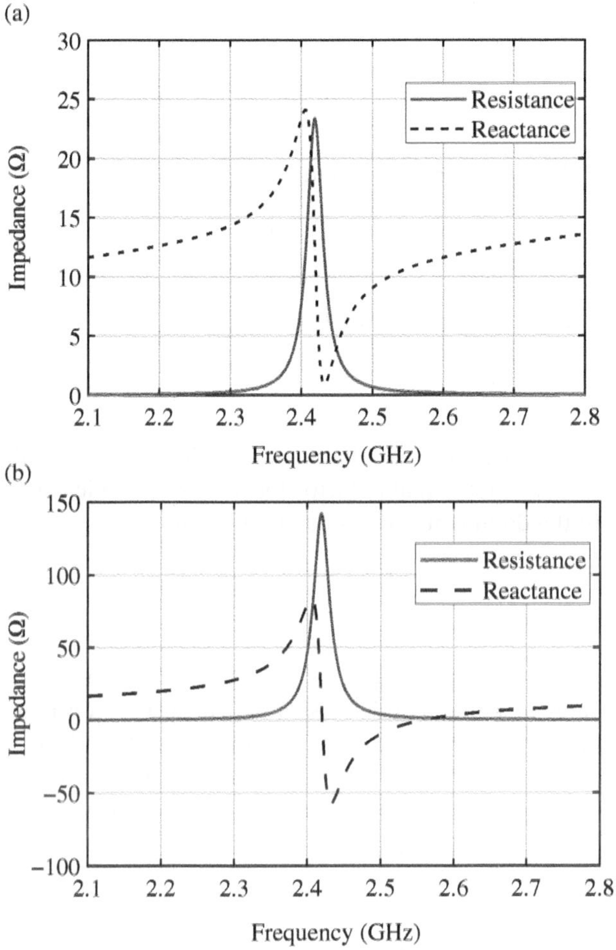

Figure 4.18 Simulated input impedance as a function of frequency for the differential circular microstrip antenna with different electrical separations of (a) 0.032 and (b) 0.08.

which reduces for very thin probes to

$$R_{d11} \;\; 4\frac{2\omega\mu h}{\pi}\frac{J_1^2(k_{11}\rho_1)}{J_1^2(k_{11}a_e)}\frac{k_{11}^2k_0^2\varepsilon_r\delta_{eff}}{\left(k_{11}^2a_e^2-1\right)} = 4\frac{2.8374h\omega^3\mu^2\varepsilon_0\varepsilon_r\delta_{eff}}{\pi a_e^2}\frac{J_1^2(k_{11}\rho_1)}{J_1^2(k_{11}a_e)}. \tag{4.161}$$

Equation (4.161) shows that the resonant differential input resistance reaches the maximum when the dual feeds are at the opposite edges of the circular microstrip patch. The resonant differential input resistance gets smaller when the dual feeds are closer.

4.4.5 Quality Factor

The quality factor due to radiation loss can be found as

$$Q_r = \frac{\omega U}{P_r} = \frac{2\omega U_e}{P_r} = \frac{\omega \epsilon_{0n} \epsilon_r \epsilon_0 \pi a_e^2}{hG_r}\left[1 - \frac{n^2}{(k_{nm}a_e)^2}\right]. \tag{4.162}$$

For the dominant TM_{11} mode, equation (4.162) reduces to

$$Q_r = \frac{1.41\omega\varepsilon_r\varepsilon_0\pi a_e^2}{hG_r}. \tag{4.163}$$

The calculation from equation (4.163) reveals that the value of Q_r decreases with increasing h/λ or reducing ε_r. The total quality factor for the circular microstrip patch antenna operating in the dominant TM_{11} mode is given by

$$Q = \left(\frac{hG_r}{1.41\omega\varepsilon_0\varepsilon_r\pi a_e^2} + \frac{1}{\pi h}\sqrt{\frac{\lambda}{120\sigma}} + \delta\right)^{-1}. \tag{4.164}$$

The differential input impedance bandwidths for various circular microstrip patch antennas on substrates of different thicknesses and dielectric constants can be more conveniently calculated using the quality factor and VSWR. The calculated results are listed in Table 4.4. Note that differential input impedance bandwidth increases with increasing substrate electrical thickness but decreases with

Table 4.4 Input impedance bandwidths of differential circular microstrip patch antennas.

	BW% for VSWR ≤ 2				
h/λ	$\varepsilon_r = 1.00$	$\varepsilon_r = 2.21$	$\varepsilon_r = 2.55$	$\varepsilon_r = 4.70$	$\varepsilon_r = 10.2$
0.005	0.93	0.86	0.92	0.84	0.63
0.010	1.36	1.20	1.22	1.02	0.65
0.015	1.85	1.61	1.61	1.28	0.78
0.020	2.35	2.04	2.43	1.87	1.08
0.025	2.84	2.47	2.41	1.67	0.95
0.030	3.32	2.88	2.83	2.16	1.25
0.035	3.79	3.30	3.22	2.46	1.41
0.040	4.25	3.70	3.61	2.77	1.58

increasing substrate dielectric constant. The differential input impedance bandwidth is typically less than a few percent.

4.4.6 Differential Versus Single-Ended Circular Microstrip Patch Antennas

There have been much less studies about differential as compared with single-ended circular microstrip patch antennas although the first microstrip patch antenna paper is about a differential circular microstrip patch antenna [6]. This section compares differential with single-ended circular microstrip patch antennas in terms of fundamental difference and performance improvement.

4.4.6.1 Excitation and Suppression of Resonant Modes

There are numerous resonant modes in a circular microstrip patch antenna cavity. The fundamental difference between differential and single-ended circular microstrip patch antennas lies in the excitation and suppression of resonant modes. Table 4.5 lists the first seven resonant modes. Note that all these modes can be generated in a single-ended circular microstrip patch antenna but only three of them can be excited in a differential circular microstrip patch antenna. The TM_{01} mode is static, which is automatically suppressed in the differential circular microstrip patch antenna but needs to be suppressed by inserting a shorting-pin in the center of the single-ended circular microstrip patch antenna. The TM_{11} mode is the dominant mode. Both differential and single-ended circular microstrip patch antennas are commonly designed to operate in this mode. The TM_{21} mode is the higher-order mode closest to the dominant TM_{11} mode, which will distort the radiation pattern in the E plane, increase the cross-polarized radiation, and reduce the AR bandwidth of the single-ended circular microstrip patch antenna. Fortunately, this undesirable higher-order mode is automatically suppressed in the differential circular microstrip patch antenna. The other higher-order modes are far away from the dominant TM_{11} mode. They will not cause serious problems and thus can be ignored.

Table 4.5 First seven resonant modes.

Resonant modes	TM_{01}	TM_{11}	TM_{21}	TM_{02}	TM_{31}	TM_{41}	TM_{12}
Differential	Suppressed	Excited	Suppressed	Suppressed	Excited	Suppressed	Excited
Single-ended	Excited	Excited	Excited	Excited	Excited	Excited	Excited

4.4.6.2 Input Impedance Ratio

For the circular microstrip patch antenna with dual feeds of coaxial probes, the differential input impedance was given by equation (4.159) and the single-ended input impedance at feed 1 with feed 2 being opened is actually Z_{11} given by equation (4.156). It is meaningful to examine the relationship between the differential and single-ended input impedances. It can be found from (4.156) and (4.157) that

$$\frac{Z_{11nm}}{Z_{12nm}} = \frac{\psi_{nm}(\rho_1, \phi_1)}{\psi_{nm}(\rho_2, \phi_2)} = \frac{J_n(k_{nm}\rho_1)\cos(n\phi_1)}{J_n(k_{nm}\rho_2)\cos(n\phi_2)}. \tag{4.165}$$

Since $\rho_1 = \rho_2$ and $\phi_2 = \pi + \phi_1$, thus

$$\frac{Z_{11nm}}{Z_{12nm}} = \frac{\psi_{nm}(\rho_1, \phi_1)}{\psi_{nm}(\rho_2, \phi_2)} = \frac{J_n(k_{nm}\rho_1)\cos(n\phi_1)}{J_n(k_{nm}\rho_2)\cos(n\phi_2)} = \frac{1}{\cos(n\pi)} = \begin{cases} 1 & for\, n = 0, 2, 4 \ldots \\ -1 & for\, n = 1, 3, 5 \ldots \end{cases}. \tag{4.166}$$

As a result, the differential input impedance Z_d is

$$Z_d = 2\sum_{m=1}^{\infty}\sum_{n=0}^{\infty}(Z_{11nm} - Z_{12nm}) = 4\sum_{m=1}^{\infty}\sum_{n=2i+1}^{\infty} Z_{11nm} \tag{4.167}$$

where $i = 0, 1, 2, 3, \ldots$. It is important to note that Z_d has no term of n being an even number. As for the single-ended input impedance Zs, it is given by

$$Z_s = \sum_{m=1}^{\infty}\sum_{n=0}^{\infty} Z_{11nm}. \tag{4.168}$$

Note that Z_s has terms regardless of n being odd or even numbers. Thus, the ratio of Z_d to Z_s is

$$\frac{Z_d}{Z_s} = 4\left(\sum_{m=1}^{\infty}\sum_{n=2i+1}^{\infty} Z_{11nm}\right) \Bigg/ \left(\sum_{m=1}^{\infty}\sum_{n=0}^{\infty} Z_{11nm}\right) \approx 4. \tag{4.169}$$

Figure 4.19 shows the calculated variation of input resistances with the feed locations at the resonant frequency of the dominant TM_{11} mode. Note that both resistances increase with the probes moving to the edge of the patch. It is interesting to note that the differential resistance is indeed about four times the single-ended resistance. This means that if the dual-feed antenna is designed to well match to a 50-Ω signal source for the single-ended excitation, it will not well match to a 100-Ω but a 200-Ω signal source for the differential excitation.

4.4.6.3 Reduction of Cross-Polarized Radiation

A differential circular microstrip patch antenna that produces much lower cross-polarized radiation than a single-ended circular microstrip patch antenna does has

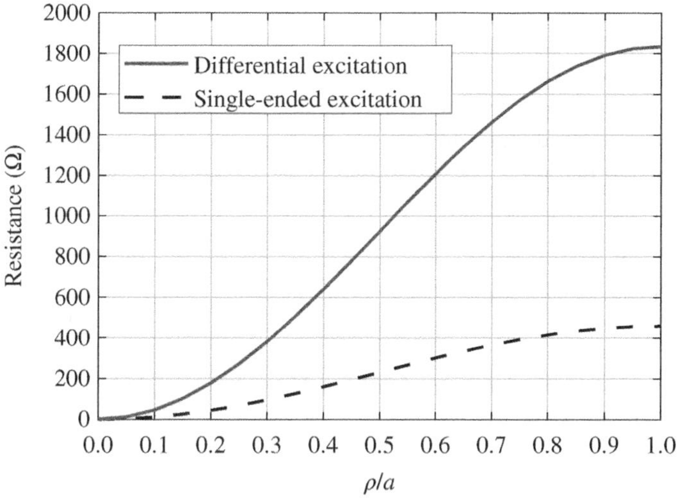

Figure 4.19 Variation of input resistances at the resonant frequency of the dominant TM_{11} mode with the feed locations.

long been known [46]. The single-ended circular microstrip patch antenna on a thick and low-permittivity substrate can achieve broadband operation. However, such a broadband single-ended circular microstrip patch antenna produces high cross-polarized radiation and pattern distortion due to unwanted probe radiation and excitation of the higher-order TM_{21} mode. In addition, the feeding from the single probe that causes the radiation with different time delays at the radiating edge further enhances cross-polarized level. The broadband differential circular microstrip patch antenna, by introducing a second probe and fed out of phase, can effectively suppress the probe radiation and the TM_{21} mode as well as balance the time delays at the radiating edge. Therefore, the broadband differential circular microstrip patch antenna has much lower cross-polarized radiation.

Vollbracht analyzed the cross-polarized radiation of both differential and single-ended circular microstrip patch antennas [38]. The circular microstrip patch antenna had radius $a = 6.335$ mm on RO6002 substrate ($\varepsilon_r = 2.94$, $\delta = 0.0012$, $h = 0.508$ mm). The probes had the same diameter of 0.4 mm and were separated by 11.67 mm. The resonance frequency was at 9.395 GHz with a bandwidth of 152 MHz for VSWR ≤ 2. The *XPL* of 14 dB for the differential circular microstrip patch antennas was established at $\pm 60°$ azimuth position in the *D* plane, which is more than 7 dB better than the *XPL* of the single-ended circular microstrip patch antenna. He concluded that the value of 14 dB somehow could reflect the theoretical maximum of *XPL* at $\pm 60°$ for microstrip patch antennas.

4.4.6.4 Axial Ratio Improvement

Unlike the rectangular microstrip patch, the circular microstrip patch is required to have the same radius for both differential and single-ended operations at the same frequency. Consequently, the differential circular microstrip patch antenna has a similar impedance bandwidth and radiation efficiency as its single-ended counterpart as indicated below

$$\frac{BW_d}{BW_s} = \frac{Q_s}{Q_{dif}} \approx \frac{a}{a} = 1, \tag{4.170}$$

$$\frac{e_d}{e_s} = \frac{Q_{dif}}{Q_s}\frac{Q_{rs}}{Q_{rdif}} = 1. \tag{4.171}$$

For the circular microstrip patch antenna with dual feeds of coaxial probes, if the probe positions are changed to

$$\rho_1 = \rho_2 \, and \, \phi_2 - \phi_1 = \pi/2 \tag{4.172}$$

and the probes are fed with two signals of the same amplitude but different phases by ±90°, the single-ended circular microstrip patch antenna radiates a right-hand or left-hand circularly polarized wave. Suzuki observed that the unwanted TM_{21} mode gives rise to some coupling between the orthogonal dual feeds, which degrades the AR as a result [46]. The degradation increases with a lower quality factor.

Since the differential circular microstrip patch antenna can suppress the unwanted TM_{21} mode, Suzuki proposed the differential schemes by adding new dual feeds mirrored to original dual feeds of (4.172) as

$$\rho_1 = \rho_2 = \rho_3 = \rho_4 \, and \, \phi_2 - \phi_1 = \phi_3 - \phi_2 = \phi_4 - \phi_3 = \pi/2 \tag{4.173}$$

and exciting all four feeds ±90° out of phase from each other with equal amplitude for LHCP and right-hand circular polarization (RHCP), respectively.

4.4.7 Design Procedure

The design procedure of a single-ended circular microstrip patch antenna has been described in [46]. The design procedure is enriched for differential circular microstrip patch antennas below with design examples for single and dual linear polarizations as well as for circular polarizations.

4.4.7.1 Design for Linear Polarization

Let us assume that the design requires the differential circular patch antenna to operate at 1268.52 MHz with a bandwidth of 20.46 MHz for VSWR ≤2.0. In this case, the resonant frequency and relative bandwidth of the differential circular patch antenna are specified as $f_r = 1268.52$ MHz and BW = 20.46/1268.52 ≈1.61%.

Step 1: RT/duroid 5880 substrate is selected for its low dielectric constant $\varepsilon_r = 2.2$ and low dielectric loss $\delta = 0.0009$.

Step 2: The patch effective radius a_e is calculated to be 46.74 mm from (4.128) for the resonant frequency and the thickness of the substrate h is $0.015\lambda = 3.54$ mm from Table 4.4 for the relative bandwidth.

Step 3: The patch radius a can be obtained by solving

$$46.7 = a\left[1 + \frac{2h}{\pi a \varepsilon_r} \ln\left(\frac{\pi a}{2h} + 1.7726\right)\right]^{1/2}, \tag{4.174}$$

which yields $a \approx 45$ mm. The substrate with the thickness $h = 3.8$ mm is selected for commercial availability.

Step 4: The locations of the dual feeds need to be determined. The typical impedance of a differential signal source is 100 Ω. Hence, the self-impedance at either feed should be 25 Ω in order to match to the 100-Ω source. The location of feed 1 can be found from equation (4.161) to be $\rho_1 = 8$ mm. The location of feed 2 can be obtained by rotating feed 1 by 180°.

Step 5: The electrical separation should be checked. If the electrical separation condition is satisfied, the resonance will be guaranteed, otherwise, the design should be modified. It is reported in [36] that the perturbation of the cavity field will affect the degree of electrical separation and may cause the occurrence of resonance of the (quasi) TM_{11} mode even if the electrical separation condition is not satisfied. It is known that the cavity field of the circular microstrip patch antenna can be perturbed by cutting a small rectangular slot between the dual feeds [36]. The slot does make the resonance of the (quasi) TM_{11} mode occur but at a lower frequency. The lowered resonant frequency can be increased simply by reducing the patch radius.

Step 6: The design should be optimized in an electromagnetic full-wave solver before sending it to a manufacturer for fabrication.

Figure 4.20 shows the simulated $|S_{d11}|$ as a function of frequency. As shown, the design satisfies the bandwidth requirement of 1.61% at 1268.52 MHz.

The above design procedure can be easily extended to dual linear polarizations just simply by adding new dual feeds orthogonal to the original dual feeds.

4.4.7.2 Design for Circular Polarization

In Section 4.4.7.1, the differential circular microstrip patch antenna that requires four feeds for circular polarization was mentioned, which improves the AR performance but makes the feeding cumbersome. The design procedure developed below explores degenerate orthogonal dominant TM_{11} modes with dual feeds of coaxial probes for circular polarization. The key to rely on dual feeds to facilitate the feeding is to introduce a small perturbation into the circular patch geometry, which separates orthogonal dominant TM_{11} modes into two independent modes of

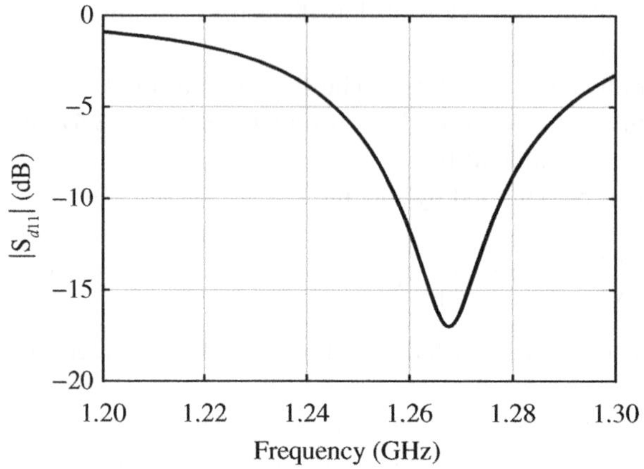

Figure 4.20 The simulated $|S_{d11}|$ as a function of frequency.

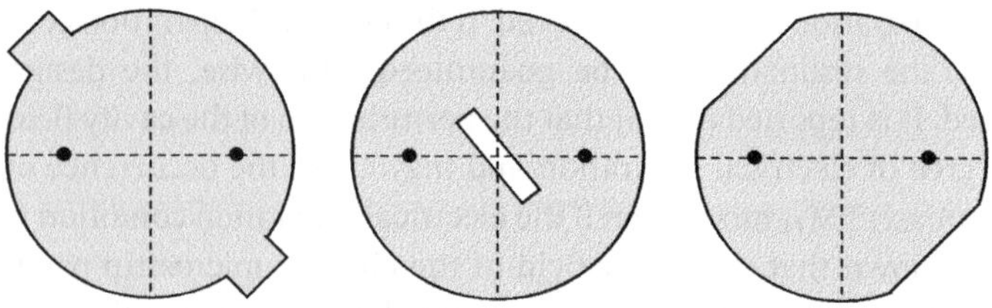

Figure 4.21 Typical perturbed circular patch geometries for circular polarization.

slightly different resonant frequencies [46]. Figure 4.21 shows typical perturbed circular patches. They radiate circularly polarized waves when excited at the dual feeds on either the horizontal or vertical dotted lines and are satisfied with the condition given by

$$\frac{\Delta A}{A} = \frac{1}{1.841Q} \tag{4.175}$$

where ΔA is the change in area from area A and Q is the quality factor of the original circular patch.

Let us assume that the design requires the differential circular patch antenna to operate at 1176.45 MHz with a bandwidth of 12 MHz for AR $\leq$3 dB. In this case, the resonant frequency and relative AR bandwidth of the differential circular patch antenna are specified as $f_r = 1176.45$ MHz and $\text{BW}_{\text{AR}} = 12/1176.45 \approx 1.02\%$.

Step 1: RT/duroid 5880 substrate is selected for its low dielectric constant $\varepsilon_r = 2.2$ and low dielectric loss $\delta = 0.0009$.

Step 2: The total Q factor is estimated by

$$Q = \frac{0.348}{BW_{AR}} = \frac{0.348}{0.0102} \approx 34.12 \tag{4.176}$$

for the relative AR bandwidth [46]. The impedance bandwidth can be calculated as

$$BW = \frac{1}{Q\sqrt{2}} = \frac{1}{34.12 \times 1.414} \approx 2.1\%. \tag{4.177}$$

Step 3: The thickness of the substrate h is $0.025\lambda = 6.3$ mm from Table 4.4 for the relative bandwidth. The patch radius a can be obtained by solving

$$50.4 = a\left[1 + \frac{2h}{\pi a \varepsilon_r} \ln\left(\frac{\pi a}{2h} + 1.7726\right)\right]^{1/2}, \tag{4.178}$$

which yields $a \approx 48$ mm.

Step 4: The locations of the dual feeds need to be determined. The typical impedance of a differential signal source is 100 Ω. Hence, the self-impedance at either feed should be 25 Ω in order to match to the 100-Ω source. The location of feed 1 can be found from equation (4.161) to be $\rho_1 = 8$ mm. The location of feed 2 can be obtained by rotating feed 1 by 180°.

Step 5: The electrical separation should be checked. If the electrical separation condition is satisfied, the resonance will be guaranteed, otherwise, the design should be modified.

Step 6: The design should be optimized in an electromagnetic full-wave solver before sending it to a manufacturer for fabrication. The optimized values are $a = 46.2$ mm, $h = 6.3$ mm, and $\Delta A = 31.5 \times 4 = 126$ mm^2.

Figure 4.22 shows the simulated AR as a function of frequency. As shown, the design satisfies the AR bandwidth requirement of 1.02% at 1176.45 MHz.

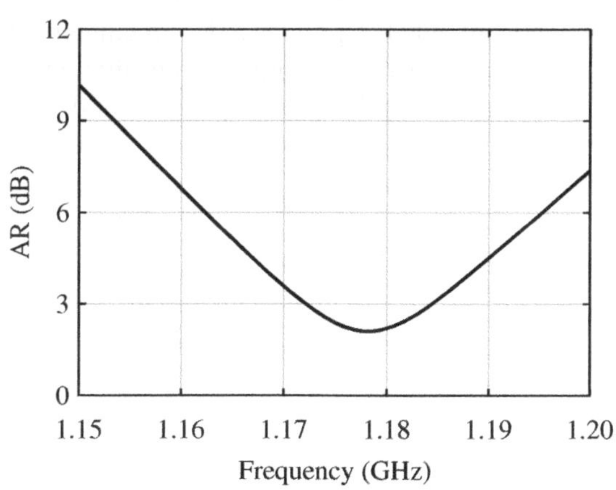

Figure 4.22 The simulated AR as a function of frequency.

4.5 Stacked Patch

A major disadvantage of basic microstrip patch antennas is narrow impedance bandwidth. The typical impedance bandwidth of a simple microstrip patch antenna is on the order of a few percent. Hence, much effort has been devoted to broadening impedance bandwidth. Some useful broadband techniques have been devised [51–56]. Among them, the technique that explores multiple resonances by stacking parasitic patches is the most popular one because it not only delivers wide impedance bandwidth but also provides additional electrical, mechanical, and thermal advantages – broad beamwidth, lower sidelobe levels, dual polarization, filter characteristics, easy-to-implement electrostatic discharge (ESD), and to meet the requirement of metal density, and beneficial to dissipate heat [57, 58].

The stacked patch technique was first reported by Hall, Wood, and Garrett in 1979 [59]. They showed that impedance bandwidth in excess of 16 times (from 1.1% to 18%) that of basic microstrip patch antennas on alumina substrate may be obtained at the expense of additional manufacturing complexity and some increase in overall height. Subsequently, Corq and Pozar demonstrated that the aperture-coupled stacked microstrip patch antenna can achieve wideband impedance bandwidth and meanwhile can naturally isolate the feed network from the patches [60]. Liu et al. presented a systematic method of designing stacked microstrip patch antennas with proximity feeding to compensate for the probe inductance [61]. Waterhouse described a strategy for creating probe-fed stacked microstrip patch antennas with bandwidths of more than 25% [62]. Gao et al. discussed how to design a stacked microstrip patch antenna for dual polarization [63].

Figure 4.23 shows the geometry and coordinate system of a stacked rectangular microstrip patch antenna. Note that the lower patch is driven directly with dual feeds of coaxial probes and the upper patch is coupled electromagnetically. This section is intended to provide a design guideline for the stacked microstrip patch antenna with particular emphasis on differential feeding [26]. It begins with a description of the principle of operation of the stacked microstrip patch antenna. This is followed by the determination of the resonant frequencies of the stacked microstrip patch antenna. Next, design and optimization issues are addressed. Finally, the simulated and measured results of the two-layer and three-layer stacked rectangular microstrip patch antennas are discussed.

4.5.1 Quasi-Even and Odd Modes

Unlike the conventional explanation of the principle of operation of a stacked microstrip patch antenna, we consider the stacked microstrip patch antenna as

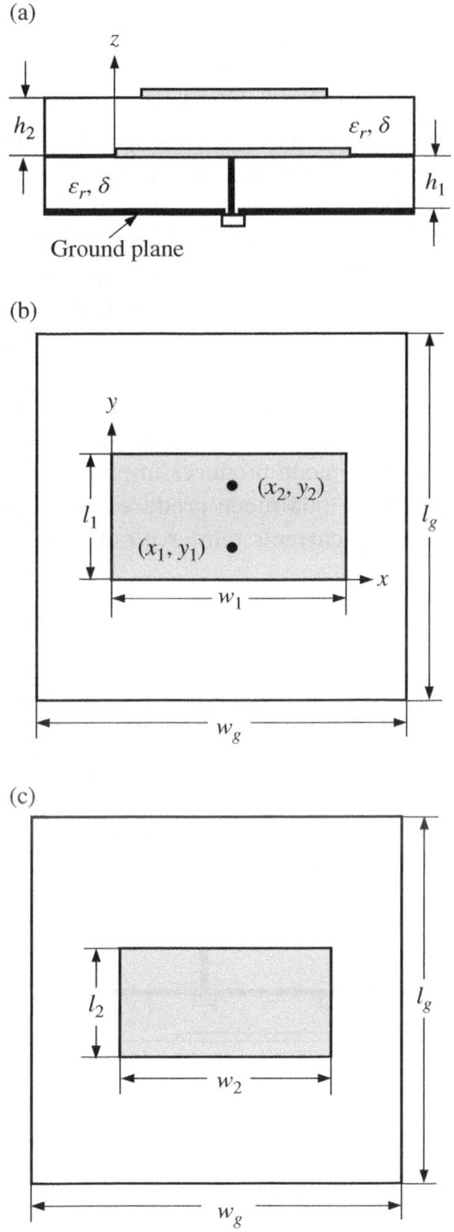

Figure 4.23 Geometry and coordinate system of the stacked rectangular microstrip patch antenna with dual feeds of coaxial probes. (a) Side view, (b) bottom view, (c) top view.

stacked two microstrip lines. Hence, we can apply quasi-even and odd modes to provide new insight into the resonance, coupling, and radiation of the stacked microstrip patch antenna.

Figure 4.24 shows the sketches of the electric field distributions on the cross sections of the stacked microstrip patch antenna for both quasi-even and odd modes. For the even mode, the charges are the same kind on both upper and lower patches, which results in a weak electric field distribution between the patches but a strong electric field distribution between the lower patch and the ground plane. On the contrary, the charges are the opposite kinds on both upper and lower patches for the odd mode, which results in a strong electric field distribution between the patches but a weak electric field distribution between the lower patch and the ground plane.

Figure 4.25 shows the sketches of the current distributions on the patches of the stacked microstrip patch antenna for the even and odd modes. It is seen that the even mode produces in-phase currents on both upper and lower patches, while the odd mode produces out-of-phase currents on the two patches. The in-phase currents reinforce radiation, while the out-of-phase currents weaken radiation.

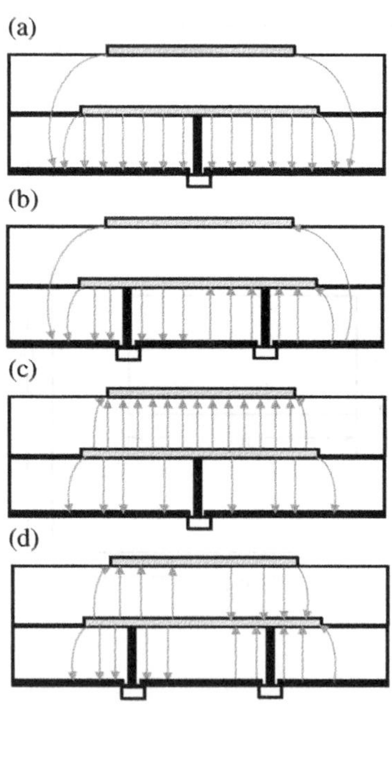

Figure 4.24 Sketched electric field distributions on the cross sections of the stacked microstrip patch antenna. (a) Even mode on the *xoz* plane, (b) even mode on the *yoz* plane, (c) odd mode on the *xoz* plane, and (d) odd mode on the *yoz* plane.

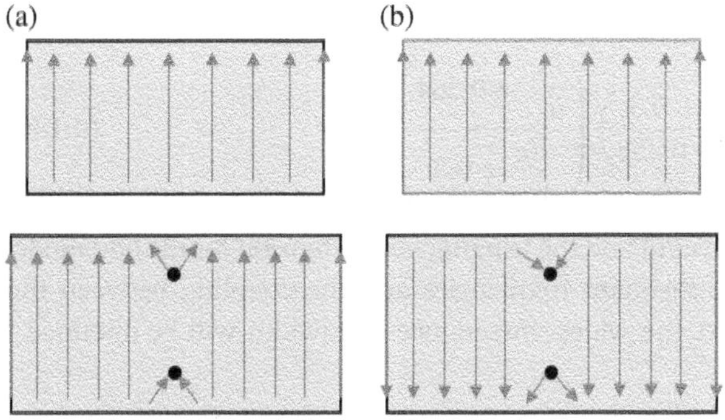

Figure 4.25 Sketched current distributions on the patches of the two-layer stacked antenna. (a) Even mode and (b) odd mode.

4.5.2 Resonant Frequencies

For the stacked rectangular microstrip patch antenna shown in Figure 4.23, it is found that the lower and upper patches with the ground plane form one resonator and the upper and lower patches form the other resonator. The lower resonant frequency is related to the even mode and can be estimated by

$$f_l = \frac{c}{2\sqrt{\varepsilon_{e1}}(l_2 + 2\Delta l_{even})} \tag{4.179}$$

where

$$\varepsilon_{e1} = \frac{\varepsilon_r + 1}{2} + \frac{\varepsilon_r - 1}{2}\left(1 + 12\frac{h_1 + h_2}{w_2}\right)^{-\frac{1}{2}} \tag{4.180}$$

and

$$\Delta l_{even} = 0.412\,(h_1 + h_2)\left(\frac{\varepsilon_{e1} + 0.3}{\varepsilon_{e1} - 0.258}\right)\left(\frac{\dfrac{h_1 + h_2}{w_2} + 0.264}{\dfrac{h_1 + h_2}{w_2} + 0.8}\right) \tag{4.181}$$

are the effective dielectric constant and the fringing field extension for the even mode. Similarly, the upper resonant frequency is related to the odd mode and can be estimated by

$$f_u = \frac{c}{2\sqrt{\varepsilon_{e2}}(l_1 + 2\Delta l_{odd})} \tag{4.182}$$

where

$$\varepsilon_{e2} = \frac{\varepsilon_r + 1}{2} + \frac{\varepsilon_r - 1}{2}\left(1 + 12\frac{h_2}{w_1}\right)^{-\frac{1}{2}} \tag{4.183}$$

and

$$\Delta l_{odd} = 0.412h_2\left(\frac{\varepsilon_{e2}+0.3}{\varepsilon_{e2}-0.258}\right)\left(\frac{\frac{h_2}{w_1}+0.264}{\frac{h_2}{w_1}+0.8}\right) \tag{4.184}$$

are the effective dielectric constant and the fringing field extension for the odd mode. If the separation of the two resonant frequencies and the coupling between the two resonators are optimized, the widest impedance bandwidth will be obtained.

4.5.3 Design and Optimization

The design rules for probe-fed single-ended stacked patch antennas can be found in [62]. The design of dual-probe-fed differential stacked patch antennas involves simulation in high-frequency structure simulator (HFSS) and optimization in Matlab. A script needs to be written to link the HFSS with Matlab. Three popular optimization methods are genetic algorithm (GA), particle swarm optimization (PSO), and bat algorithm (BA). The GA is a metaheuristic inspired by the process of natural selection, which has been used for optimization of probe-fed single-ended stacked antennas [64]. The PSO is a heuristic global optimization approach based on the movement and intelligence of swarms. The BA is a procedure for global optimization, which was inspired by the bat behavior of using echolocation to detect prey and avoid obstacles [65]. It was found that BA is potentially more powerful than GA and PSO, due to combining major advantages of both GA and PSO algorithms. We used the GA, PSO, and BA to optimize stacked microstrip patch antennas and confirmed that the BA achieves convergence faster than the GA and PSO [26].

During the BA process, each potential solution is represented as a bat in the population P. The position vector and moving velocity of bat groups in a P-dimensional search space are updated for every iteration. The new solution from every iteration is accepted if the fitness function is satisfied. To achieve the desired impedance bandwidth, the fitness function is defined as the magnitude of S_{d11} being less than or equal to − 10 dB.

Based on the BA process and the given information on the dielectric constants, loss tangents, and thicknesses of the multilayer substrates as well as the targeted impedance bandwidth, a design procedure is outlined as follows.

Specify: Target the impedance bandwidth from f_l to f_u. Select the substrate to know ε_r, δ, and the one-layer thicknesses t. The thicknesses of h_1, h_2, or h_3 are multiples of the one-layer thickness t.

Determine: The dimensions of the lower and upper patches, the thicknesses of substrates, and the separation of the dual feeds (w_1, l_1, w_2, l_2, h_1, h_2, d). To speed up the design and optimization, the thicknesses of h_1, h_2 can be first fixed due to the fact that they are multiples of the one-layer thickness.

Procedure: (1) The initial dimensional values of the lower and upper patches (w_1, l_1, w_2, l_2) are estimated by equations from (4.179) to (4.184). for the targeted frequency range with the given substrate. The initial value for the separation of the dual feeds can be chosen to be smaller than l_1. (2) The minimum and maximum values of (w_1, l_1, w_2, l_2, d) are selected. (3) Run the HFSS simulation with the initial values. (4) Get new values for (w_1, l_1, w_2, l_2, d) from the BA. (5) Run the HFSS simulation with the new values. (6) Repeat steps (4) and (5) until the targeted frequency range is covered.

For the three-layer or more layered design, the above procedure needs to be modified slightly. First, the minimum and maximum dimensional values are selected. Then, the initial dimensional values can be generated randomly between the minimum and maximum values. Next, run the HFSS simulation with the initial values. After that, get new values from the BA and run the HFSS simulation with the new values. This is repeated until the targeted frequency range is covered.

4.5.4 Design Examples

Two stacked rectangular microstrip patch antennas are designed and optimized for linear polarization using a standard FR4 substrate ($\varepsilon_r = 4.4$, $\delta = 0.02$, $t = 2$ mm) with a size of 50×50 mm^2. For the two-later design, the thicknesses h_1 and h_2 are 4 mm. For the three-layer design, one more dielectric layer with the thickness h_3 and one more patch with the width w_3 and the length l_3 are added on the top surface of the two-layer design. The thicknesses h_1 and h_3 are 4 mm and the thickness h_2 is 2 mm. The impedance bandwidth is targeted from 3.3 to 4.2 GHz.

Table 4.6 shows the description of the parameters to be optimized from the initial values within the minimum and maximum values. Table 4.7 shows the final

Table 4.6 Optimization parameters and description.

Parameters	Min	Max	Initial value	Description
l_1	5	20	12.5	Length of the first layer patch
l_2	5	20	12.5	Length of the second layer patch
l_3	5	20	12.5	Length of the third layer patch
w_1	5	35	20	Width of the first layer patch
w_2	5	35	20	Width of the second layer patch
w_3	5	35	20	Width of the third layer patch
d	5	20	12.5	Distance between the two probes

Table 4.7 Optimized dimensions (mm).

Dimensions	l_1	w_1	l_2	w_2	l_3	w_3	d
Two layers	15.74	22.48	13.92	20.07			12.01
Three layers	16.33	22.07	11.42	33.40	8.78	7.80	13.49

optimized values. For the two-layer design, it is seen that the upper patch is smaller than the lower patch, which is different from the suggestion that the upper patch be equal to or slightly larger than the lower patch in order to obtain a suitable coupling between them for a wider impedance bandwidth [53].

Both two-layer and three-layer design prototypes were first fabricated in a single-layer PCB process and then assembled together by nylon screws. Figure 4.26 shows the photo of the assembled samples.

Figure 4.27 shows the simulated electric field distributions on the cross-section of the final optimized two-layer design. Figure 4.28 shows the simulated current distributions on the upper and lower rectangular patches of the final optimized two-layer design. They have validated the explanation of the principle of operation of the two-layer stacked rectangular microstrip patch antennas based on the quasi-even and odd modes.

Figure 4.26 Photo of assembled two-layer and three-layer samples.

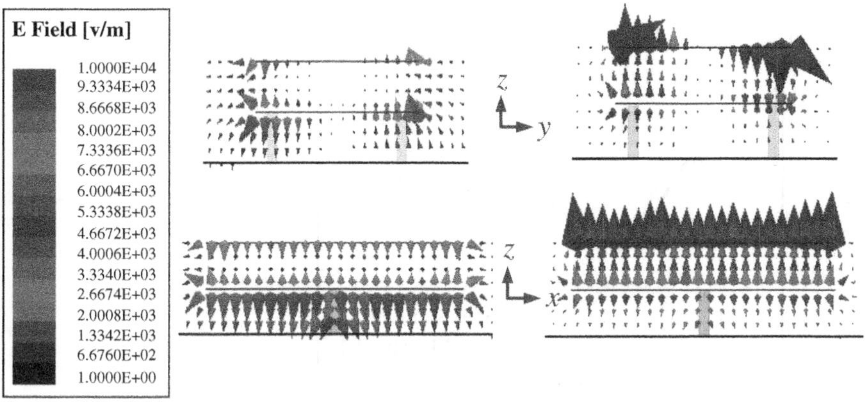

Figure 4.27 Simulated electric field distributions of the two-layer stacked antenna where the dielectric substrate is not shown for clarity.

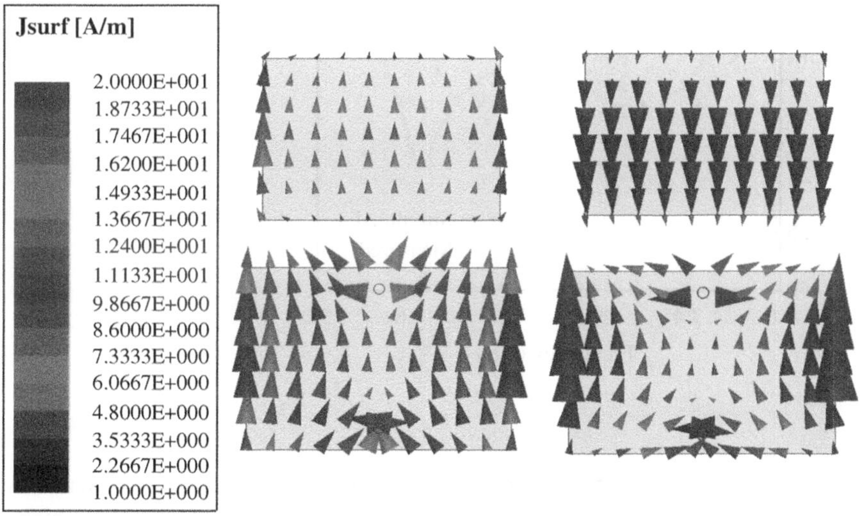

Figure 4.28 Simulated current distributions on the patches of the two-layer stacked antenna.

4.5.5 Results and Discussion

The Agilent network analyzer E5071C was employed for measuring S_{d11}. The two differential ports of the stacked patch antenna were directly connected to the network analyzer in the measurement. Figure 4.29 compares the simulated and measured

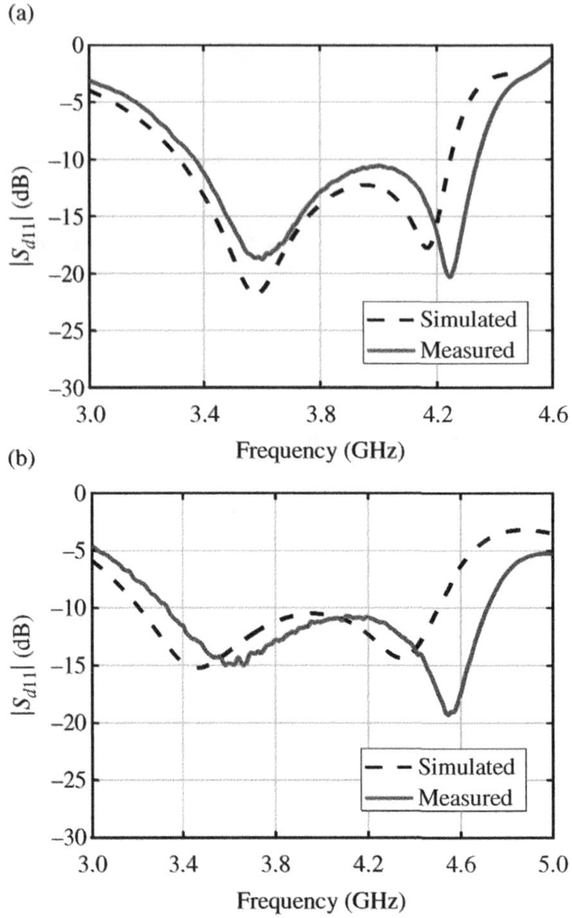

Figure 4.29 Simulated and measured $|S_{d11}|$. (a) Two-layer and (b) three-layer structures.

$|S_{d11}|$ for the stacked patch antennas as a function of frequency [49]. Generally, they are in good agreement. However, the measured curves shift slightly to higher frequencies for both two-layer and three-layer prototypes, which is due to the smaller dielectric constant possessed by the substrate than that used in the simulation. The two-layer design obtained 24.6% simulated and 25.5% measured impedance bandwidths. As expected, the three-layer design has a wider impedance bandwidth. It achieved 33.6% simulated and 34.9% measured impedance bandwidths.

The ET Industries J-212-180 hybrid junction was employed as a balun for measuring the radiation patterns of the differential stacked patch antennas and the effects of the balun and additional cables were calibrated to obtain the gain values. Figure 4.30 shows the simulated and measured gain and radiation efficiency over the frequency

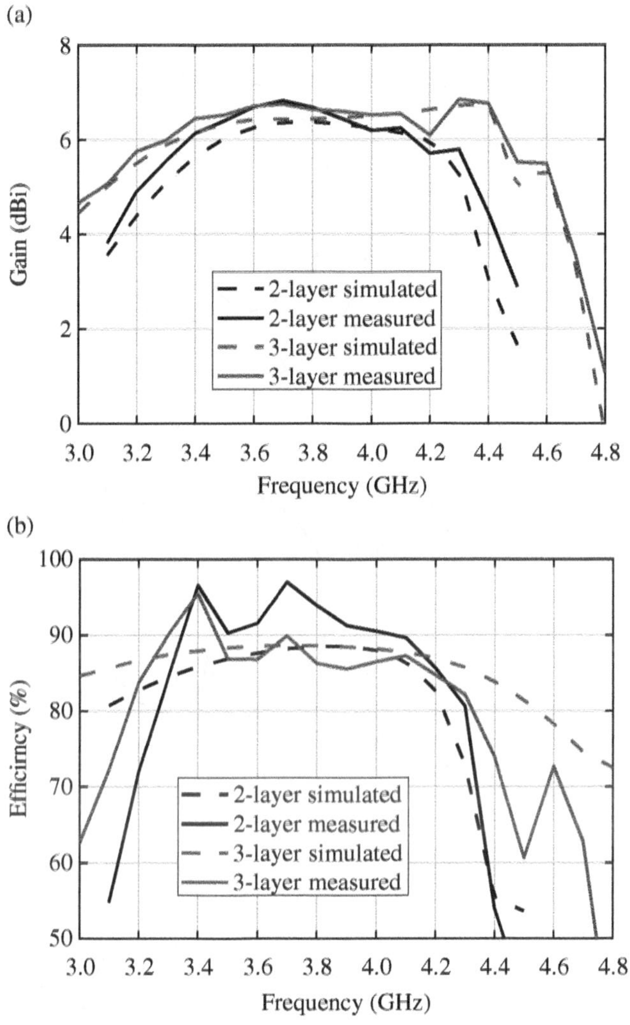

Figure 4.30 Simulated and measured gains and efficiencies. (a) Two-layer and (b) three-layer structures.

range from 3 to 4.8 GHz. Note that the simulated and measured gain curves agree well. The three-layer design has a wider 3-dB gain bandwidth than the two-layer design. However, there are large differences between the simulated and measured efficiency values at some frequencies. It is known that the efficiency of a microstrip patch antenna is hard to be simulated and measured accurately [48].

Figures 4.31 and 4.32 show the simulated and measured radiation patterns for the two-layer and three-layer prototypes, respectively. For the two-layer prototype,

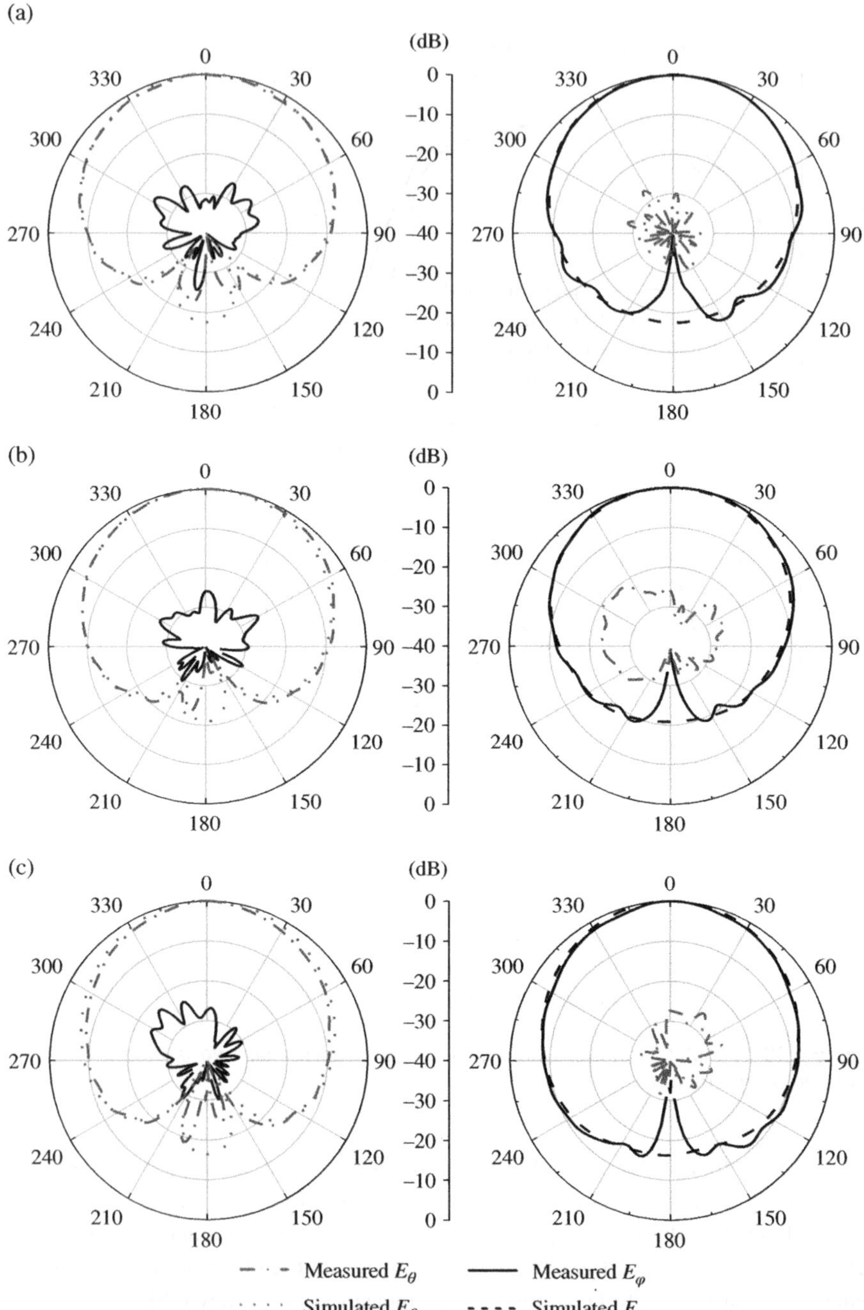

Figure 4.31 Simulated and measured radiation patterns of the two-layer design at (a) 3.3 GHz (b) 3.8 GHz, and (c) 4.3 GHz.

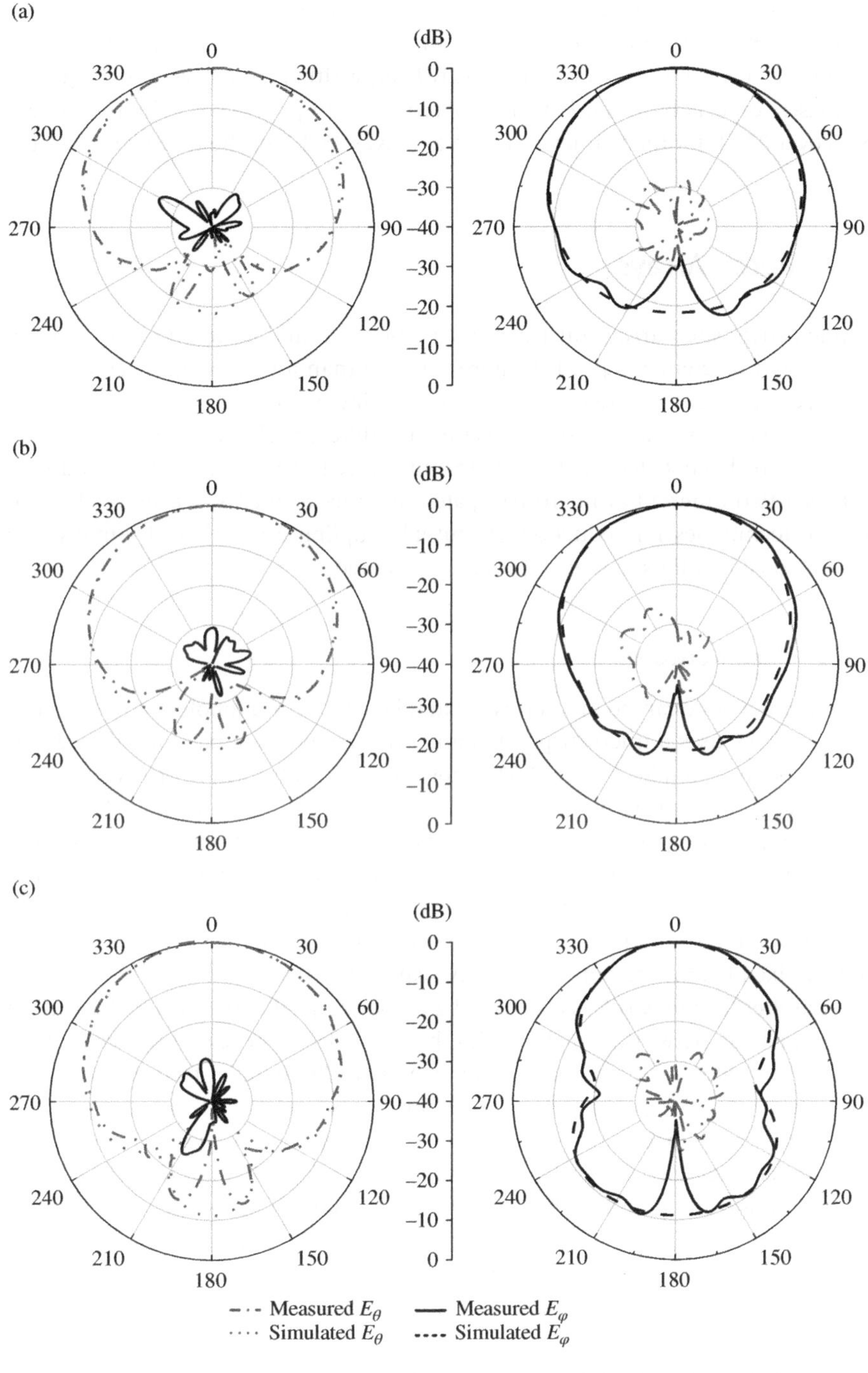

Figure 4.32 Simulated and measured radiation patterns of the three-layer design at (a) 3.3 GHz, (b) 4.2 GHz, and (c) 4.6 GHz.

it radiates in the broadside direction with a 3-dB beamwidth of 93° ± 1° in the *E* plane and 87° ± 3° in the *H* plane. The three-layer prototype also radiates in the broadside direction. However, the 3-dB beamwidth gets narrow as frequency increases, especially in the *H* plane. Moreover, both prototypes exhibit excellent XPL over 22 dB for the two-layer design and over 25 dB for the three-layer design.

4.6 Patch Arrays

A microstrip patch array consists of a feed network and a number of microstrip patch antenna elements [9]. It is more versatile than a microstrip patch antenna because it can realize a required pattern or an increased directivity that a microstrip patch antenna cannot. In this section, the differential microstrip patch arrays are described. First, feed networks are introduced. Then, the mutual coupling between two differential microstrip patch antenna elements is examined. Next, several techniques for reduction of mutual coupling are reviewed. Finally, the design considerations and examples are presented.

4.6.1 Feed Networks

The feed network can be in a series, a parallel, or a hybrid topology [66]. A series feed network uses a microstrip patch antenna element as a part of the feed network. As an example, Figure 4.33a shows the series-feed network with a five-element array featuring how an amplitude taper can be implemented. As shown, the five microstrip patch antenna elements are connected in series along two uniform coplanar strip (CPS) lines. The first to fourth elements play the roles of both transmission and radiation, while the last or the fifth element act the roles of both radiation and termination. The series feed is simple to realize but limits the array to be a fixed-beam array or a frequency-scanning array.

A parallel-feed network is also called a corporate-feed network. It is basically a power division network, which divides the power from the signal source sequentially and excites each microstrip patch element. The parallel-feed network uses the quarter-wavelength impedance transformer extensively. Hence, the parallel-feed network is lengthy, and the insertion loss may be significant. However, the parallel-feed network has wideband characteristics and offers flexibility to adjust the amplitude and phase of the signal to an individual microstrip patch antenna element. Therefore, it is widely adopted in array design. Figure 4.33b shows two parallel-feed networks to excite the microstrip patch array differentially. Note that the array has four microstrip patch elements and thus four pairs of differential signals are needed to excite them.

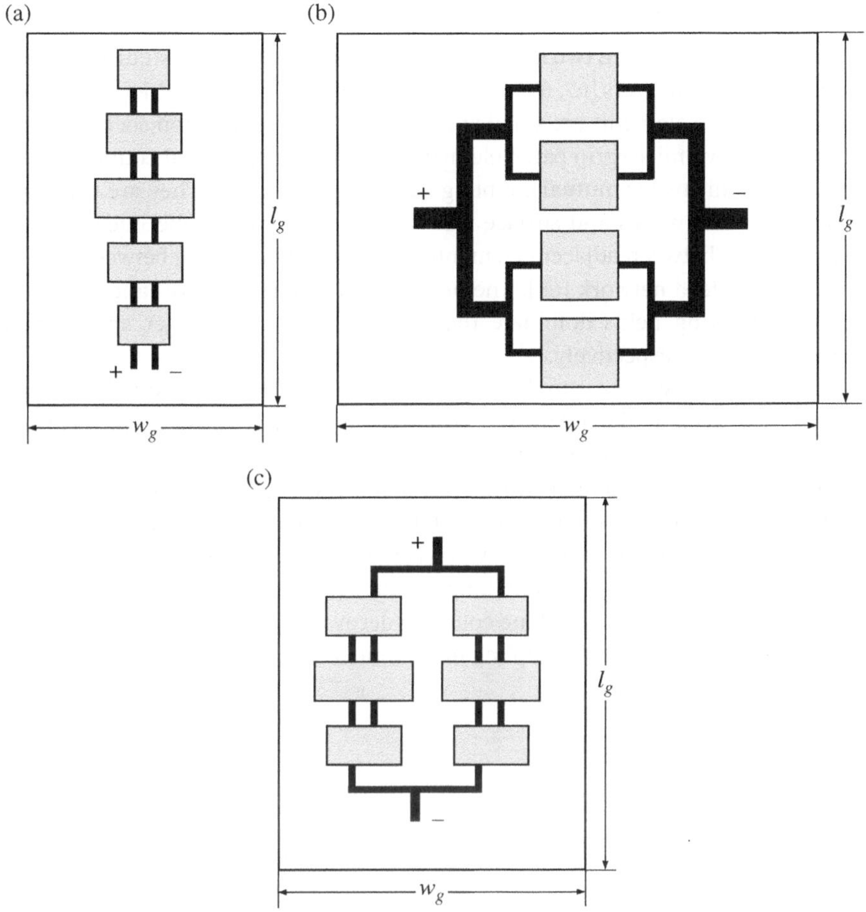

Figure 4.33 Feed networks for differential microstrip patch array. (a) Series, (b) parallel, (c) hybrid.

A hybrid feed network combines the series- and parallel-feed networks. It is commonly used due to its ability to provide power division, amplitude taper, and phase progression. Figure 4.33c shows the hybrid-feed network with a two-by-three-element array. Note that it is formed by a pair of parallel-feed networks and CPS lines. It should be mentioned that the hybrid feed of microstrip patch array is limited, particularly with respect to amplitude taper of the series-feed elements. The limitation in magnitude of the amplitude distribution results directly from the limited sensitivity of the single-element microstrip patch to variation in patch width.

4.6.2 Mutual Coupling

Mutual coupling between two microstrip patch elements has been studied in terms of standard S-parameters [67, 68]. Assuming that the reflection coefficients of both microstrip patch elements are small at resonance, the transmission coefficient S_{21} between the two microstrip patch elements represents the mutual coupling level. Three mechanisms for mutual coupling have been identified. They are the propagation of space waves and surface waves between elements, the interaction of fringing fields between adjacent elements, and the poor shielding between an element and the feed network [68]. The propagation of surface waves and the interaction of fringing fields dominate the mutual coupling at larger and smaller element spacing, respectively.

Figure 4.34 shows measured data for the E- and H-plane mutual coupling between two rectangular microstrip patch elements on a substrate of dielectric constant $\varepsilon_r = 2.5$ and thickness $h = 1.575$ mm [67]. The dimensions of the rectangular microstrip patch element are $w = 10.57$ cm and $l = 6.55$ cm. The feed location is at ($w/2$, 0.2l). The frequency is 1.41 GHz. Note that the E-plane coupling is greater than the H-plane coupling, which can be readily explained by the fact that the radiating edge is considerably longer than the nonradiating edge. Also note that the magnitude of the H-plane coupling decays much faster than the E-plane coupling, while the E-plane coupling decays much slower owning to the surface-wave effect of TM_0 mode.

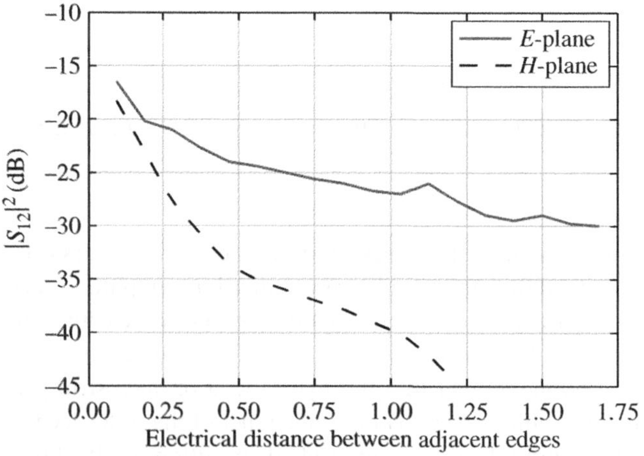

Figure 4.34 Measured mutual coupling between two single-ended rectangular microstrip patch antenna elements.

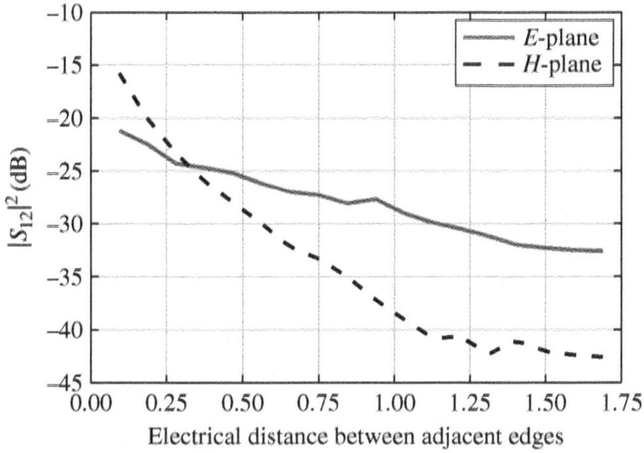

Figure 4.35 Measured mutual coupling between two single-ended circular microstrip patch antenna elements.

Figure 4.35 shows measured data for the *E*- and *H*-plane mutual coupling between two circular microstrip patch elements on a substrate of dielectric constant $\varepsilon_r = 2.5$ and thickness $h = 1.575$ mm [67]. The radius of the circular microstrip patch element is $a = 3.85$ cm. The feed location is at 1.1 cm radius. The frequency is 1.44 GHz. Note that there is a crossover in the *E*- and *H*-plane curves. Before the crossover, the *H*-plane coupling is greater than the *E*-plane coupling, while after the crossover, the *E*-plane coupling is greater than the *H*-plane coupling. It is interesting to note that the *E*-plane coupling shows a single decay rate of 3 dB per doubled spacing due to the dominant surface-wave effect of TM_0 mode. It is more important to note that the *H*-plane coupling shows three decay rates per doubled spacing. The decay rates are 12, 6, and 3 dB per doubled spacing for the element spacing shorter than 0.25λ, between 0.25λ and 1.0λ, and longer than 1.0λ, respectively. The three decay rates can be attributed to the coupling due to fringing fields, space wave, and surface wave, correspondingly.

All the above studies, however, dealt with mutual coupling between two single-ended microstrip patch elements. An investigation of mutual coupling between two differential microstrip patch elements in terms of mixed-mode *S*-parameters has been recently conducted. We deliberately choose the same substrate and patch dimensions as those used for Figures 4.34 and 4.35. We add one more feed at the symmetric position of the original feed. Figure 4.36 shows simulated data for the *E*- and *H*-plane mutual coupling between two differential rectangular microstrip patch elements at 1.41 GHz. Figure 4.37 shows simulated data for the *E*- and *H*-plane mutual coupling between two differential circular microstrip patch

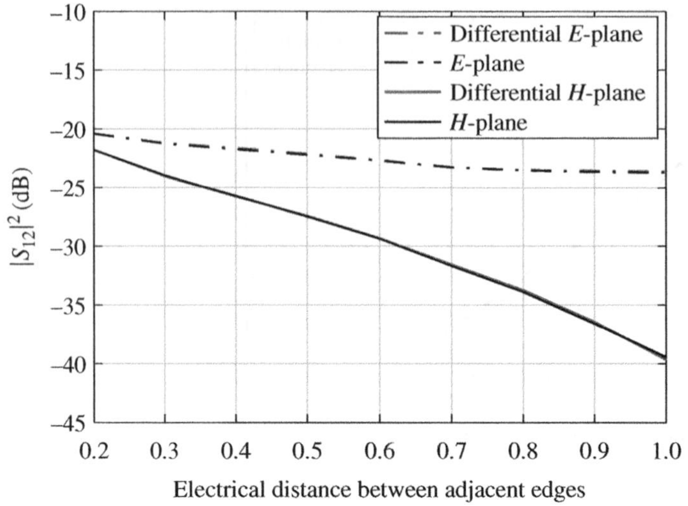

Figure 4.36 Simulated mutual coupling between two differential rectangular microstrip patch antenna elements.

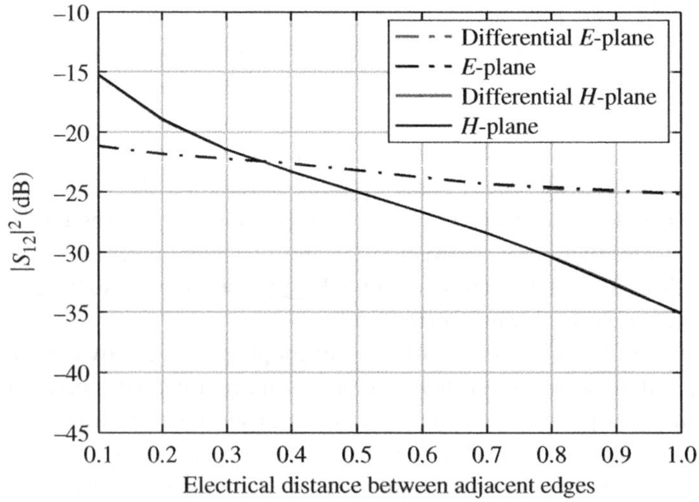

Figure 4.37 Simulated mutual coupling between two differential circular microstrip patch antenna elements.

elements at 1.44 GHz. It should be noted that the mutual coupling levels are comparable for the single-ended cases in Figures 4.34 and 4.35 and differential cases in Figures 4.36 and 4.37.

In addition to the element spacing, the mutual coupling also depends on the substrate thickness and dielectric constant. It is found that the mutual coupling increases gradually with the substrate thickness and dielectric constant.

4.6.3 Reduction of Mutual Coupling

Generally, mutual coupling is undesirable. It can change the input impedance of an individual element in an array. Especially, the input-matching condition may get worse as the scanned angle increases in a phased array. At some scanned angles, all the power may be reflected to the feed network and scan blindness may occur. It can also degrade the radiation pattern of the array. Sidelobe levels may increase, ripples may occur, nulls may be shifted, and grating lobes may appear. It reduces the channel capacity in a multiple-input and multiple-output (MIMO) array and increases the response time of an adaptive array. Several techniques to reduce mutual coupling between microstrip patch elements have been devised and are reviewed as follows.

4.6.3.1 Modification to the Patch, Substrate, or Ground Plane

A circular microstrip patch antenna can be designed not to excite the dominant TM_0 surface wave. This requires the radius of the circular microstrip patch to be a particular critical value. However, a standard circular microstrip patch having this critical radius will not be resonant. To make it resonant, two different modifications can be made. One has a central core removed and replaced by a material with a different permittivity as a cored patch. The other has shorting pins placed as a shorted ring patch. Both modified patch designs enjoy reduced mutual coupling, due to the reduced surface-wave excitation. However, a larger radius than a standard circular microstrip patch could impose restrictions in an array environment [69, 70].

A rectangular microstrip patch antenna loaded with shorting pins can realize an effective dielectric constant either close to that of air to suppress the mutual coupling of fringing field and space wave for arrays on thin substrates [71] or equal to the specific value not to excite the dominant TM_0 surface wave for arrays on thick substrates [72]. However, this requires increasing the patch size considerably. For example, the differential rectangular microstrip patch element loaded with shorting pins has much weaker mutual coupling than the conventional rectangular microstrip patch element does. However, the dimensions of the former are 63 mm wide and 45 mm long, while the dimensions of the latter are 22.4 mm wide and 13.75 mm long. The resonance frequency in both cases is around 2.14 GHz [72].

Micromachining enables a selective removal of substrate material under the microstrip patch area to realize an equivalent substrate of a much lower effective dielectric constant [73]. Hence, the micromachined substrate can suppress unwanted surface wave modes, resulting in improved radiation efficiency and reduced mutual coupling [74]. In addition, micromachining can etch grooves or trenches between the microstrip patches, which also helps reduce mutual coupling due to the space wave. The drawbacks of microstrip patches on a micromachined substrate are less robust and large electrical size.

By placing microstrip patch elements in their individual metal cavities, the substrate can no longer support surface waves [75]. As a result, the mutual coupling due to surface waves can be reduced. Those metal cavities can be formed with pins or plating through holes in the ground plane and dielectric substrate [27]. The cavity-backed microstrip patch retains the patch size, although some modest increase in weight and cost can probably be expected.

4.6.3.2 Integration of Electromagnetic Bandgap Structures

High-impedance surfaces with a forbidden frequency band have aroused great interest among antenna designers for more than two decades [76]. A high-impedance surface was first built with subwavelength mushroom-like metal cells, which function as parallel LC circuits. The capacitance C is provided by the gap of the metal caps and the inductance L is related to the metal stems or vias. At the LC resonance, the surface shows the high-impedance characteristics, which reflect incident waves with a phase shift of 0°, rather than 180° as with an electric conductor. Hence, it is also known as an artificial magnetic conductor (AMC), which can lead to low-profile antenna designs [77]. In addition, the high-impedance surface has a bandgap within which no surface wave of either transverse magnetic or transverse electric polarization can propagate. Therefore, it may be considered an electromagnetic bandgap (EBG) structure for surface waves. Yang and Rahmat-Samii investigated the reduction of mutual coupling in both the E and H planes when the mushroom-like EBG structure was inserted between two microstrip patch elements on substrates of different thicknesses and various dielectric constants [78]. They demonstrated that the mushroom-like EBG structure had a better capability in lowering the mutual coupling than other approaches. For example, the microstrip patch elements were on a substrate with $h = 2$ mm and dielectric constant $\varepsilon_r = 10.2$. The size of the microstrip patch was 7 mm × 4 mm. The frequency of operation was 5.8 GHz. The distance between the microstrip patch elements was 38.8 mm (0.75λ) in the E plane. As expected, the mutual coupling was the highest for the normal case, a 2-dB reduction was observed for the cavity-backed case, and the lowest mutual coupling was obtained in the EBG case as an 8.8-dB reduction was achieved.

Probably inspired by the mushroom-like EBG structure, a uniplanar compact EBG (UC-EBG) structure was soon developed [79]. The UC-EBG structure gets rid of metal vias, which facilitates fabrication with low-cost printed-circuit board technology. Subsequently, several UC-EBG structures have been proposed and integrated in microstrip patch arrays [80, 81]. The effects of different arrangements of the UC-EBG structure along with two adjacent microstrip patch elements on a substrate with thickness $h = 1.6$ mm and dielectric constant $\varepsilon_r = 4.8$ have been investigated. The size of the microstrip patch was 28.45 mm × 28.45 mm. The frequency of operation was 5.59 GHz. The dimensions of the unit cell of the UC EBG structure were 5.05 mm × 6.52 mm or approximately $\lambda/8 \times \lambda/10$. The distances between the adjacent edges of the microstrip patch elements were 0.3λ, 0.4λ, and 0.5λ, respectively in the E plane. Table 4.8 compares the simulated and measured results. The results indicate that mutual coupling is reduced for all different distances between the microstrip patch elements [82].

Note that the above EBG structures were demonstrated at microwave frequencies, Lamminen et al. integrated the first UC-EBG structure in an array of 16 microstrip patch elements in the Ferro A6-S LTCC technology at millimeter-wave frequencies [83]. A reduction of up to 11.8 dB at 60 GHz was observed in the mutual coupling between two microstrip patch elements in the E plane. Nevertheless, the implementation of UC-EBG for 60-GHz array applications was very challenging due to its highly frequency-dependent resonance structure. It was concluded that the small bandwidth and large size make the current UC-EBG design quite impractical to be used in the 60-GHz microstrip patch phased array applications. The same improvements could be achieved by using simpler and more

Table 4.8 Mutual coupling between microstrip patch elements with and without UC-EBG structure.

Status	**0.3λ**		**0.4λ**		**0.5λ**	
S_{21} (dB)	**Simulated**	**Measured**	**Simulated**	**Measured**	**Simulated**	**Measured**
Without UC-EBG	−18.66	−18.79	−21.21	−21.79	−24.83	−25.18
With UC-EBG	−30.06	−36.41	−40.07	−47.14	−49.07	−55.34
Mutual coupling reduction (dB)	14.4	17.62	18.86	25.35	24.24	30.16

robust methods, such as processing of air cavities inside LTCC substrates [84]. Furthermore, the tolerances required to create the gaps between conducting areas are very tight at millimeter-wave frequencies. There is clearly a need to consider structures that can be manufactured using well-established technologies that comply with the dimensional requirements. One example of such an EBG structure is the woodpile or layer-by-layer structure [85].

The defected ground structure (DGS) was proposed for microstrip line [86]. It was found that DGS can provide the bandgap characteristic in some frequency bands with only one or more cells. Hence, the DGS is much more compact and easier to be fabricated than an EBG structure. Xiao et al. investigated the reduction of mutual coupling between microstrip patch elements using DGS [87]. Two kinds of DGSs, namely back-to-back U-shaped and dumb bell-shaped DGSs, were analyzed and compared. It was found that the back-to-back U-shaped DGS performed better at suppressing the propagation of surface waves and the degree of the mutual coupling reduction was increased when the element spacing was reduced. The scan blindness of an infinite microstrip patch phased array in the E plane was simulated, indicating that the scan blindness can be eliminated by applying a back-to-back U-shaped DGS to the infinite phased array. The main disadvantage of the DGS is that it radiates, which degrades the array of radiation patterns and leads to a poor front-to-back ratio.

4.6.3.3 Utilization of Additional Structures

An additional structure such as a cover is often utilized to protect a microstrip patch array from environmental hazards. Alexopoulos and Jackson investigated the fundamental effects of cover (superstrate) materials on printed circuit antennas. They derived the conditions for the substrate-superstrate resonance to maximize antenna gain, radiation resistance, and radiation efficiency. They also determined the criteria for the material properties and dimensions for which surface waves are eliminated, which require that the substrate must be electrically thin for a nonmagnetic superstrate and can be much thicker for a magnetic superstrate [88].

A metamaterial consists of artificial structural elements, designed to achieve advantageous and unusual electromagnetic properties [89]. An early metamaterial superstrate was developed to reduce the mutual coupling between dipole elements in an array environment [90]. For the spacing between dipoles around 0.25λ, low coupling of -20 dB was obtained. A recent metamaterial superstrate was developed to suppress the mutual coupling and to achieve high isolation between the microstrip patch elements [91]. The superstrate was made of capacitively loaded loop cells, which can be designed separately. It provided a unique solution to improve the isolation of the previously fabricated microstrip patch array without replacing them with the new ones. To validate the effect of the superstrate, the fabricated structure with a size of 56.4 mm × 14.3 mm × 90 mm was tested on two microstrip patch elements at 3.3 GHz. About a 55 dB reduction was obtained in the measurement for an edge-to-edge spacing of 0.12λ.

A metasurface is a special class of planar metamaterials, which is composed of thin and densely packed resonant or nearly resonant subwavelength elements. An early metasurface superstrate was developed to reduce the mutual coupling between dipole elements. The metasurface was based on grids of short metal strips and continuous wires. On average, between 3 and 14 dB reduction of the mutual coupling was achieved when the metasurface superstrate was used [92]. A recent metasurface superstrate was proposed to reduce the mutual coupling of two closely placed dual-band microstrip patch elements. The metasurface superstrate was made of pairs of nonuniform strips with two different lengths. It was compact in size and effective in reducing the mutual coupling between two dual-band microstrip patch elements with the edge-to-edge spacing of only 0.008λ at low-frequency band. The results showed that the isolation between two microstrip patch elements can be improved to more than 25 dB at both 2.5–2.7 GHz and 3.4–3.6 GHz bands, while their reflection coefficients remained below −10 dB after the metasurface superstrate was introduced [93].

Wu et al. proposed an interesting surface cover for reducing the mutual coupling between antenna elements in a large-scale antenna array [94]. The surface was a plurality of electrically small metal patches printed on the substrate with a dielectric constant of 2.6, loss tangent of 0.001, and thickness of 1.0 mm. Two array examples were given to illustrate the design process and considerations. For the example of the phased array, there were 8 air patch elements in a linear arrangement, and the center-to-center spacing was 55 mm or 0.45λ at 2.45 GHz. Measured results showed that the surface reduced the mutual coupling more than 15 dB between any two adjacent patch elements but maintained the mutual coupling between nonadjacent patch elements at 2.45 GHz. The surface did not affect the matching condition of the original array significantly and reduced the mutual coupling by more than 9 dB over a bandwidth wider than that of 10-dB return loss, showing its potential for a wide band operation.

4.6.4 Design Considerations

Currently, there is increasing interest in differential microstrip patch arrays for millimeter-wave radio and radar applications [28, 30, 95–103]. This section will consider the design of differential microstrip patch arrays. As will be seen, design challenges come from the layout and integration of the differential feed network with differential microstrip patch elements. The feed network often requires use of baluns. A balun has amplitude and phase imbalance, which imply that a differential microstrip patch element cannot be simply treated as the element of a single differential port but as that of two separate single-ended ports with the port imbalance included. Since the port asymmetry may result in scan blindness for differential microstrip patch phased arrays, it must be taken properly into account in the design.

4.6.4.1 Frequency Scanning Array

A frequency scanning array steers the beam by changing the frequency. The frequency scanning array is typically a linear series-fed differential microstrip patch array. The number of differential microstrip patch elements is N and the spacing between the adjacent elements is d. An approximate equation to estimate the directivity of such an array D_a is given in dBi by

$$D_a(\text{dBi}) = 10\log\left(\frac{2Nd}{\lambda}\right) + D_e(\text{dBi}) \tag{4.185}$$

where D_e is the element gain, which is typically 5–7 dBi. Taking $d/\lambda = 1, N = 6$, and $D_e = 6$ dBi as an example, one can find that the directivity in the broadside direction is 16.8 dBi. The gain of the array will be lower due to the mismatch and insertion loss of the CPS line that guides the differential signal to the nonradiating edges of all rectangular microstrip patches. The design of a rectangular differential microstrip patch element in the array follows the procedure in Section 4.3. The operating frequency sets the patch length. The patch width has an effect on the input impedance and the bandwidth. The input impedance also depends strongly on the CPS line separation. The larger the separation, the larger the input impedance. If the separation is too small, the patch will not resonate. It is found that the input impedance of the series-fed rectangular differential microstrip patch array tends to be inductive. Hence, an impedance-matching network needs to be inserted between the first patch and the CPS line. The impedance-matching network is usually implemented with an inset on the first patch and a quarter wavelength CPS section, which transforms the array inductive impedance to a 100-Ω differential resistance [97].

The linear series-fed differential rectangular microstrip patch array can act as the frequency scanning array because the radiation from each patch adds in phase in a certain special direction, depending on the frequency (or array dimensions). The beam direction is broadside at the frequency for which the total length of the patch and the following CPS line is exactly one wavelength. At lower and higher frequencies, the beam will be steered to the directions of the first and last patches, respectively.

Ahmad et al. reported the linear series-fed differential rectangular microstrip patch array with six patch elements on Rogers RO3003 substrate with 127-μm thickness for frequency scanning in the 60-GHz band [98]. The dimensions of patches were deliberately perturbed and symmetrically paired for impedance bandwidth enhancement. The spacing between the adjacent patches was fixed and optimized to be 0.89λ. The measured impedance bandwidth was from 59 to 67 GHz. The beam was able to be directed to 1° at 59 GHz, to 8° at 61 GHz and to 23° at 67 GHz with an average gain of 13 dBi.

4.6.4.2 Phased Array

A phased array features multiple antennas and phase shifters. Basically, the design of a phased array can be started from uniformly spaced antennas with reference to the diagram in Figure 4.38. We first set the EIRP within the target beam coverage angle as

$$20\log N + P_{1E} + G_{1E} \geq EIRP\,\text{dBm} \tag{4.186}$$

where N is the number of antennas in the array, P_{1E} is the input power to each antenna in dBm, and G_{1E} is the gain of each antenna in dBi. For a fixed EIRP, as N increases (i.e., the antenna aperture is increased), P_{1E} decreases by a factor of N^2, the total input power to all antennas decreases by a factor of N, and the overall DC power consumption in the array also decreases by a factor of N. These benefits are obtained at the cost of increased circuit and array size, more complex power and signal distribution, and increased testing difficulty. Balancing these power/area/complexity considerations and knowing P_{1E} will lead to a proper choice of the number of elements N for the array.

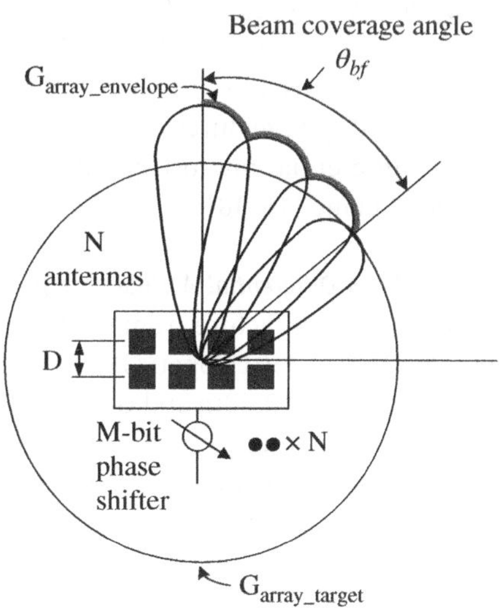

Figure 4.38 The diagram for the design of a phased array.

To ensure the array factor has only one maximum at broadside, the maximum scan angle θ_{max} can be calculated as

$$\theta_{max} = \sin^{-1}\left(\frac{\lambda}{d} - 1\right) \tag{4.187}$$

where λ is wavelength at the center frequency and d is the spacing between adjacent antennas. It should be noted that (4.187) only links θ_{max} with d. In an actual design, however, one also needs to consider other factors such as the array size, the sidelobe level, the effect of mutual coupling, etc., to choose d.

To define the resolution in the phase shifter, the main concern is to guarantee that the pointing error is minimal (<1 dB) when the beam is steered in different directions. For a given scan angle step $\Delta\psi$, the phase shifter step $\Delta\phi$ can be calculated as

$$\Delta\phi = k_0 d[\sin\theta - \sin(\theta + \Delta\psi)] \approx k_0 d \Delta\psi \cos\theta. \tag{4.188}$$

Accordingly, the minimum phase step or the resolution of the phase shifter is

$$\Delta\phi_{min} \approx k_0 d \Delta\psi \cos\theta_{max}. \tag{4.189}$$

The phased array envelope gains $G_{array\text{-}envelop}$ should be larger than that of the target gain $G_{array\text{-}target}$ within the target beam coverage angle θ_{max}. The pointing error ($G_{max} - G_{min}$) decreases as the number of phase shifter bits increases, but the power efficiency of the transmitter (Tx) and the noise figure (NF) of the receiver (Rx) degrade due to phase shifter loss. Therefore, there is a value of M for an optimal Tx efficiency and Rx signal-to-noise ratio (SNR).

The phased array using differential microstrip patch antennas can be classified into one of two types. The first type features a differential architecture from circuits to antennas. The use of differential microstrip patch antennas is to get rid of bulky and lossy baluns and to reduce the cross-polarization in the H plane [101–103]. The second type features the use of baluns in the feeding network. The reason to use differential microstrip patch antennas is to reduce the cross-polarization in the H plane [15–18].

4.6.4.2.1 A Differential Architecture from Circuits to Antennas A phased array in a differential architecture from circuits to microstrip patch antennas was reported by Alberto et al. [101] for 60-GHz wireless communications. Assuming realistically that P_{1E} was 10–12 dBm and G_{1E} was 4.0 dBi, they calculated the number of microstrip patch antennas N in the array from (4.125) to be 16 in order to achieve the EIRP of 40 dBm. They chose $d = 0.628\lambda$ at 60 GHz to ensure the mutual coupling between any two microstrip patch antennas to be higher than 17 dB. They found that the phase resolution of $M = 3$ (or 45°) was sufficient. However, the higher resolution of $M = 5$ (or 11.25°) was chosen to compensate for potential digital-analogue conversion inaccuracies and to enable the reduction of side lobes and the formation of nulls in particular directions, when desired.

For this phased array, the circuits were integrated in a 0.12-μm SiGe BiCMOS process as a single die. The microstrip patch antennas were implemented in an advanced packaging technology. The codesign of the die, patches, and package made it possible to realize this phased array as a compact antenna-in-package (AiP) module [28].

Figure 4.39 shows the cross-sectional view of the AiP module [28]. As shown, the AiP module consists of five dielectric layers (D1–D5) with adhesive layers (A1–A4) in between. There are four buried layers and one bottom-layer metallization (M1–M5) in the construction. A trench is cut in D2, which forms an embedded air cavity with D1 as the ceiling and M2 as the ground floor. The microstrip patches are realized with M1 and can be viewed as the ceiling decorates of the embedded air cavity. There are slots in M2 that allow the microstrip patches to be excited by the feed network on M3 beneath D3. Portions of D4 and D5 are removed locally to create an open cavity to house the die. M4 is used for interconnects and reflectors. M5 provides the metallization for pads for the AiP to be assembled on the application board. Table 4.9 lists the thicknesses and properties of each layer. To improve the feed network efficiency, a low-loss adhesive is used close to D3, while relatively lossy adhesive material is used elsewhere to reduce cost and enhance the AiP manufacturability. These higher-loss adhesives are placed further away from the microstrip patch antennas so that they cannot affect the performance significantly.

An aperture-coupled microstrip patch antenna has significant advantages compared to a probe-fed microstrip patch antenna. The main advantage is that the feed line and the patch can be located on different substrates. This allows one to choose the substrates separately to optimize the performances of both the feed network and the microstrip patch antenna. Figure 4.40 shows the geometry of the

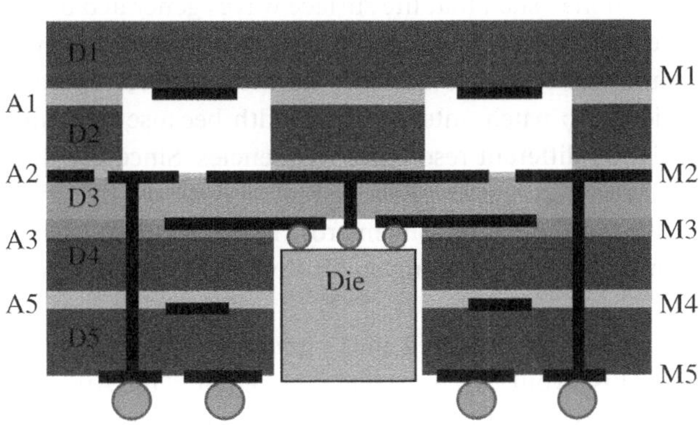

Figure 4.39 Stack-up of the AiP module.

Table 4.9 Layer thickness and property.

Layers	Thickness	Dielectric constant	Loss tangent	Conductivity
D1, D2, D4, D5	254 μm	3.74	0.0044	Not applicable
D3	50.8 μm	3.28	0.0018	Not applicable
A1, A2, A4	25.4 μm	Not available	Not available	Not applicable
A3	25.4 μm	Not available	Not available	Not applicable
M1, M4, M5	18 μm	Not applicable	Not applicable	3×10^7 S/m
M2, M3	9 μm	Not applicable	Not applicable	3×10^7 S/m

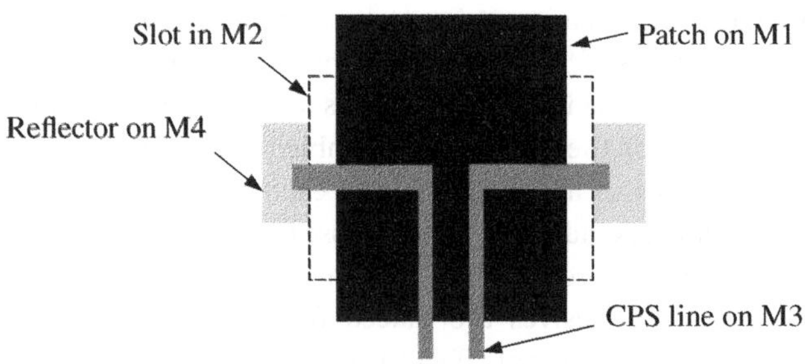

Figure 4.40 Geometry of the aperture-coupled microstrip patch antenna.

aperture-coupled microstrip patch antenna. The sizes for the patch, slot, and reflector are 1.6×1.32, 1.1×0.2, and 1.95×1.0 mm^2, respectively. The central separation between the two slots is 1.43 mm. The two slots play an important role in the design. First, they are positioned such that the surface waves generated by the slots and the patch interfere destructively and therefore the radiation efficiency of the microstrip patch antenna can be improved. Second, the slots are resonant, which can improve the microstrip patch antenna bandwidth because the patch and the slots, each have slightly different resonance frequencies. Since the resonant slots radiate, back radiation becomes an issue. To reduce this back radiation, a reflector is placed behind the ground plane. The microstrip patch antenna bandwidth can be further enhanced by implementing an air cavity between the patch and the ground plane.

The microstrip patch antennas are arranged in a circular configuration and share the annular air cavity. This layout creates room in the center of the AiP module to place the die and facilitates the routing of CPS lines to each of the microstrip patch antennas. The annular air cavity had a 12-mm inner diameter and 20-mm outer diameter, and the overall package dimension was $28 \times 28 \times 1.2\,\text{mm}^3$. The patches are almost uniformly located on a circle with a 16-mm diameter; therefore, the spacing between the patches is 0.628λ at 60 GHz.

The locations of the microstrip patch antennas dictate the optimal port locations of the power and low noise amplifiers on the die for minimum loss routing. All the CPS lines of the microstrip patch antennas should have an equal length, which is important to ensure equal loss and phase responses across the operational frequency band. Moreover, it also simplifies or even eliminates the need for calibration. Figure 4.41 shows the layout of the feeding network.

Figure 4.42 shows four different 16-element radiation patterns measured in an antenna chamber; this measurement demonstrates beam-steering along the θ spherical coordinate. The ideal radiation patterns employed for comparison to these measurements assume 16 identical elements and isotropic antennas. It is also important to mention that these exemplary beams are general and were designed without any special considerations for sidelobe suppression or notch in a particular direction. The phased array achieved − 6 dB side lobe levels which are 7 dB higher than the −13 dB levels for the 16-element linear case. This difference is expected and is due to the circular configuration which does not include antenna elements in the center. Despite the relatively higher side lobe levels, the circular configuration makes it easy to design equal-length feed lines in the package environment. A relatively narrow beam width of element radiation patterns will also increase the array side lobe levels when the main beam is steered far away from the broadside direction.

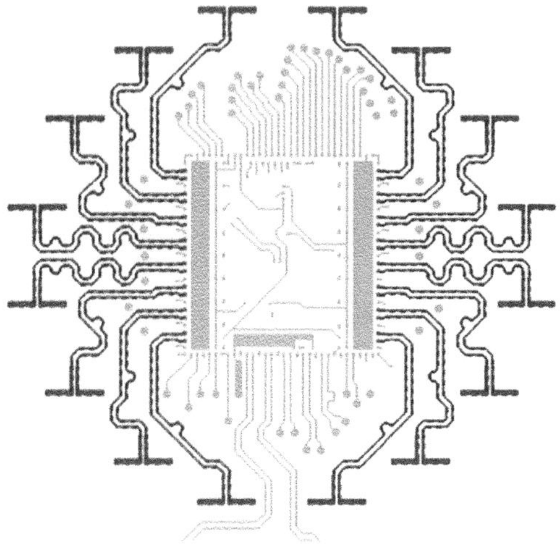

Figure 4.41 Symmetric layout of the feeding network with CPS lines. *Source:* Kam et al. [28]. (© 2011 IEEE).

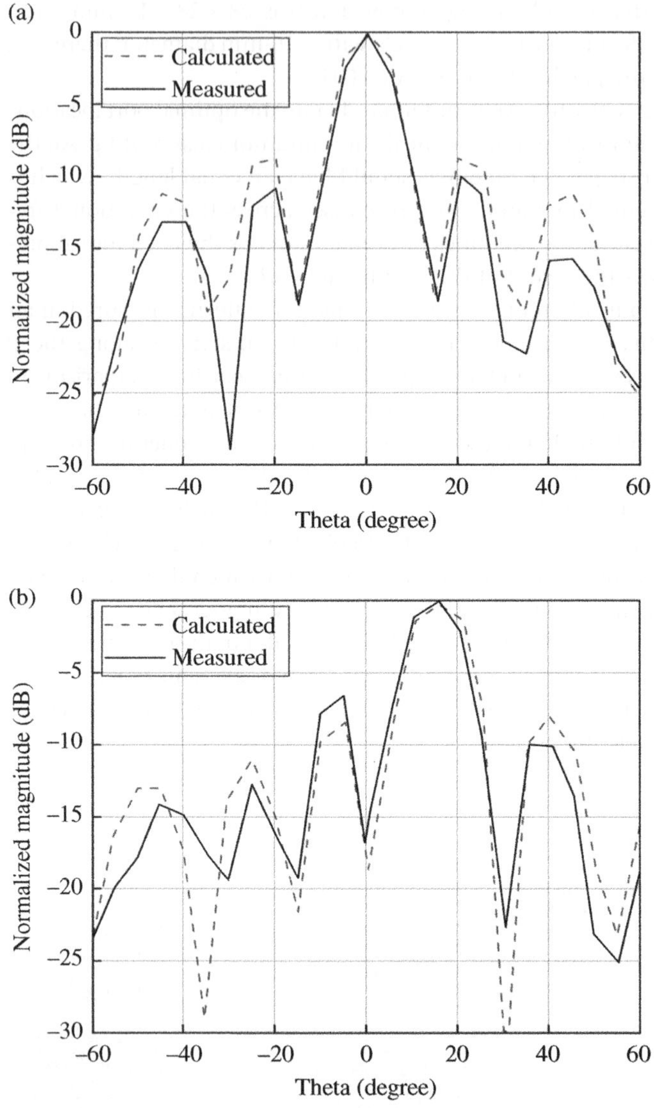

Figure 4.42 The phased array radiation patterns at 60.48 GHz for (a) 0°, (b) 15°, (c) 30°, and (d) 45°.

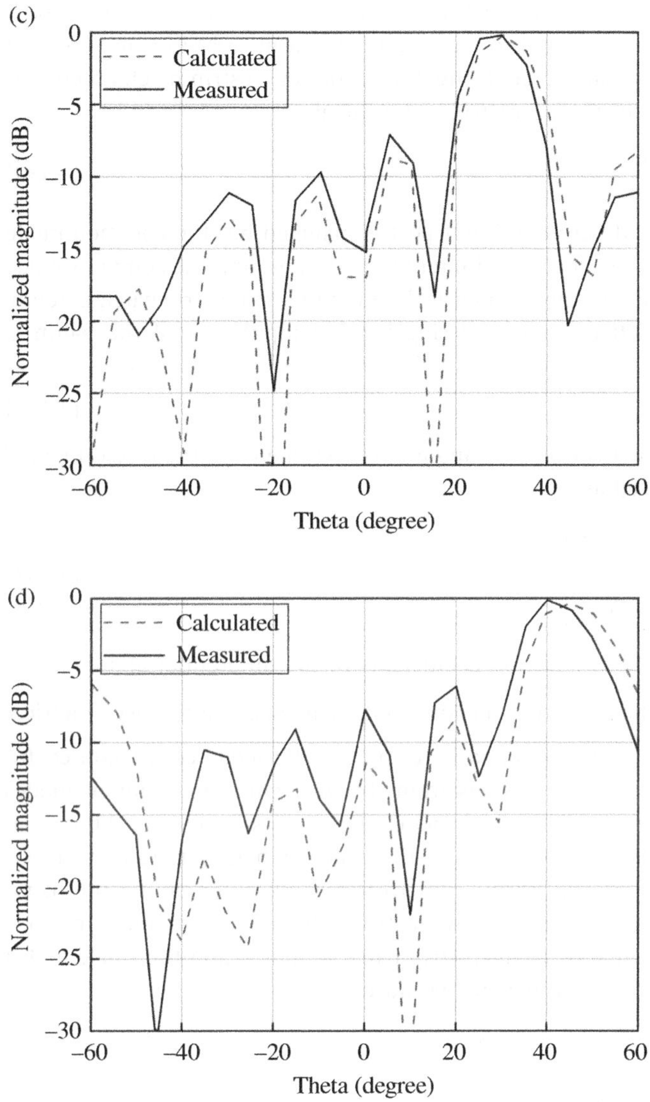

Figure 4.42 (Continued)

4.6.4.2.2 The Use of Baluns in the Feed Network A reactive balun is a simple broadband structure. It is widely used to feed a differential microstrip patch antenna from a single-ended signal source. If the differential microstrip patch antenna is symmetrical, the mixed-mode S matrix is thus given by

$$S^{s}_{mm} = \begin{bmatrix} 0 & 0 \\ 0 & 1 \end{bmatrix}, \tag{4.190}$$

which shows that the differential input port is matched but the common-mode input port is not matched, and the differential input port and the common-mode input port are isolated. For the symmetrical differential microstrip patch antenna connected to the reactive balun, the single-ended input port of the balun is matched because

$$S^{s}_{11} = 0. \tag{4.191}$$

If the differential microstrip patch antenna is asymmetrical, the mixed-mode S matrix can be expressed as

$$S^{as}_{mm} = \begin{bmatrix} \Gamma_a & -\dfrac{\Delta}{2} \\ -\dfrac{\Delta}{2} & \Gamma_b \end{bmatrix} \tag{4.192}$$

where Γ_a and Γ_b are the reflection coefficients of the differential and common modes and $-\dfrac{\Delta}{2}$ is the mode conversion coefficient. Because the asymmetrical differential microstrip patch antenna is reciprocal, the mode conversion coefficient is the same for the conversions from the differential to common modes and from the common to differential modes. For the asymmetrical differential microstrip patch antenna connected to the reactive balun, reflections occur between the asymmetrical differential microstrip patch antenna and the balun junction, mode conversions exist between the differential and common-mode input ports of the asymmetrical microstrip patch antenna, and the single-ended input port of the balun is not matched because

$$S^{as}_{11} = e^{-j\frac{4\pi l}{\lambda}} \left[\Gamma_a - \frac{\dfrac{\Delta^2}{4} e^{-j\frac{6\pi l}{\lambda}}}{1 + \Gamma_b e^{-j\frac{4\pi l}{\lambda}}} \right]. \tag{4.193}$$

If the differential microstrip patch antenna is structurally symmetrical but somehow becomes asymmetrical and nonreciprocal, the mixed-mode S matrix can be written as

$$S^{an}_{mm} = \begin{bmatrix} \Gamma_d & -\dfrac{\Delta_d}{2} \\ -\dfrac{\Delta_c}{2} & \Gamma_c \end{bmatrix} \tag{4.194}$$

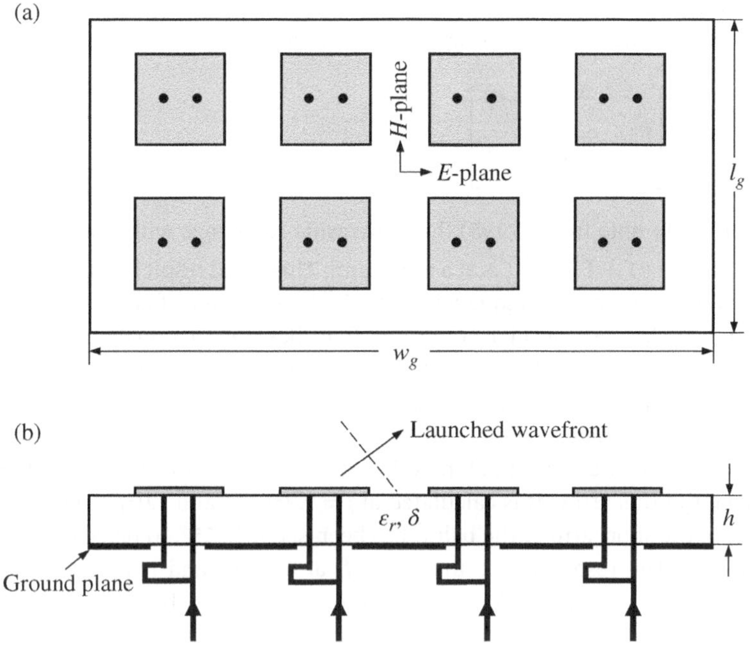

Figure 4.43 Illustration of the differential microstrip patch phased array using reactive baluns in the feed network. (a) Top view and (b) side view.

where Γ_d and Γ_c are the reflection coefficients of the differential and common modes, $-\frac{\Delta_d}{2}$ and $-\frac{\Delta_c}{2}$ are the mode conversion coefficients. It should be clear that the mode conversion coefficients are different for the conversions from the differential to common modes and from the common to differential modes, due to nonreciprocity. Figure 4.43 shows the structurally symmetrical differential microstrip patch antenna used as an element in the array. As can be seen, any component of the outgoing wave vector in the array *E*-plane breaks the symmetry between the two halves of the structurally symmetrical differential microstrip patch antenna. Thus, when the array launches an oblique wave having an *E*-plane component of its wave vector, the two halves will each respond differently to the differential excitation. Consequently, the structurally symmetrical differential microstrip patch antenna becomes electronically asymmetrical when the array scans in the *E* plane. The reason that the structurally symmetrical differential microstrip patch antenna can be nonreciprocal in the array is that it does not correspond to a real two-port device composed of reciprocal components. For the asymmetrical and nonreciprocal differential microstrip patch antenna connected to the reactive balun, the single-ended input port of the balun is neither matched because

$$S_{11}^{as} = e^{-j\frac{4\pi l}{\lambda}} \left[\Gamma_d - \frac{\frac{\Delta_d \Delta_c}{4} e^{-j\frac{6\pi l}{\lambda}}}{1 + \Gamma_c e^{-j\frac{4\pi l}{\lambda}}} \right]. \tag{4.195}$$

It is important to note from (4.195) that a parasitic resonance will occur at the frequency at which $| 1 + \Gamma_c e^{-j\frac{4\pi l}{\lambda}} |$ is at a minimum. This could result in scan blindness within a narrow range of frequencies [16]. To illustrate the effect of this parasitic resonance, let us consider an array that has the parameters: lattice dimensions in the E and H planes are 0.51λ and 0.50λ, the width and length of the square microstrip patch are 0.363λ, the substrate thickness is 0.07λ, probe offsets in the E plane from patch center is $\pm 0.102\lambda$, width of feed probes in the H plane is 0.07λ, and the dielectric constant of the substrate ε_r is 1.13. Figure 4.44 shows the global behavior of this array. It is calculated at $f = 0.95f_o$, f_o, and $1.05f_o$. The electrical length of the short arm of the balun has been set at 0.455λ, to place the resonance at $f = f_o$. The resonance occurs at $f = f_o$ at $\theta = 30°$ and results in scan blindness; however, at $f = (1 \pm 0.05) f_o$, the resonance is nearly absent. This parasitic resonance depends on the probe impedance and the balun line length. The safest way of avoiding it is to employ resistive baluns.

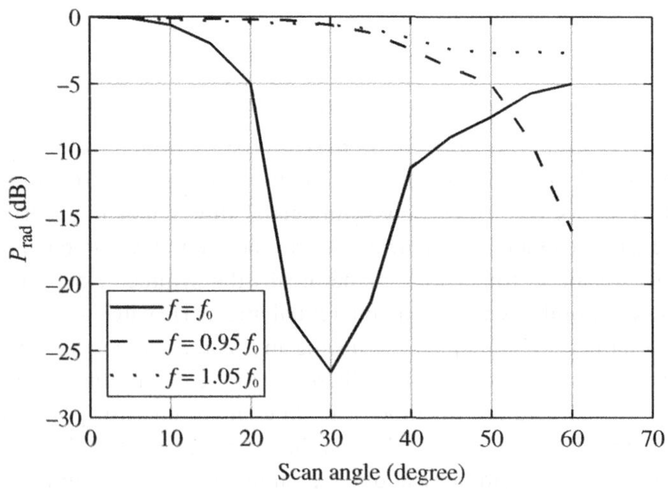

Figure 4.44 Scan blindness due to parasitic resonance of the reactive baluns in the feed network.

4.7 Applications

Highly integrated radios, radars, and imagers in complementary metal–oxide–semiconductor (CMOS) technology have been demonstrated in various frequency bands for many applications. The high level of integration has the benefit of cost reduction and system reliability and therefore represents the trend for implementing wireless systems. As stated in Chapter 1, a fully differential architecture from the antenna to the chip is favored for wireless SoC [104–106].

Figure 4.45 shows a chip package in a cavity-down ceramic ball grid array format [27, 104]. It consists of three layers, with an open cavity formed in the middle. The top layer with the patch and guard ring printed on its surface is 0.8 mm thick, the middle layer is 0.4 mm thick, and the bottom layer where solder balls are attached is also 0.4 mm thick. There are two buried layers and one top-layer metallization in the construction. The lower buried layer provides the metallization for the signal traces, while the upper buried layer provides the metallization for the ground plane. The differential rectangular microstrip patch antenna and the guard ring are realized with top-layer metallization. The package was realized in low-temperature co-fired ceramic (LTCC) technology with a size of $15 \times 15 \times 1.6\ \text{mm}^3$. The rectangular patch measured $9.5 \times 9.8\ \text{mm}^2$ and achieved impedance bandwidth of 2%, radiation efficiency of 84%, and gain of 3.2 dBi at 5.425 GHz.

Figure 4.46 shows another chip package in a cavity-down ball grid array format. This package integrates 16 differential rectangular microstrip patch antennas [28]. To improve antenna bandwidth and radiation efficiency, an air cavity is used, resulting in a superstrate microstrip patch antenna structure. Additionally, aperture-coupled instead of probe-fed techniques are used to further increase the

Figure 4.45 Differential rectangular microstrip patch antenna for RF single-chip radio.

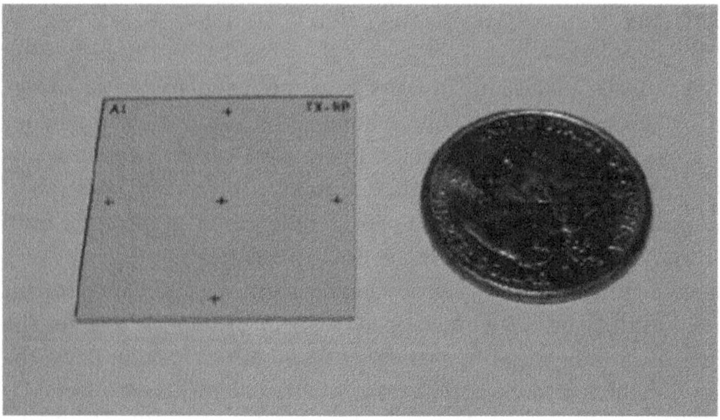

Figure 4.46 Differential rectangular microstrip patch antenna for single-chip 60-GHz radio. *Source:* Kam et al. [28]/with permission of IEEE.

antenna bandwidth. The package was realized in a multilayer organic (MLO) technology with a size of $15 \times 15 \times 1.6\ \mathrm{mm}^3$. The rectangular patch measured $1.32 \times 1.6\ \mathrm{mm}^2$ and achieved good performance at 60 GHz in terms of peak gain of about 8 dBi for a single element and 17 dBi for a 16-element array, bandwidth of more than 10 GHz, and radiation efficiency of 80% for single-element from simulation.

Figure 4.47 shows a third chip package in a ball grid array format [29]. The package was developed on the embedded wafer-level ball grid array (eWLB) technology

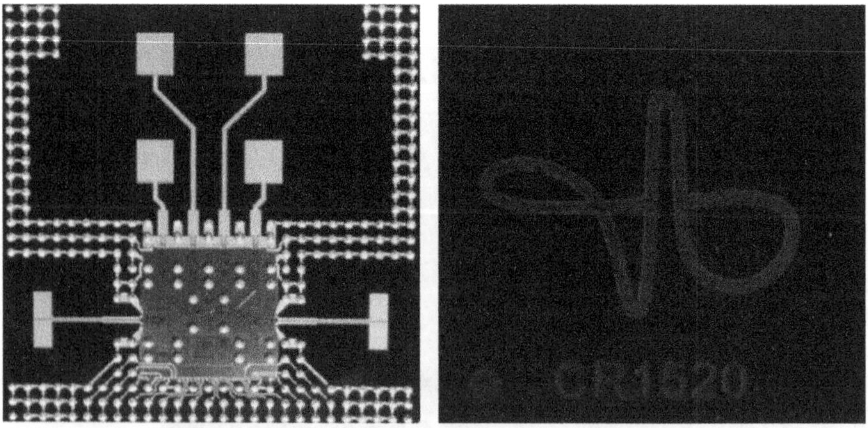

Figure 4.47 Differential rectangular microstrip patch antenna for single-chip 60-GHz radar. *Source:* Nasr et al. [29]/with permission of IEEE.

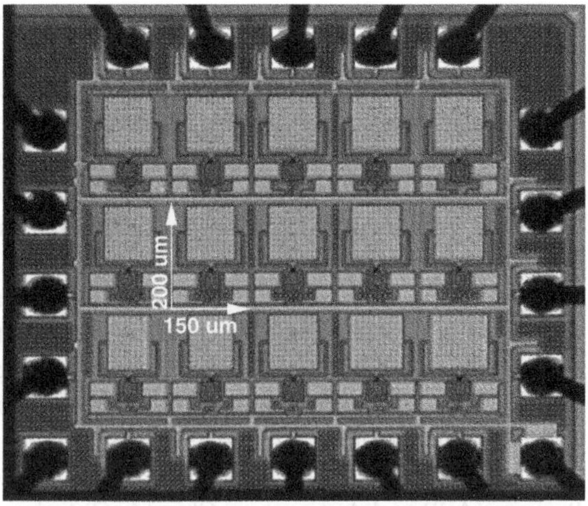

Figure 4.48 Differential rectangular microstrip patch antenna for single-chip 0.65-THz imager. *Source:* Ojefors et al. [30]/with permission of IEEE.

with a size of $14 \times 14 \times 0.8\ \text{mm}^3$. It integrates two differential rectangular microstrip patch antennas for transmission and four single-ended rectangular microstrip patch antennas for reception. The dimensions of the differential rectangular microstrip patch antenna were not given. The differential rectangular microstrip patch antenna was fed by two microstrip transmission lines. To reduce the electrical separation, a slit was cut on the patch between the two feeding lines. Simulated results showed a gain of 4 dBi and a bandwidth of about 15 GHz at 60 GHz.

Figure 4.48 shows a chip in 0.25 μm CMOS technology [30]. It integrates 15 differential rectangular patch antennas for imaging at 0.65 THz. The rectangular patch of size $88 \times 100\ \mu\text{m}^2$ was separated 8 μm from the ground plane and connected to the detector through two 130-μm-long microstrip transmission lines. Simulated results showed a gain of 2.1 dBi and a bandwidth of about 15 GHz.

4.8 Concluding Remarks

This chapter first extended the cavity model of microstrip patch antennas and then applied it to the analysis and design of differential rectangular and circular microstrip patch antennas. It was shown that the resonance of a differential microstrip patch antenna depends on the electrical separation between the dual feeds. When the dual feeds are located far from each other, the resonance occurs. However, when the dual feeds are located near to each other, the resonance does not occur, and the input impedance is inductive. It was found that there exist simple quantitative relations, which state that the ratio of differential to single-ended input impedances of the differential microstrip patch antenna are 4 or 0, depending on the relationship between the self-impedance and mutual impedance. The differential microstrip patch antennas have similar co-polar radiation patterns and lower cross-polar radiation components as compared with the single-ended microstrip patch antennas. Next, a new approach to the design and optimization of stacked microstrip patch antennas based on the BA for wide bandwidth was developed. A stacked microstrip patch antenna was treated as stacked two microstrip lines. A new insight into the resonance, coupling, and radiation of the stacked microstrip patch antenna was provided in terms of quasi-even and odd modes. After that, the differential microstrip patch arrays are described. It was emphasized that the use of the reactive balun to feed a differential microstrip patch antenna element in an array may result in scan blindness, due to the parasitic resonance. The safest way of avoiding it is to employ resistive baluns. Finally, the applications of differential microstrip patch antennas were highlighted with some examples of Antenna-in-Package (AiP) and Antenna-on-Chip (AoC) technologies.

References

1 D. D. Grieg, H. F. Engelmann, "Microstrip-A new transmission technique for the kilomegacycle range," *Proc. IRE*, vol. 40, no. 12, pp. 1644–1650, Dec. 1952.

2 G. Deschamps, W. Sichak, "Microstrip microwave antennas," in *Proc. Third Symp. USAF Antenna Research and Development Program*, Monticello, Illinois, USA, 18–22 Oct. 1953.

3 H. Gutton, G. Baissinot, "Flat aerial for ultra-high frequencies," French patent No. 703113 (1955).

4 P. S. Hall, "Keeping a low profile: A personal history of the microstrip antenna," in *Proc. Loughborough Antennas & Propagation Conf.*, Loughborough, UK, 16–17 Nov. 2009.

5 E. J. Denlinger, "Radiation from microstrip radiators," *IEEE Trans. Microw. Theory Technol.*, vol. 17, no. 4, pp. 235–236, April 1969.

6 E. V. Byron, "A new flush mounted antenna for phased-array applications," in *Phased-Array Antenna Symp. Digest*, Polytechnic Institute of Brooklyn, June 1970.

7 T. C. Cheston, E. V. Byron, G. J. Laughlin, "Very wide-band phased arrays," *IEEE AP-S Intl. Symp. Digest*, pp. 226–232, Columbus, OH, USA, 14–16 Sep. 1970.

8 R. E. Munson, "Microstrip phased array antennas," in *Proc. Twenty-Second Symp. USAF Antenna Research and Development Program*, Oct. 1972.

9 R. E. Munson, "Conformal microstrip antennas and microstrip phased arrays," *IEEE Trans. Antennas Propagat.*, vol. 22, no. 1, pp. 74–78, Jan. 1974.

10 J. Q. Howell, "Microstrip antennas," in *IEEE AP-S Intl. Symp. Digest*, pp. 177–180, Williamsburg, VA, USA, 11–14 Dec. 1972.

11 J. Q. Howell, "Microstrip antennas," *IEEE Trans. Antennas Propagat.*, vol. 23, no. 1, pp. 90–93, Jan. 1975.

12 D. R. Jackson, "Introduction to microstrip antennas," Short course at *IEEE AP-S Intl. Symp.*, Orlando, FL, USA, 7–13 Jul. 2013.

13 K. F. Lee, "A personal overview of the development of microstrip patch antennas," in *IEEE AP-S Intl. Symp.*, Puerto Rico, 26 Jun.–1 Jul. 2016.

14 T. Chiba, Y. Suzuki, N. Miyano, "Suppression of higher modes and cross polarized component for microstrip antenna," *IEEE AP-S Intl. Symp. Digest*, pp. 485–288, Albuquerque, NM, USA, 24–28 May 1982.

15 J. D. Hanfling, J. J. Schuss, "Experimental results illustrating performance limitations and design tradeoffs in probe-fed patch-element phased arrays," in *IEEE AP-S Intl. Symp. Digest*, pp. 802–805, Philadelphia, PA, USA, 8–13 Jun. 1986.

16 J. J. Schuss, J. D. Hanfling, "Non-reciprocity and scan blindness in phased arrays using balanced radiators," *IEEE Trans. Antennas Propagat.*, vol. 35, no. 2, pp. 134–138, Feb. 1987.

17 R. L. Bauer, J. J. Schuss, "Axial ratio of balanced and unbalanced fed circularly polarized patch radiator arrays," in *IEEE AP-S Intl. Symp. Digest*, pp. 286–289, Blacksburg, VA, USA, 15–19 Jun. 1987.

18 J. J. Schuss, J. D. Hanfling, R. E. Morrow, "Observation of scan blindness due to surface wave resonance in an array of printed circuit patch radiators," in *IEEE AP-S Intl. Symp. Digest*, pp. 802–805, Blacksburg, VA, USA, 15–19 Jun. 1987.

19 K. R. Carver, J. W. Mink, "Microstrip antenna technology," *IEEE Trans. Antennas Propagat.*, vol. 27, no. 1, pp. 2–24, Jan. 1981.

20 W. F. Richards, Y. T. Lo, D. D. Harrison, "An improved theory for microstrip antennas and applications," *IEEE Trans. Antennas Propagat.*, vol. 27, no. 1, pp. 38–46, Jan. 1981.

21 Y. T. Lo, D. Solomon, W. F. Richards, "Theory and experiment on microstrip antennas," *IEEE Trans. Antennas Propagat.*, vol. 27, no. 2, pp. 137–145, Mar. 1979.

22 Y. P. Zhang, J. J. Wang, "Theory and analysis of differentially-driven microstrip antenna," *IEEE Trans. Antennas Propagat.*, vol. 54, no. 4, pp. 1092–1099, Apr. 2006.

23 Y. P. Zhang, "Design and experiment on differentially-driven microstrip antennas," *IEEE Trans. Antennas Propagat.*, vol. 55, no. 10, pp. 2701–2708, Oct. 2007.

24 H. L. Peng, Z. Tang, Y. P. Zhang, J. F. Mao, "Cavity model analysis of a dual-probe-feed circular microstrip patch antenna," *IEEE Antennas Wireless Propagat. Lett.*, vol. 15, pp. 44–47, Apr. 2015.

25 G. Xu, et al., "Differential probe fed liquid crystal-based frequency tunable circular ring patch antenna," *IEEE Access*, vol. 6, pp. 3051–3058, 2017.

26 Z. J. Shao, L. F. Qiu, Y. P. Zhang, "Design of wideband differentially-fed multilayer stacked patch antennas based on bat algorithm," *IEEE Antennas Wireless Propagat. Lett.*, vol. 19, no. 7, pp. 1172–1176, Jul. 2019.

27 Y. P. Zhang, "Enrichment of package antenna approach with dual feeds, guard ring, and fences of vias," *IEEE Trans. Adv. Packag.*, vol. 32, no. 3, pp. 612–618, Aug. 2009.

28 D. G. Kam, D. Liu, A. Natarajan, S. K. Reynolds, R. A. Floyd, "Organic packages with embedded phased-array antennas for 60-GHz wireless chipsets," *IEEE Trans. Compon. Packag. Manufact. Technol.*, vol. 1, no. 11, pp. 1806–1814, Nov. 2011.

29 I. Nasr, et al., "A highly integrated 60 GHz 6-channel transceiver with antenna in package for smart sensing and short-range communications, " *IEEE J. Solid-State Circuits*, vol. 51, no. 9, pp. 2066–2076, Sep. 2016.

30 E. Ojefors, U. R. Pfeiffer, A. Lisauskas, H. G. Roskos, "A 0.65 THz focal-plane array in a quarter-micron CMOS process technology," *IEEE J. Solid-State Circuits*, vol. 44, no. 7, pp. 1968–1976, Jul. 2009.

31 A. C. Ludwig, "The definition of cross polarization," *IEEE Trans. Antennas Prapagat.*, vol. 21, no. 1, pp. 116–119, Jan. 1973.

32 B. Nauwelaers, A. V. Decapelle, "Surface wave losses of rectangular microstrip antennas," *Electron. Lett.*, vol. 25, no. 11, pp. 696–697, May 1989.

33 Y. L. Fang, Y. P. Zhang, "On surface-wave suppression of differential circular microstrip antennas," *IEEE Antennas Wirel. Propagat. Lett.*, vol. 20, no. 9, pp. 1691–1695, Sep. 2021.

34 K. F. Lee, S. R. Chebolu, W. Chen, R. Q. Lee, "On the role of substrate loss tangent in the cavity model theory of microstrip patch antennas," *IEEE Trans. Antennas Propagat.*, vol. 42, no. 1, pp. 110–112, Jan. 1994.

35 Y. P. Zhang, "Impedance relations for differential antennas and single-ended counterparts," *IEEE Trans. Antennas Propagat.*, vol. 70, no. 2, pp. 953–959, Feb. 2022.

36 Y. P. Zhang, "Electrical separation and fundamental resonance of differentially-driven microstrip antennas," *IEEE Trans. Antennas Propagat.*, vol. 59, no. 4, pp. 1078–1084, Apr. 2011.

37 F. Abboud, J. P. Damiano, A. Papiernik, "Simple model for the input impedance of coax-fed rectangular microstrip patch antenna for CAD," *Proc. Inst. Elect. Eng. Microw., Antennas Propagat.*, vol. 135, no. 5, pp. 323–326, Oct. 1988.

38 D. Vollbracht, "Understanding and optimizing microstrip patch antenna cross polarization radiation on element level for demanding phased array antennas in weather radar applications," *Adv. Radio Sci.*, vol. 13, pp. 251–268, Nov. 2015.

39 D. Thouroude, M. Himdi, J. P. Daniel, "CAD-oriented cavity model for rectangular patches," *Electron. Lett.*, vol. 26, no. 13, pp. 842–844, Jun. 1990.

40 A. Petosa, A. Ittipiboon, N. Gagnon, "Suppression of unwanted probe radiation in wideband probe-fed microstrip patches," *Electron. Lett.*, vol. 35, no. 5, pp. 355–357, Mar. 1999.

41 G. Dubost, "Far field radiated by rectangular patch microstrip antenna," *Electron. Lett.*, vol. 18, no. 23, pp. 991–993, Nov. 1982.

42 M. L. Oberhart, Y. T. Lo, R. Q. H. Lee, "New simple feed network for an array module of four microstrip elements," *Electron. Lett.*, vol. 23, no. 9, pp. 436–437, Apr. 1987.
43 R. C. Hansen, "Cross polarization of microstrip patch antennas," *IEEE Trans. Antennas Propagat.*, vol. 35, no. 6, pp. 731–732, Jun. 1987.
44 T. Huynh, K. F. Lee, R. Q. Lee, "Crosspolarisation characteristics of rectangular patch antennas," *Electron. Lett.*, vol. 24, no. 8, pp. 463–464, Apr. 1988.
45 S. Bhardwaj, Y. Rahmat-Samii, "Revisiting the generation of cross-polarization in rectangular patch antennas: A near-field approach," *IEEE Antennas Propagat. Mag.*, vol. 56, no. 1, pp. 14–38, Feb. 2014.
46 Y. Suyuki, "Key points in the design and measurement of microstrip antennas," *Analysis, Design, and Measurement of Small and Low-Profile Antennas*, K. Hirasawa, M. Haneishi, Eds. Norwood, MA: Artech House, 1991, pp. 83–159.
47 Z. H. Tu, Y. P. Zhang, "Comparison of three methods for the measurement of the differential impedance of a balanced device," *IEEE Antennas Propagat. Mag.*, vol. 55, no. 5, pp. 142–151, Oct. 2013.
48 Y. P. Zhang, Z. H. Chen, "The Wheeler method for the measurement of the efficiency of differentially-driven microstrip antennas," *IEEE Trans. Antennas Propagat.*, vol. 62, no. 6, pp. 3436–3439, Jun. 2014.
49 Y. P. Zhang, Z. H. Tu, "Measuring the impedance of differentially-driven microstrip antennas by two balun methods," *IEEE Trans. Antennas Propagat.*, vol. 61, no. 11, pp. 1246–1252, Nov. 2014.
50 L. C. Shen, S. A. Long, M. R. Allerding, M. D. Walton, "Resonant frequency of a circular disc, printed-circuit antenna," *IEEE Trans. Antennas Propagat.*, vol. 25, no. 4, pp. 595–596, Jul. 1977.
51 K. F. Lee, J. S. Dahele, "Characteristics of microstrip patch antennas and some methods of improving frequency agility and bandwidth," *Handbook of Microstrip Antennas*, J. R. James, P. S. Hall, Eds. London, UK: Peter Peregrinus Ltd, 1989, pp. 111–217.
52 K. F. Lee, K. M. Luk, *Microstrip Patch Antenna*. London, UK: Imperial College Press, 2011.
53 D. H. Schaubert, "Multilayer and parasitic configurations," *Handbook of Microstrip Antennas*, J. R. James, P. S. Hall, Eds. London, UK: Peter Peregrinus Ltd, 1989, pp. 311–351.
54 T. Huynh, K. F. Lee, "Single-layer single-patch wideband microstrip antenna," *Electron. Lett.*, vol. 31, no. 16, pp. 1310–1312, Aug. 1995.
55 B. L. Ooi, Q. Shen, "Novel E-shaped broadband microstrip patch antenna," *Microw. Opt. Technol. Lett.*, vol. 27, no. 5, pp. 348–352, Oct. 2000.
56 C. L. Mak, K. M. Luk, K. F. Lee, Y. L. Chow, "Experimental study of a microstrip patch antenna with an L-shaped probe," *IEEE Trans. Antennas Propagat.* vol. 48, no. 5, pp. 777–783, May 2000.

57 D. Liu, X. Gu, C. Baks, A. Valdes-Garcia, "Antenna-in-package design considerations for Ka-band 5G communication applications," *IEEE Trans. Antennas Propagat.*, vol. 65, no. 12, pp. 6372–6379, Dec. 2017.

58 Y. P. Zhang, "Antennas," *Antenna-in-Package Technology and Applications*, D. X. Liu, Y. P. Zhang, Eds. Hoboken, NJ: Wiley-IEEE, 2020, pp. 17–56.

59 P. S. Hall, C. Wood, "Wide bandwidth microstrip antennas for circuit integration," *Electron. Lett.*, vol. 15, no. 15, pp. 458–460, Jul. 1979.

60 F. Croq, D. M. Pozar, "Millimeter wave design of wide-band aperture coupled stacked microstrip antennas," *IEEE Trans. Antennas Propagat.*, vol. 39, no. 12, pp. 1770–1776, Dec. 1991.

61 Z. F. Liu, P. S. Kooi, L. W. Li, M. S. Leong, "A method for designing broad-band microstrip antennas in multilayered planar structures," *IEEE Trans. Antennas Propagat.*, vol. 47, no. 9, pp. 1416–1420, Sep. 1999.

62 R. B. Waterhouse, "Design of probe-fed stacked patches," *IEEE Trans. Antennas Propagat.*, vol. 47, no. 12, pp. 1780–1784, Dec. 1999.

63 S. Gao, L. W. Li, M. S. Leong, T. S. Yeo, "A broad-band dual-polarized microstrip patch antenna with aperture coupling," *IEEE Trans. Antennas Propagat.*, vol. 51, no. 4, pp. 898–900, Apr. 2003.

64 A. Mitchell, M. Lech, D. M. Kokotoff, R. B. Waterhouse, "Search for high-performance probe-fed stacked patches using optimization," *IEEE Trans. Antennas Propagat.*, vol. 51, no. 2, pp. 249–255, Feb. 2003.

65 X. S. Yang, A. H. Gandomi, "Bat algorithm: A novel approach for global engineering optimization", *Int. J. Comput. Aided Eng. Softw.*, vol. 29, no. 5, pp. 464–483, Jul. 2012.

66 R. P. Owens, "Microstrip antenna feeds," *Handbook of Microstrip Antennas*, J. R. James, P. S. Hall, Eds. London, UK: Peter Peregrinus Ltd, 1989, pp. 815–870.

67 R. P. Jedlicka, M. T. Poe, K. R. Carver, "Measured mutual coupling between microstrip antennas," *IEEE Trans. Antennas Propagat.*, vol. 29, no. 1, pp. 147–149, Jan. 1981.

68 D. M. Pozar, "Input impedance and mutual coupling of rectangular microstrip antennas," *IEEE Trans. Antennas Propagat.*, vol. 30, no. 6, pp. 1191–1196, Nov. 1982.

69 D. R. Jackson, et al., "Microstrip patch designs that do not excite surface waves," *IEEE Trans. Antennas Propagat.*, vol. 41, no. 8, pp. 1026–1037, Aug. 1993.

70 M. A. Khayat, J. T. Williams, D. R. Jackson, S. A. Long, "Mutual coupling between reduced surface-wave microstrip antennas," *IEEE Trans. Antennas Propagat.*, vol. 48, no. 10, pp. 1581–1593, Oct. 2000.

71 M. M. Nikolic, A. R. Djordjevic, A. Nehorai, "Microstrip antennas with suppressed radiation in horizontal directions and reduced coupling," *IEEE Trans. Antennas Propagat.*, vol. 53, no. 11, pp. 3469–3476, Nov. 2005.

72 V. R. Komanduri, D. R. Jackson, J. T. Williams, A. R. Mehrotra, "A general method for designing reduced surface wave microstrip antennas," *IEEE Trans. Antennas Propagat.*, vol. 61, no. 6, pp. 2887–2894, Jun. 2013.

73 G. P. Gauthier, J. P. Raskin, L. P. B. Katehi, G. M. Rebeiz, "A 94-GHz aperture-coupled micromachined microstrip antenna," *IEEE Trans. Antennas Propagat.*, vol. 47, no. 12, pp. 1761–1766, Dec. 1999.

74 J. G. Yook, L. P. B. Katehi, "Micromachined microstrip patch antenna with controlled mutual coupling and surface waves," *IEEE Trans. Antennas Propagat.*, vol. 49, no. 9, pp. 1282–1289, Sep. 2001.

75 J. T. Aberle, F. Zavosh, "Infinite phased arrays of cavity-backed patches," *IEEE Trans. Antennas Propagat.*, vol. 42, no. 3, pp. 390–398, Mar. 1994.

76 D. Sievenpiper, L. J. Zhang, R. F. J. Broas, N. G. Alexopolous, E. Yablonovitch, "High-impedance electromagnetic surfaces with a forbidden frequency band," *IEEE Trans Microw. Theory Technol.*, vol. 47, no. 11, pp. 2059–2074, Nov. 1999.

77 A. P. Feresidis, G. Goussetis, S. H. Wang, J. C. Vardaxoglou, "Artificial magnetic conductor surfaces and their application to low-profile high-gain planar antennas," *IEEE Trans. Antennas Propagat.*, vol. 53, no. 1, pp. 209–215, Jan. 2005.

78 F. Yang, Y. Rahmat-Samii, "Microstrip antennas integrated with electromagnetic band-gap (EBG) structures: a low mutual coupling design for array applications," *IEEE Trans. Antennas Propagat.*, vol. 51, no. 10, pp. 2936–2946, Oct. 2003.

79 F. R. Yang, K. P. Ma, Y. Qian, T. Itoh, "A uniplanar compact photonic-bandgap (UC-PBG) structure and its applications for microwave circuits," *IEEE Trans Microw. Theory Technol.*, vol. 47, no. 8, pp. 1509–1514, Aug. 1999.

80 A. Yu, X. Z. Zhang, "A novel method to improve the performance of microstrip antenna arrays using a dumbbell EBG structure," *IEEE Antennas Wirel. Propagat. Lett.*, vol. 2, pp. 170–172, 2003.

81 B. Mohamadzade, A. Lalbakhsh, R. B. V. B. Simorangkir, A. Rezaee, R. M. Hashmi, "Mutual coupling reduction in microstrip array antenna by employing cut side patches and EBG structures," *Prog. Electromagn. Res. M*, vol. 89, pp. 179–187, 2020.

82 B. Mohamadzade, M. Afsahi, "Mutual coupling reduction and gain enhancement in patch array antenna using a planar compact electromagnetic bandgap structure," *IET Microw. Antennas Propagat.*, vol. 11, no. 12, pp. 1719–1725, Sep. 2017.

83 A. E. I. Lamminen, A. R. Vimpari, J. Saily, "UC-EBG on LTCC for 60-GHz frequency band antenna applications," *IEEE Trans. Antennas Propagat.*, vol. 57, no. 10, pp. 2904–2912, Oct. 2009.

84 D. G. Kam, et al., "LTCC packages with embedded phased-array antennas for 60 GHz communications," *IEEE Microw. Wirel. Comp. Lett.*, vol. 21, no. 3, pp. 142–144, Mar. 2011.

85 I. Ederra, et al., "Experimental verification of the reduction of coupling between dipole antennas by using a woodpile substrate," *IEEE Trans. Antennas Propagat.*, vol. 54, no. 7, pp. 2105–2112, Jul. 2007.

86 D. Ahn, et al., "A design of the low-pass filter using the novel microstrip defected ground structure," *IEEE Trans Microw. Theory Technol.*, vol. 49, no. 1, pp. 86–93, Jan. 2001.

87 S. Xiao, M. C. Tang, Y. Y. Bai, S. Gao, B. Z. Wang, "Mutual coupling suppression in microstrip array using defected ground structure," *IET Microw. Antennas Propagat.*, vol. 5, no. 12, pp. 1488–1494, Sep. 2011.

88 N. Alexopoulos, D. Jackson, "Fundamental superstrate (cover) effects on printed circuit antennas," *IEEE Trans. Antennas Propagat.*, vol. 32, no. 8, pp. 807–816, Aug. 1984.

89 J. B. Pendry, A. J. Holden, D. J. Robbins, W. J. Stewart, "Magnetism from conductors and enhanced nonlinear phenomena," *IEEE Trans Microw. Theory Technol.*, vol. 47, no. 11, pp. 2075–2084, Nov. 1999.

90 E. Saenz, R. Gonzalo, I. Ederra, P. Maagt, "Radiation performances of a dipole array configuration inserted in a left-handed media," in *Proc. IEEE Workshop Antenna Technol.*, pp. 187–190, Singapore, 7–9 Mar. 2005.

91 A. Jafargholi, A. Jafargholi, J. H. Choi, "Mutual coupling reduction in an array of patch antennas using CLL metamaterial superstrate for MIMO applications," *IEEE Trans. Antennas Propagat.*, vol. 67, no. 1, pp. 179–189, Jan. 2019.

92 E. Saenz, et al., "Coupling reduction between dipole antenna elements by using a planar meta-surface," *IEEE Trans. Antennas Propagat.*, vol. 57, no. 2, pp. 383–394, Feb. 2009.

93 F. Liu, et al., "Dual-band metasurface-based decoupling method for two closely packed dual-band antennas," *IEEE Trans. Antennas Propagat.*, vol. 68, no. 1, pp. 552–557, Jan. 2020.

94 K. L. Wu, C. N. Wei, X. D. Mei, Z. Y. Zhang, "Array-antenna decoupling surface," *IEEE Trans. Antennas Propagat.*, vol. 65, no. 12, pp. 6728–6738, Dec. 2017.

95 H. Y. Jin, et al., "60-GHz LTCC differential-fed patch antenna array with high gain by using soft-surface structures," *IEEE Trans. Antennas Propagat.*, vol. 65, no. 1, pp. 206–216, Jan. 2017.

96 H. Y. Jin, et al., "Differential-fed patch antenna arrays with low cross polarization and wide bandwidths," *IEEE Antennas Wirel. Propagat. Lett.*, vol. 13, pp. 1069–1072, 2004.

97 Z. Tong, A. Stelzer, C. Wagner, R. Feger, E. Kolmhofer, "A novel differential microstrip patch antenna and array at 79 GHz," in *Proc. Intl. Symp. Antennas Propagat.*, pp. 276–280, Taipei, Taiwan, China, 27–30 Oct. 2008.

98 W. A. Ahmad, J. H. Lu, D. Kissinger, H. J. Ng, "Beam squinting in wideband 60 GHz on-board series-fed differential patch arrays," in *Proc. Asia Pacific Microw. Conf.*, Kuala Lumpur, Malaysia, 13–16 Nov. 2017.

99 R. Hasan, W. A. Ahmad, J. H. Lu, H. J. Ng, D. Kissinger, "F-band differential microstrip patch antenna array and waveguide to differential microstrip line transition for FMCW radar sensor," *IEEE Sensors J.*, vol. 19, no. 15, pp. 6486–6496, Aug. 2019.

100 H. Shahi, N. Masoumi, S. Safavi-Naeini, M. Mohammad-Taheri, "N-way planar balanced-to-balanced series feeding network for balanced frequency-scanning phased array," *IEEE Trans. Antennas Propagat.*, vol. 68, no.7, pp. 5334–5343, Jul. 2020.

101 Alberto Valdes-Garcia, S. T., et al., "A fully integrated 16-element phased-array transmitter in SiGe BiCMOS for 60-GHz communications," *IEEE J. Solid-State Circuits*, vol. 45, no. 12, pp. 2757–2773, Dec. 2011.

102 A. Natarajan, et al., "A fully-integrated 16-element phased-array receiver in SiGe BiCMOS for 60-GHz communications," *IEEE J. Solid-State Circuits*, vol. 46, no. 5, pp. 1059–1075, Dec. 2011.

103 T. Zhang, L. M. Li, H. Y. Xia, X. J. Ma, T. J. Cui, "A low-cost and high-gain 60-GHz differential phased array antenna in PCB process," *IEEE Trans. Compon. Packag. Manufact. Technol.*, vol. 8, no. 7, pp. 1281–1291, Jul. 2018.

104 Y. P. Zhang, J. J. Wang, Q. Li, X. J. Li, "Antenna-in-package and transmit-receive switch for single-chip radio transceivers of differential architecture," *IEEE Trans. Circuits Syst. I, Regular Papers*, vol. 55, no. 11, pp. 3564–3570, Dec. 2008.

105 Y. P. Zhang, "Differential antennas: Fundamentals and applications," *Electromagn. Sci.*, vol. 1, no. 1, pp. 1–17, Mar. 2023.

106 D. Liu, Y. P. Zhang, *Antenna-in-Package Technology and Applications*, IEEE-Wiley, 2020.

5

Differential Microstrip Shorted Patch Antennas

5.1 Introduction

A half-wavelength microstrip patch antenna operates in the fundamental mode. A virtual electric wall exists along the center line between the two radiating edges. Hence, a physical metal wall was proposed to replace the virtual electric wall to create a quarter-wavelength microstrip patch antenna or a shorted patch antenna (SPA) [1]. A comparative study of the SPA with the patch antenna showed that the dielectric plays a fundamental role in the far-field pattern, the input impedance, and the bandwidth [2]. When $\varepsilon_r = 1$, the impedance of the SPA and the patch antenna has almost the same value, whereas the SPA bandwidth is twice as wide. On the other hand, with usual dielectric $\varepsilon_r = 2.2$, the bandwidth retains the same value for the two structures, but the impedance of the SPA is twice that of the patch antenna. These differences are attributed to the polarization currents whose effect should be more significant for the SPA than for the patch antenna because of the cosine distribution of the resonant mode, which has no change of phase along the length of the SPA. The maximum gain of the SPA is about half that of the patch antenna [3].

Although the SPA preserves the advantages of a half-wavelength patch antenna with a reduced size, it has a major issue of lower *H*-plane XPL, which limits the SPA from many applications. This chapter focuses on the SPA, differential SPA (DSPA), and their variations, aiming to improve *H*-plane XPL. Section 5.2 presents the basic SPA and DSPA [4]. It will show that the DSPA achieves a symmetrical *E*-plane radiation pattern and an improved *H*-plane XPL as compared with the SPA. More importantly, the DSPA has a smaller size than the SPA for the same resonant frequency. Section 5.3 describes the modified SPA (mSPA) and DSPA (mDSPA) for further enhanced XPL [5–7].

The planar inverted-F antenna (PIFA) can be considered a special SPA. The PIFA finds wide application in portable wireless devices, due to its compact size [8–10]. There have been few attempts to design balanced-fed PIFAs. For example,

Differential Antennas: Theory and Practice, First Edition. Yueping Zhang.

Ogawa et al. developed a built-in balanced-fed PIFA [11] and Chan et al. devised an aperture-coupled differentially fed PIFA [12]. The former consists of two largely separated PIFAs and a U-balun through coaxial cables and thus appears bulky. The latter is an asymmetrical structure, which results in tilted radiation pattern and large cross-polarization components. Section 5.4 presents a basic differential PIFA (DPIFA) [13], compares its performance with that of the PIFA, and shows how to miniaturize it.

The magnetoelectric (ME) dipole consists of an SPA as a magnetic dipole and a planar electric dipole [14]. The ME dipole has many advantages such as broad bandwidth, low cross-polarization, symmetrical radiation pattern, and specially, very low back radiation [15]. The ME dipole was first invented as a single-ended antenna [14] and then was modified for differential operation [16, 17]. Section 5.5 demonstrates how to design ME dipole and array for millimeter-wave 5G mobile communications [18–22]. Finally, Section 5.6 provides concluding remarks.

5.2 SPA and DSPA

The SPA was devised from a microstrip patch antenna [1]. Figure 5.1 shows the geometry and the coordinate system of the SPA. It is formed with a rectangular patch above a ground plane, a shorting wall (or a fence of metal vias), and a coaxial probe feed. The patch has a size of width w and length l and is kept at a height of h above the ground plane of dimensions $w_g \times l_g$. The substrate between the patch and the ground plane has a dielectric constant ε_r and a loss tangent δ. The feed is located at (x_1, y_1).

The DSPA is shown in Figure 5.2. It is created from the SPA in three steps. First, a narrow gap g is cut along the centerline perpendicular to the shorting edge. Then, one half shorting wall is moved to the opposite edge. Finally, another coaxial probe feed is added at (x_2, y_2) to the half patch [3].

5.2.1 Resonant Modes

The SPA, like the patch antenna, can be modelled as a resonant cavity. However, it is surrounded not by four magnetic walls at the side periphery but by three magnetic and one electric walls. Thus, according to the cavity model in Section 4.3.1, the interior electric field with a z-directed current ribbon at the feed (x_1, y_1) is given by

$$\begin{aligned} E_z &= j\omega\mu \sum_{m=0}^{\infty}\sum_{n=1,3,5\ldots}^{\infty} \frac{1}{k_e^2 - k_{mn}^2} \frac{\langle J_z(x_1,y_1)\psi_{mn}(x,y)\rangle}{\langle \psi_{mn}\psi_{mn}\rangle}\psi_{mn}(x,y) \\ &= j\omega\mu I \sum_{m=0}^{\infty}\sum_{n=1,3,5\ldots}^{\infty} \frac{\psi_{mn}(x_1,y_1)}{k_e^2 - k_{mn}^2}\psi_{mn}(x,y) j_0\left(\frac{m\pi d}{2w}\right)\frac{\epsilon_{0m}\epsilon_{0n}}{wl} \\ &= \sum_{m=0}^{\infty}\sum_{n=1,3,5\ldots}^{\infty} E_{mn}\psi_{mn}(x,y) \end{aligned} \tag{5.1}$$

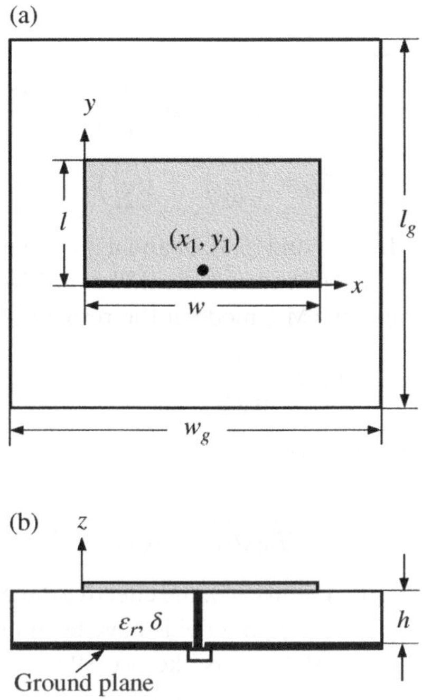

Figure 5.1 Geometry and coordinate system of the SPA. (a) Top view and (b) side view.

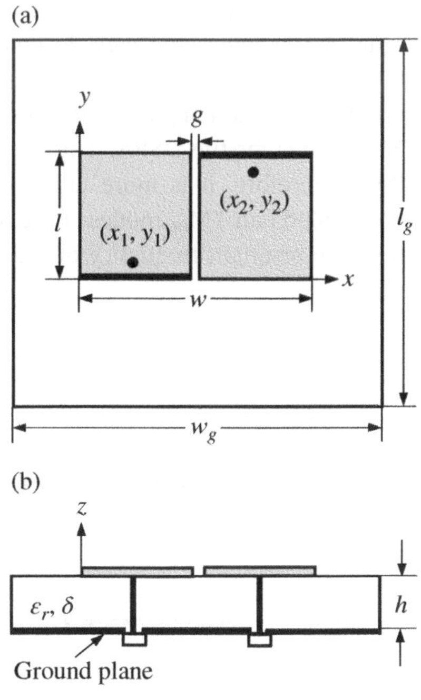

Figure 5.2 Geometry and coordinate system of DSPA. (a) Top view and (b) side view.

where the eigenfunctions and eigenvalues of the cavity can be written, respectively, as

$$\psi_{mn}(x,y) = \cos\left(\frac{m\pi x}{w}\right)\sin\left(\frac{n\pi y}{2l}\right), \tag{5.2}$$

$$k_{mn}^2 = \left(\frac{m\pi}{w}\right)^2 + \left(\frac{n\pi}{2l}\right)^2. \tag{5.3}$$

In (5.2) and (5.3), m and n are the mode indices in the x and y directions, respectively. The index n should be positive odd integer. Typically, the SPA operates in the half TM_{01} mode at the resonant frequency

$$f_{01} = \frac{c}{4l\sqrt{\varepsilon_e}} \tag{5.4}$$

with the size

$$l = \frac{c}{4f_{01}\sqrt{\varepsilon_e}} = \frac{\lambda}{4\sqrt{\varepsilon_e}} = \frac{\lambda_g}{4}. \tag{5.5}$$

In (5.4) and (5.5), the effective dielectric constant is used, which implies that the effects of fringing fields have been already taken into account.

The DSPA seems like two SPAs and thus can be modelled as two resonant cavities. It is found that the interior electric field within each cavity can be approximately expressed as (5.1) with the proper changes in coordinates, dimensions, and excitations. Also, (5.2)–(5.5) must be modified to account for the two cavities and the interaction of the fringing fields due to the gap between the two SPAs. Figure 5.3 sketches the electric field distributions on the cross sections of the SPA and the DSPA. As can be seen, the electric field distribution confirms the dominance of the TM_{01} mode for the SPA and the resemblance of two out-of-phase TM_{01} modes for the DSPA. The gap causes electrical coupling between the two SPAs. Therefore, it is more appropriate to say that the DSPA operates in the two coupled half TM_{01} modes at the resonant frequency. Due to the electrical coupling, the resonant frequency of the DSPA should be lower than that of the SPA.

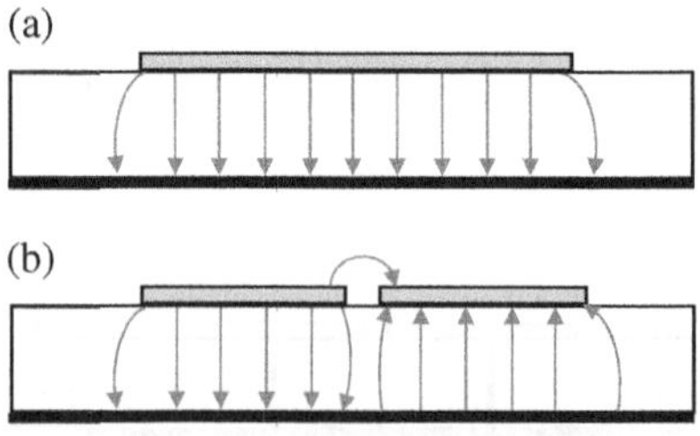

Figure 5.3 Electric field distributions at the resonant frequencies on the cross sections. (a) SPA and (b) DSPA.

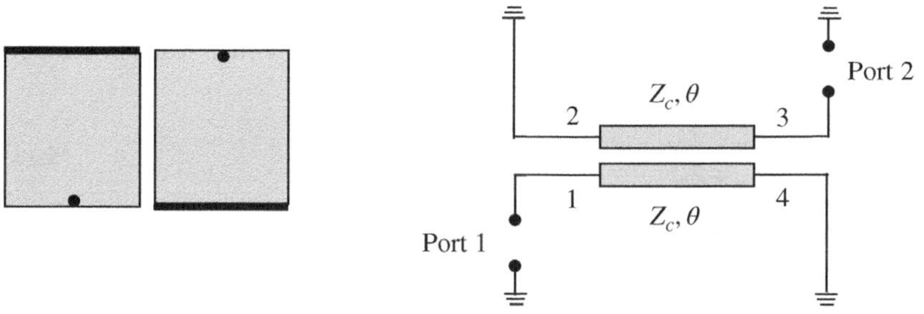

Figure 5.4 Coupled transmission-line model for the DSPA resonance.

5.2.2 Coupling and Resonance

The coupling introduced by the gap in the DSPA will affect its resonance. For a patch antenna, the resonance mainly depends on the patch length l and weakly on the feed location. Hence, a simple coupled transmission-line model shown in Figure 5.4 is proposed to determine the DSPA resonance. Note that the coupling effect is treated as a parallel-coupled transmission-line section with a four-port network.

After considering the shorted edges, the four-port network reduces to a two-port network [4]. Then, the input impedance of the DSPA can be derived as

$$Z_d = 2(Z'_{11} - Z'_{13}) \tag{5.6}$$

where

$$Z'_{11} = Z_{11} + \frac{Z_{11}(Z_{12}^2 + Z_{14}^2) - 2Z_{12}Z_{13}Z_{14}}{Z_{13}^2 - Z_{11}^2}, \tag{5.7}$$

$$Z'_{13} = Z_{13} - \frac{Z_{11}(Z_{12}^2 + Z_{14}^2) - 2Z_{11}Z_{13}Z_{14}}{Z_{13}^2 - Z_{11}^2}. \tag{5.8}$$

Substituting (5.7) and (5.8) into (5.6) yields

$$Z_d = 2\frac{(Z_{12} - Z_{14})^2 - (Z_{11} - Z_{13})^2}{Z_{13} - Z_{11}} \tag{5.9}$$

where Z_{11}, Z_{12}, Z_{13}, and Z_{14} are the self and mutual impedances at port 1 and between port 1 and other ports, respectively. Using the expressions for the self-impedance and mutual impedance in [23], one can rewrite (5.9) as

$$Z_d = \frac{2Z_{0e}Z_{0o}}{Z_{11} - Z_{13}} \tag{5.10}$$

where Z_{0e} and Z_{0o} are the characteristic impedances of the even and odd modes, respectively. When Z_{11} is equal to Z_{13}, Z_d goes to infinity, which leads to the parallel resonance and

$$\cosh\theta = \frac{Z_{0e} - Z_{0o}}{Z_{0e} + Z_{0o}}. \tag{5.11}$$

Neglecting the losses, (5.11) reduces to

$$\cos\theta = \frac{Z_{0e} - Z_{0o}}{Z_{0e} + Z_{0o}}. \tag{5.12}$$

If there were no coupling between the two halves, the DSPA would degenerate to the SPA loaded with the gap and one would get

$$Z_{0e} = Z_{0o} \tag{5.13}$$

or

$$\theta = \frac{\pi}{2}, \tag{5.14}$$

which results in the resonance at $\lambda_g/4$ for the SPA. While if there is coupling between the two halves, one will get

$$Z_{0e} > Z_{0o} \tag{5.15}$$

or

$$\theta < \frac{\pi}{2}, \tag{5.16}$$

which results in the resonance at less than $\lambda_g/4$, indicating that the DSPA has a smaller size than the SPA for a given resonant frequency.

5.2.3 Radiation Characteristics

The radiation of the SPA can be thought mainly from the magnetic currents on the three walls at $x = 0$, w and $y = l$. The magnetic current is given by

$$\boldsymbol{M} = \hat{n} \times E_z = \hat{n} \times \sum_{m=0}^{\infty} \sum_{n=1,3,5\ldots}^{\infty} E_{mn} \cos\left(\frac{m\pi x}{w}\right) \sin\left(\frac{n\pi y}{2l}\right). \tag{5.17}$$

Since the SPA operates in the half TM_{01} mode, (5.17) simplifies to

$$\boldsymbol{M} = \hat{n} \times E_z \approx \hat{n} \times E_{01} \sin\left(\frac{\pi y}{2l}\right). \tag{5.18}$$

At $x = 0$,

$$\boldsymbol{M}_1 = \hat{y}\, E_{01} \sin\left(\frac{\pi y}{2l}\right), \tag{5.19}$$

at $y = l$,

$$\boldsymbol{M_2} = \hat{x}E_{01}, \tag{5.20}$$

and at $x = w$,

$$\boldsymbol{M_3} = -\hat{y}E_{01}\sin\left(\frac{\pi y}{2l}\right). \tag{5.21}$$

The electric vector potential of the magnetic current can be calculated by evaluating the resultant integral of (4.10). The radiation fields in the far zone at r are evaluated from equations (4.11) and (4.12) as

$$E_\theta = j\frac{wh}{\lambda}E_{01}\frac{e^{-jk_0 r}e^{jk_0 l\sin\theta}}{r}, \tag{5.22}$$

$$E_\phi = 0 \tag{5.23}$$

for the E plane ($\phi = 90°$) and

$$E_\phi = j\frac{h}{\tan\theta}E_{01}\frac{e^{-jk_0 r}}{\pi r}\sin\left(\frac{\pi w}{\lambda}\sin\theta\right), \tag{5.24}$$

$$E_\theta = -\frac{4lh}{\lambda}E_{01}\frac{e^{-jk_0 r}}{\pi r}\sin\left(\frac{\pi w}{\lambda}\sin\theta\right) \tag{5.25}$$

for the H plane ($\phi = 0°$). Therefore, for an infinite ground plane, the ratio of the cross-polarization component of the SPA at any elevation angle of the H-plane pattern to the co-polarization component can be expressed as

$$\left|\frac{E_\theta}{E_\phi}\right| = \frac{4l\tan\theta}{\lambda}. \tag{5.26}$$

As the co-polarization component E_ϕ of the H plane reaches the maximum in the broadside direction ($\theta = 0°$), the H-plane XPL of the SPA can be expressed as

$$XPL = \left|\frac{E_\theta|_{max}}{E_\phi|_{\theta=0°}}\right| = \frac{4l}{\pi w}\sin\left(\frac{\pi w}{\lambda}\sin\theta\right)\Big|_{max}. \tag{5.27}$$

Let ε_e be the effective dielectric constant, (5.27) can be written as

$$XPL = \left|\frac{E_\theta|_{max}}{E_\phi|_{\theta=0°}}\right| = \frac{4l}{w\pi}\sin\left(\frac{w\pi}{4l\sqrt{\varepsilon_e}}\sin\theta\right)\Big|_{max}, \tag{5.28}$$

which can be further simplified as

$$XPL = \left|\frac{E_\theta|_{max}}{E_\phi|_{\theta=0°}}\right| = \begin{cases} \dfrac{4l}{w\pi} & w/l \geq 2\sqrt{\varepsilon_e} \\ \dfrac{4l}{w\pi}\sin\left(\dfrac{w\pi}{4l\sqrt{\varepsilon_e}}\right) & w/l < 2\sqrt{\varepsilon_e} \end{cases}. \tag{5.29}$$

It can be seen from (5.29) that the H-plane XPL of the SPA will decrease as w/l increases. In particular, when w/l is much smaller than 1, the H-plane XPL of the SPA is approximately $1/\sqrt{\varepsilon_e}$. Therefore, the H-plane XPL of the SPA can be reduced to a certain extent by increasing the dielectric constant of the substrate.

Consider an SPA with $\varepsilon_r = 2.2$, $\delta = 0.001$, $l = 26$ mm, $l_g \times w_g \times h = 150 \times 150 \times 3\ \text{mm}^3$, $x_1 = w/2$, $y_1 = 5.5$ mm. Figure 5.5 shows the simulated H-plane pattern of the SPA with different w/l on an infinite ground plane. It can be seen that by increasing w/l from 1 to 4, the H-plane XPL of the SPA can be increased from 3.2 to 8.6 dB. This result is in good agreement with the calculation from (5.29). Therefore, the improvement of the H-plane XPL of the SPA is always accompanied by an increase in the patch size.

Similarly, the radiation of the DSPA can be thought mainly from the magnetic currents on the six walls at $x = 0, 0.5(w - g), 0.5(w + g)$, and w as well as $y = 0$ and l. For the DSPA operating in the two coupled half TM_{01} modes, the expressions for the magnetic currents on the three walls at $x = 0, 0.5(w - g)$ and $y = l$ are the same as (5.19)–(5.21), while the expressions for the magnetic currents on the other three walls at $x = 0.5(w + g)$, w, and $y = l$ are given by

$$\boldsymbol{M}_4 = -\hat{y}E_{01}\cos\left(\frac{\pi y}{2l}\right), \tag{5.30}$$

$$\boldsymbol{M}_5 = \hat{x}E_{01}, \tag{5.31}$$

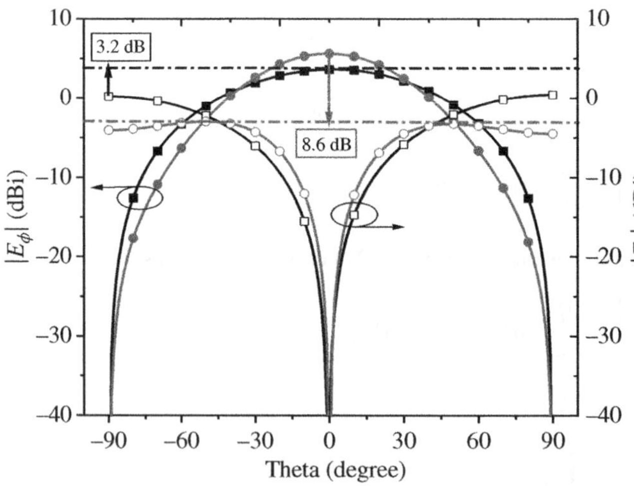

Figure 5.5 Simulated H-plane pattern of the SPA with $w/l = 1$ in gray with circles and 4 in black with squares.

and

$$\boldsymbol{M_6} = \hat{y}E_{01}\cos\left(\frac{\pi y}{2l}\right), \tag{5.32}$$

respectively.

The electric vector potential of the magnetic current can be calculated by evaluating the resultant integral of (4.10). The radiation fields in the far zone at r are evaluated from equations (4.11) and (4.12) as

$$E_\theta = j\frac{(w-g)h}{2\lambda}E_{01}\frac{e^{-jk_0 r}\left[1 + e^{jk_0 l\sin\theta}\right]}{r}, \tag{5.33}$$

$$E_\phi = 0 \tag{5.34}$$

for the E plane ($\phi = 90°$) and

$$E_\phi = j\frac{2h}{\tan\theta}E_{01}\frac{e^{-jk_0 r}}{\pi r}\cos\left(\frac{\pi(w+g)}{2\lambda}\sin\theta\right)\sin\left(\frac{\pi(w-g)}{2\lambda}\sin\theta\right), \tag{5.35}$$

$$E_\theta = j\frac{8lh}{\lambda}E_{01}\frac{e^{-jk_0 r}}{\pi r}\sin\left(\frac{\pi(w-g)}{2\lambda}\sin\theta\right)\sin\left(\frac{\pi(w+g)}{2\lambda}\sin\theta\right) \tag{5.36}$$

for the H plane ($\phi = 0°$). Therefore, for an infinite ground plane, the ratio of the cross-polarization component of the DSPA at any elevation angle of the H-plane pattern to the co-polarization component can be expressed as

$$\left|\frac{E_\theta}{E_\phi}\right| = \frac{4l\tan\theta}{\lambda}\tan\left(\frac{\pi(w+g)}{2\lambda}\sin\theta\right). \tag{5.37}$$

As the co-polarization component E_ϕ of the H plane reaches the maximum in the broadside direction ($\theta = 0°$), the H-plane XPL of the SPA can be expressed as

$$XPL = \left|\frac{E_\theta|_{max}}{E_\phi|_{\theta=0°}}\right| = \frac{8l}{\pi(w-g)}\sin\left(\frac{\pi(w-g)}{2\lambda}\sin\theta\right)\sin\left(\frac{\pi(w+g)}{2\lambda}\sin\theta\right)\Bigg|_{max}. \tag{5.38}$$

For $g \ll w$, (5.38) can be written as

$$XPL = \left|\frac{E_\theta|_{max}}{E_\phi|_{\theta=0°}}\right| = \begin{cases} \dfrac{8l}{w\pi} & w/l \geq 4\sqrt{\varepsilon_e} \\[2ex] \dfrac{8l}{w\pi}\sin^2\left(\dfrac{w\pi}{8l\sqrt{\varepsilon_e}}\right) & w/l < 4\sqrt{\varepsilon_e} \end{cases} \tag{5.39}$$

where ε_e is the effective dielectric constant for the patch width $(w-g)$.

Consider a DSPA with $\varepsilon_r = 2.2$, $\delta = 0.001$, $l = 26$ mm, $l_g \times w_g \times h = 150 \times 150 \times 3\ \text{mm}^3$, $x_1 = 0.25(w-g)$, $y_1 = 5.5$ mm, $x_2 = 0.75(w-g)$, $y_2 = 21.5$ mm, and $g = 1.0$ mm.

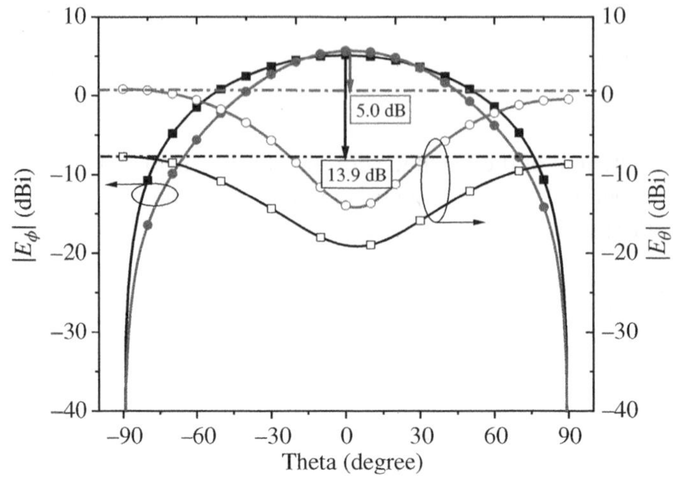

Figure 5.6 Simulated H-plane pattern of the DSPA with w/l = 1 in gray with circles and 4 in black with squares.

Figure 5.6 shows the simulated H-plane pattern of the DSPA with different w/l on an infinite ground plane. It can be seen that by increasing the ratio of the width to length w/l from 1 to 4, the H-plane XPL of the DSPA can be decreased from 13.9 to 5.0 dB. This result is in good agreement with the calculation from (5.39). Therefore, the improvement of the H-plane XPL of the DSPA is always accompanied by a decrease in the patch size.

The XPL ratio of the DSPA to the SPA can be deduced from (5.29) and (5.39) as

$$\left|\frac{XPL|_{DSPA}}{XPL|_{SPA}}\right| = \begin{cases} 2 & w/l \geq 4\sqrt{\varepsilon_e} \\ 2\sin^2\left(\dfrac{w\pi}{8l\sqrt{\varepsilon_e}}\right) & 2\sqrt{\varepsilon_e} < w/l < 4\sqrt{\varepsilon_e}, \\ \tan\left(\dfrac{w\pi}{8l\sqrt{\varepsilon_e}}\right) & w/l < 2\sqrt{\varepsilon_e} \end{cases} \tag{5.40}$$

which reveals that for the most common condition, the DSPA always exhibits a better H-plane XPL than that of the SPA.

5.2.4 Frequency Ratio and Size Reduction

According to (5.12), the ratio of the resonant frequency of the DSPA to the SPA with the same physical size can be calculated as

$$a_f = \frac{f_d}{f_s} = \frac{2}{\pi}\cos^{-1}\left(\frac{Z_{0e} - Z_{0o}}{Z_{0e} + Z_{0o}}\right). \tag{5.41}$$

It is known that the resonant frequency is inversely proportional to the patch length l. Thus, equation (5.41) can also be interpreted as the size reduction ratio of the DSPA to the SPA for the same resonant frequency.

Let us consider that the DSPA and SPA are built on FR4 substrates with $\varepsilon_r = 4.4$, $\delta = 0.02$, $l \times w = 13.5\,\text{mm} \times 6.75\,\text{mm}$, $l_g \times w_g \times h = 50\,\text{mm} \times 50\,\text{mm} \times 0.8\,\text{mm}$, and $g = 0.1\,\text{mm}$. Figure 5.7 shows the calculated and simulated a_f as a function of height h. It is seen that a_f decreases as h increases because the resonant frequency of the DSPA reduces faster than that of the SPA. Figure 5.8 shows the calculated and simulated a_f as a function of width w.

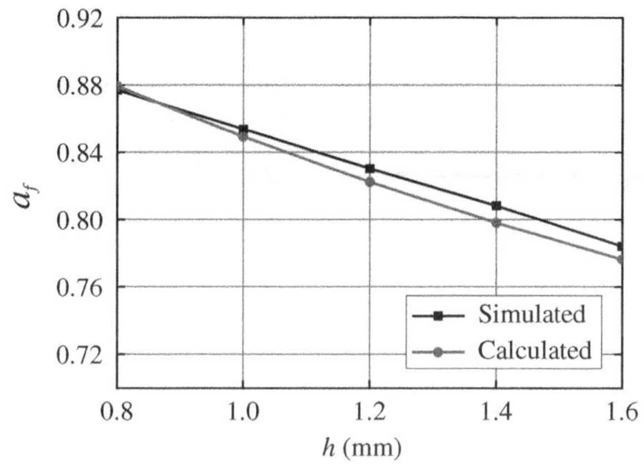

Figure 5.7 Calculated and simulated a_f as a function of h. ($l \times w = 13.5 \times 6.75\ \text{mm}^2$).

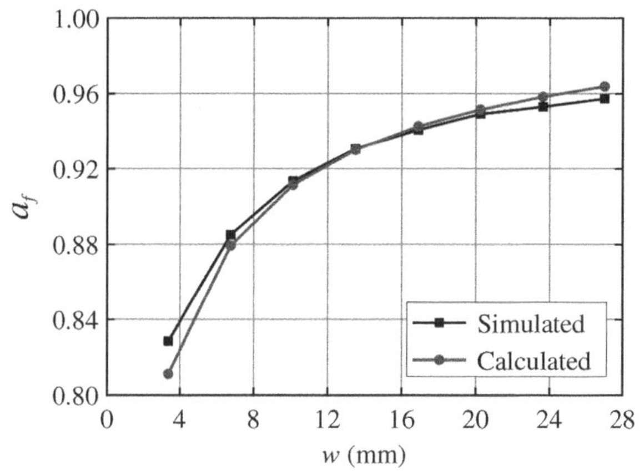

Figure 5.8 Calculated and simulated a_f as a function of w. ($l = 13.5$ mm).

It is seen that a_f increases as w increases. This is because the resonant frequency of the DSPA increases as w increases, while the resonant frequency of the SPA decreases as w increases. Figure 5.9 shows the effect of the relative dielectric constant ε_r on reduction factor a_f. As expected, it is seen that a_f is almost independent of dielectric constant. The effect of the patch ratio w/l on a_f for different substrate thickness h is shown in Figure 5.10. As the patch ratio w/l increases, a_f first remains a stable value of about 1 for w smaller than h. As w/l continues to increase, a_f declines sharply and then rises gradually, producing a minimum. The minimum decreases with an increase in h. Besides, the patch ratio corresponding to the minimum point also decreases as h increases.

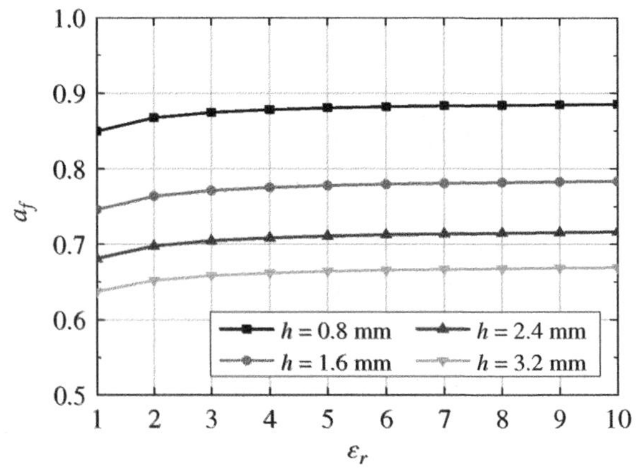

Figure 5.9 Effect of the relative dielectric constant ε_r on a_f. ($l \times w = 13.5 \times 6.75$ mm^2).

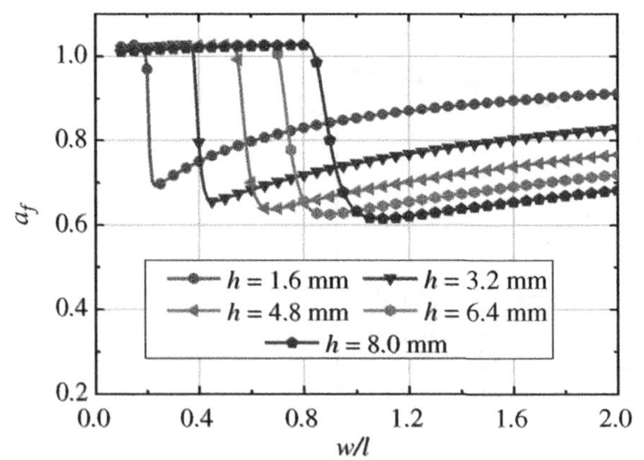

Figure 5.10 Effect of the patch ratio w/l on the frequency ratio a_f. ($l = 13.5$ mm).

5.2.5 Experiment Results

Figure 5.11 shows the photo of the fabricated samples of SPAs and DSPAs on Rogers 5880 substrates of $l_g \times w_g \times h = 50 \times 50 \times 1.575\,\text{mm}^3$ with $\varepsilon_r = 2.2$ and $\delta = 0.0009$ for three aspect ratios $l/w = 2$, 1, and 1/2, where $l = 20\,\text{mm}$. The gap width g for DSPAs is 0.2 mm. The feed probe has a radius of 0.45 mm. The shorting pins have a radius of 0.2 mm, and they are spaced 0.8 mm apart. For the SPAs, the feeding distance y_1 was chosen as 2, 2.9, and 4.5 mm for the patch ratio $l/w = 2$, 1, and 1/2, respectively. For the DSPAs, the corresponding y_2 was chosen as 18, 17.9, and 16.8 mm, respectively.

Figure 5.12 shows the simulated and measured $|S_{11}|$ and $|S_{d11}|$ as a function of frequency. Note that the SPAs show excellent matching in the simulated and measured results, while the DSPAs show acceptable shifts in the resonant frequencies. The measured resonant frequency shifts up to 0.75%, 0.54%, and 0.66% for the DSPAs with respect to the simulated resonant frequencies of 2.137, 2.214, and 2.274 GHz, respectively. The simulated and measured bandwidths are listed in Table 5.1. As can be seen, the impedance bandwidths of the DSPAs are narrower than those of the SPAs on Rogers 5880 substrates. Considering that the radiation quality factor dominated the total quality factor for low-loss Roger 5880 substrate, this can be attributed to the enhanced radiation quality factor of the DSPA due to the smaller electrical size of the resonant cavity.

Figure 5.13 shows the simulated and measured radiation patterns of the SPAs and DSPAs on the Roger 5880 substrate. They are in good agreement. As can be seen, symmetrical radiation patterns and generally better cross-polarization

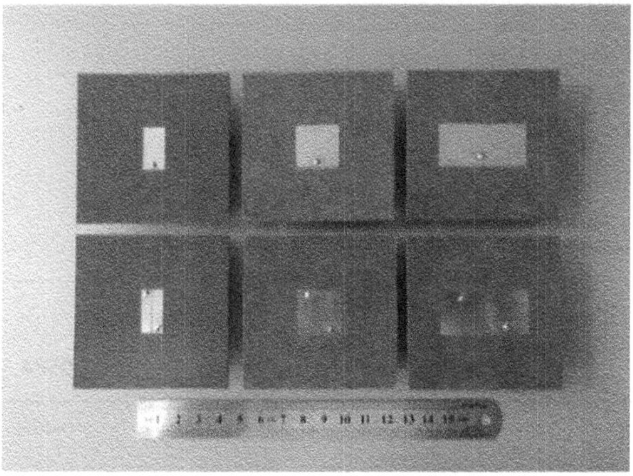

Figure 5.11 Photo of fabricated SPA and DSPA samples on the Roger 5880 substrate.

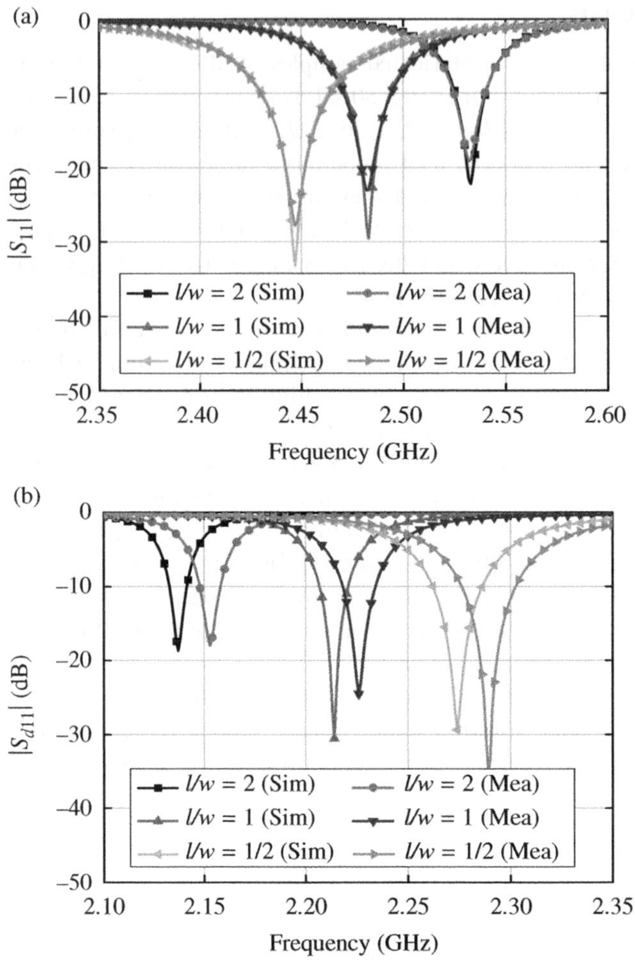

Figure 5.12 Simulated and measured $|S_{11}|$ and $|S_{d11}|$. (a) SPAs and (b) DSPAs.

Table 5.1 Comparison of fractional bandwidths.

	Simulated		Measured	
FBW	**SPA**	**DSPA**	**SPA**	**DSPA**
$l/w = 2$	0.59%	0.42%	0.59%	0.46%
$l/w = 1$	0.89%	0.63%	0.93%	0.72%
$l/w = 1/2$	1.31%	1.05%	1.39%	1.13%

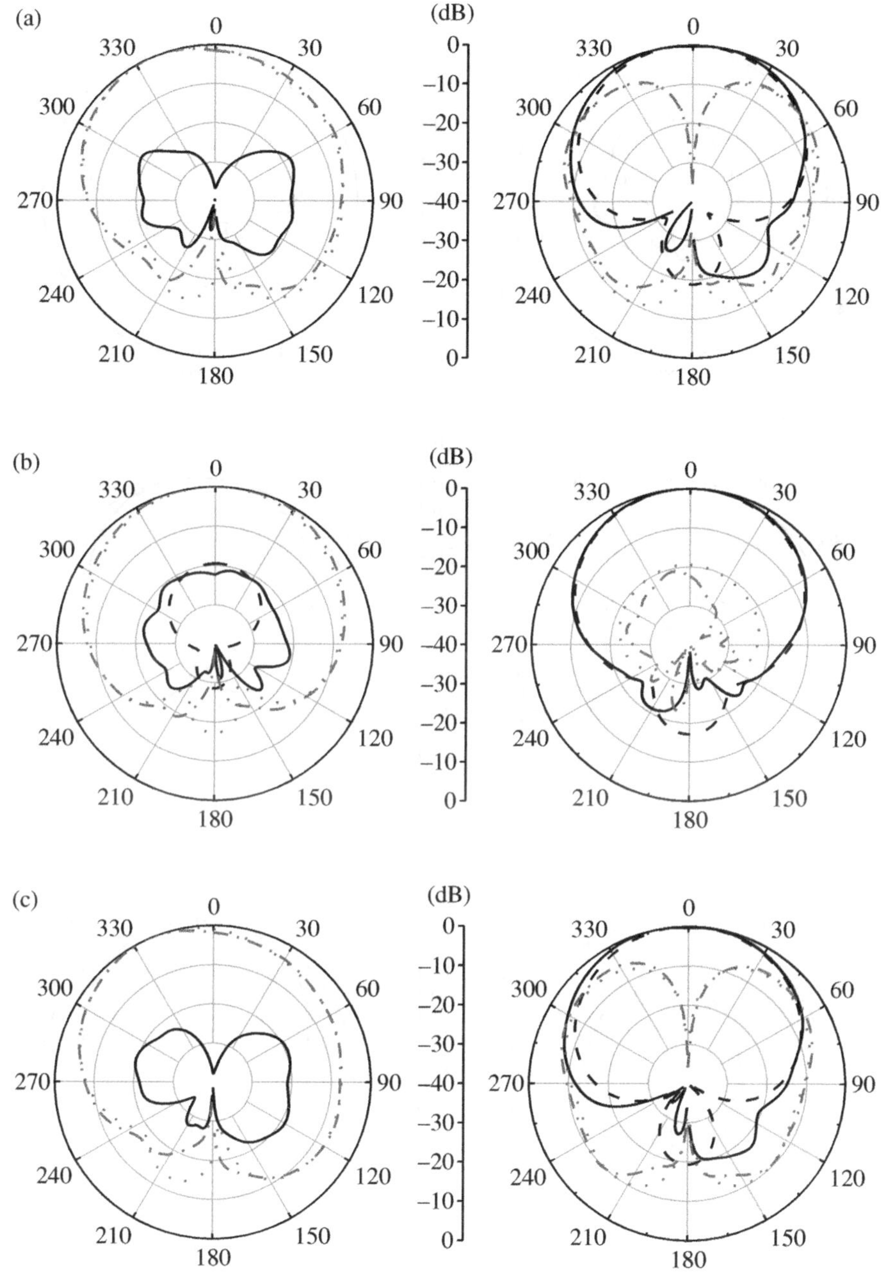

Figure 5.13 Simulated and measured radiation patterns for the cases of $l/w = 2$, 1, and 0.5, respectively. (a), (c), (e) for the SPAs and (b), (d), (f) for the DSPAs.

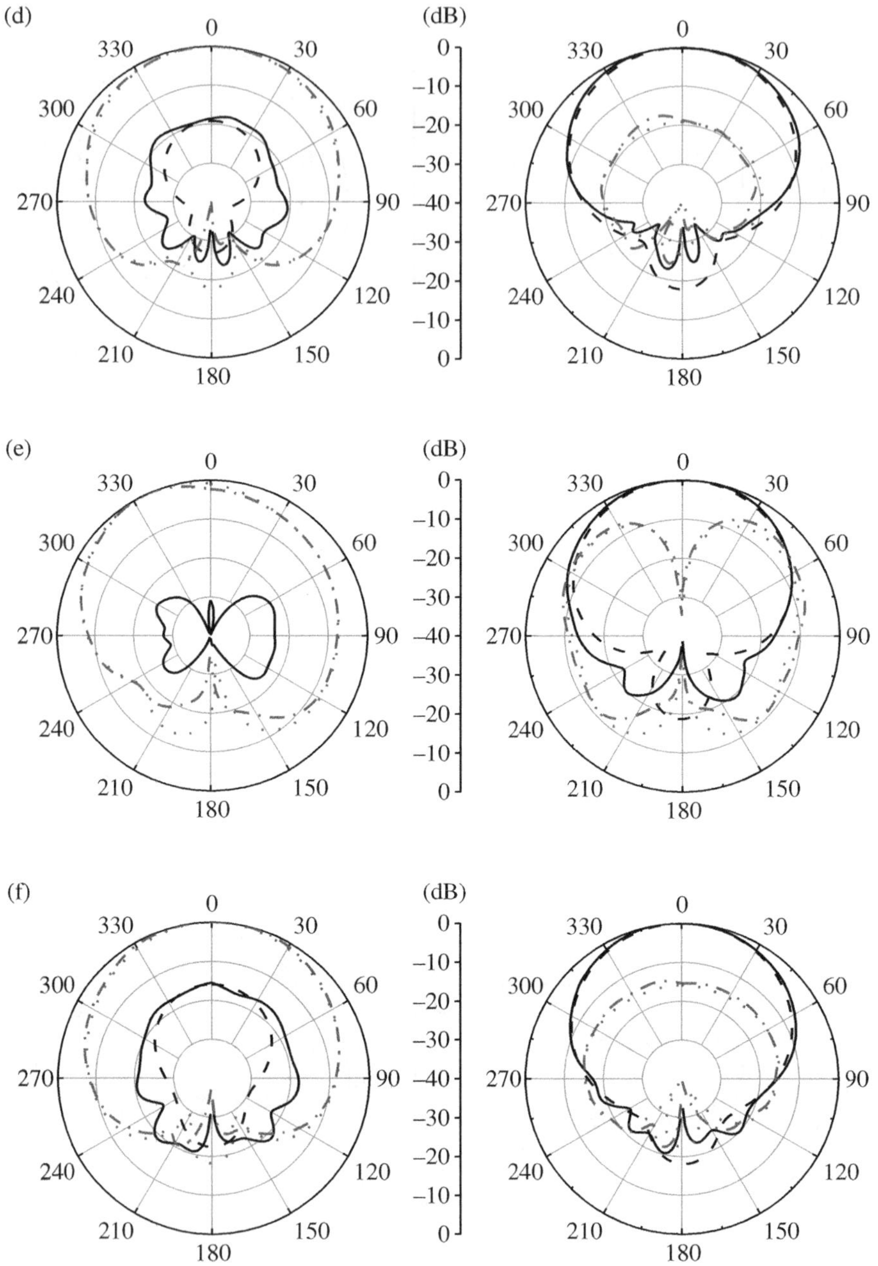

Figure 5.13 (Continued)

suppression are witnessed for the DSPAs. Besides, it can be noted that the DSPAs achieve peak radiation in the boresight direction while the peak radiation of the SPAs deviates from the boresight direction in the *E* plane due to the leakage radiation of probe and shorting pins. It should also be noted that SPAs have better cross-polarization suppression in the *E* plane, while the DSPAs have relatively larger cross-polarization with a peak along the boresight.

The different radiation characteristics between the SPA and the DSPA can be explained by their current distributions. For the SPA, the current on the patch is almost longitudinal and symmetrical, which results in an extremely low E_ϕ component in the *E* plane. The leakage radiation of probe and shorting pins is unbalanced, which causes a large E_θ component in the *H* plane and inclined pattern in the *E* plane. For the DSPA, except for the longitudinal currents, there are also transverse currents at the open-ended edges on the patches, which cause a relatively larger E_ϕ component in the *E* plane. Moreover, one can observe an increase in XPLs with reduced patch ratios for the DSPA, which can result from the enhanced transverse parasitic mode as l/w decreases. Nevertheless, the cross-polarization suppression for the DSPA is acceptable from a practical application viewpoint. Moreover, the radiation currents on the probe and shorting pins of the DSPA are naturally out of phase for the differential excitation, which leads to better cross-polarization performance in the *H* plane and peak radiation in the boresight direction.

In addition, the current distribution on the DSPA is relatively more uniform than that on the SPA, leading to narrower half-power beamwidth in the *H* plane compared with the SPA of the same physical dimension. Simulated and measured gain values as well as radiation efficiencies are listed in Table 5.2. As seen in the table, the DSPA generally shows similar gain and a little lower radiation efficiency with a reduced electrical size compared to the SPA.

Table 5.2 Comparison of gain and efficiency.

		$l/w = 2$		$l/w = 1$		$l/w = 1/2$	
Parameter		**SPA**	**DSPA**	**SPA**	**DSPA**	**SPA**	**DSPA**
Gain (dBi)	Sim	5.17	4.81	5.65	5.58	6.23	6.27
	Mea	4.93	5.23	5.89	5.71	6.52	6.68
Efficiency (%)	Sim	81.96	72.78	88.56	76.89	92.58	87.59
	Mea	81.67	77.17	84.50	84.44	93.40	92.78

5.3 Modified SPA and DSPA

The basic DSPA has an improved *H*-plane XPL, but a slightly degraded *E*-plane XPL as compared with those of the basic SPA. This section describes the mSPA and mDSPA for further enhanced XPL [5]. Figure 5.14 shows the geometry and the coordinate system of the mSPA. Note that the mSPA consists of three shorted patches separated by a pair of gaps along the *y*-axis direction. The shorting fences of metal vias are alternately arranged on the edges of the patches along the *x*-axis direction. The mSPA is excited with a coaxial probe feed at the middle-shorted patch. Let w_1 be the width of the middle-shorted patch, one can define a width $w_2 = w_1 + 2g$. Figure 5.15 shows the geometry and the coordinate system of the mDSPA. Note that the mDSPA consists of four shorted patches separated by a triple of gaps along the *y*-axis direction. The shorting fences of metal vias are alternately arranged on the edges of the patches along the *x*-axis direction. The mDSPA is excited with dual coaxial probe feeds at the inner shorted patches. Let w_1 be the summation of the width of the inner shorted patches and the gap *g*, one can define a width $w_2 = w_1 + 2s$.

5.3.1 Radiation Characteristics

For the mSPA, it operates in the coupled half TM_{01} modes. Figure 5.16 shows the equivalent magnetic current density distributions on the magnetic walls of the mSPA.

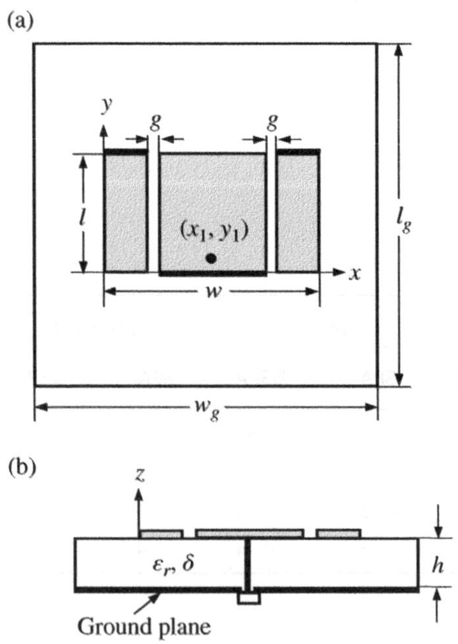

Figure 5.14 Geometry and the coordinate system of mSPA. (a) Top view and (b) side view.

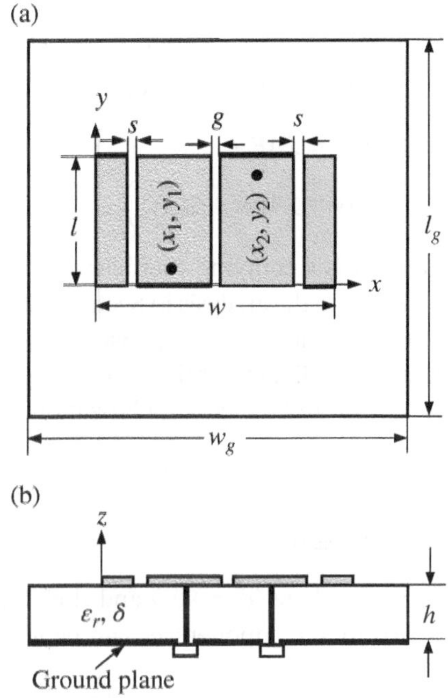

Figure 5.15 Geometry and the coordinate system of mDSPA. (a) Top view and (b) side view.

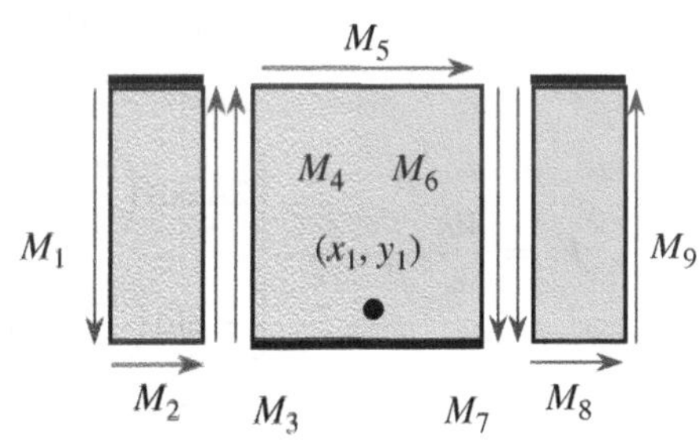

Figure 5.16 Illustration of equivalent magnetic current density distributions on the magnetic walls of the mSPA.

The magnetic currents can be expressed as

$$\boldsymbol{M_1} = -\hat{y}AE_{01}\cos\left(\frac{\pi}{2l}y\right), \tag{5.42}$$

$$\boldsymbol{M_2} = \hat{x}AE_{01}, \tag{5.43}$$

$$\boldsymbol{M_3} = \hat{y}AE_{01}\cos\left(\frac{\pi}{2l}y\right), \tag{5.44}$$

$$\boldsymbol{M_4} = \hat{y}E_{01}\sin\left(\frac{\pi}{2l}y\right), \tag{5.45}$$

$$\boldsymbol{M_5} = \hat{x}E_{01}, \tag{5.46}$$

$$\boldsymbol{M}_6 = -\hat{y}E_{01}\sin\left(\frac{\pi}{2l}y\right), \tag{5.47}$$

$$\boldsymbol{M}_7 = \hat{y}AE_{01}\sin\left(\frac{\pi}{2l}y\right), \tag{5.48}$$

$$\boldsymbol{M}_8 = \hat{x}AE_{01}, \tag{5.49}$$

$$\boldsymbol{M}_9 = \hat{y}AE_{01}\cos\left(\frac{\pi}{2l}y\right) \tag{5.50}$$

where A is a coupling coefficient.

Similar to the analysis in Section 5.2.3, the magnetic currents along the x-axis and y-axis directions are responsible for the co-polarized radiation and the cross-polarized radiation for the mSPA, respectively. As a result, the co-polarization component E_θ and the cross-polarization component E_ϕ can be found as

$$E_\theta = j\frac{h}{\lambda}E_0\frac{e^{-jk_0 r}\left[w_1 e^{jk_0 l\sin\theta} + 2A(w - w_2)\right]}{r}, \tag{5.51}$$

$$E_\phi = 0 \tag{5.52}$$

for the E plane ($\phi = 90°$) and the co-polarization component E_ϕ and the cross-polarization component E_θ can be found as

$$E_\phi = j\frac{h}{\tan\theta}E_{01}\frac{e^{-jk_0 r}}{\pi r}\left[\begin{array}{l} 2A\cos\left(\dfrac{\pi(w + w_2)}{2\lambda}\sin\theta\right)\sin\left(\dfrac{\pi(w - w_2)}{2\lambda}\sin\theta\right) \\ + \sin\left(\dfrac{\pi w_1}{\lambda}\sin\theta\right) \end{array}\right], \tag{5.53}$$

$$E_\theta = \frac{4lh}{\lambda}E_0\frac{e^{-jk_0 r}}{\pi r}\left[\begin{array}{l} 2A\cos\left(\dfrac{\pi(w + w_2)}{2\lambda}\sin\theta\right)\sin\left(\dfrac{\pi(w - w_2)}{2\lambda}\sin\theta\right) \\ - \sin\left(\dfrac{\pi w_1}{\lambda}\sin\theta\right) \end{array}\right] \tag{5.54}$$

for the H plane ($\phi = 0°$). For a small gap width, the coupling coefficient A can approximately be 1. Hence, (5.53) and (5.54) can be rewritten as

$$E_\phi = j\frac{h}{\tan\theta}E_{01}\frac{e^{-jk_0 r}}{\pi r}\left[\begin{array}{l} \sin\left(\dfrac{\pi w}{\lambda}\sin\theta\right) \\ + \sin\left(\dfrac{\pi w_1}{\lambda}\sin\theta\right) - \sin\left(\dfrac{\pi w_2}{\lambda}\sin\theta\right) \end{array}\right], \tag{5.55}$$

$$E_\theta = \frac{4lh}{\lambda}E_{01}\frac{e^{-jk_0 r}}{\pi r}\left[\sin\left(\frac{\pi w}{\lambda}\sin\theta\right) - \sin\left(\frac{\pi w_1}{\lambda}\sin\theta\right) - \sin\left(\frac{\pi w_2}{\lambda}\sin\theta\right)\right]. \tag{5.56}$$

From (5.56), it can be found that the cross-polarization component E_θ can always be canceled out by choosing proper w_1 and w_2. One can write $w_1 = \chi w - g$ and $w_2 = \chi w + g$ with $0 < \chi < 1$.

For a small gap width g, one can rewrite (5.56) as

$$E_\theta = \frac{4lh}{\lambda} E_{01} \frac{e^{-jk_0 r}}{\pi r} \left[\sin\left(\frac{\pi w}{\lambda} \sin\theta\right) - 2\sin\left(\frac{\pi w_1}{\lambda} \sin\theta\right) \right]. \tag{5.57}$$

Let $E_\theta = 0$, which leads to

$$\sin\left(\frac{\pi w}{\lambda} \sin\theta\right) = 2\sin\left(\frac{\pi \chi w}{\lambda} \sin\theta\right). \tag{5.58}$$

Therefore, for an infinite ground plane, the ratio of the cross-polarization component of the mSPA at any elevation angle of the H-plane pattern to the co-polarization component can be expressed as

$$\left|\frac{E_\theta}{E_\phi}\right| = \frac{4l\tan\theta}{\lambda} \left[\frac{\sin\left(\frac{\pi w}{\lambda}\sin\theta\right) - \sin\left(\frac{\pi u w}{\lambda}\sin\theta\right)}{\sin\left(\frac{\pi w}{\lambda}\sin\theta\right) + \sin\left(\frac{\pi w_1}{\lambda}\sin\theta\right) - \sin\left(\frac{\pi w_2}{\lambda}\sin\theta\right)} \right]. \tag{5.59}$$

As the co-polarization component E_ϕ of the H plane reaches the maximum in the broadside direction ($\theta = 0°$), the H-plane XPL of the mSPA can be expressed as

$$XPL = \left| \frac{E_\theta|_{max}}{E_\phi|_{\theta=0°}} \right| = \left\{ \frac{4l}{\pi(w-2g)} \left[\sin\left(\frac{\pi w}{\lambda}\sin\theta\right) - \sin\left(\frac{\pi \chi w}{\lambda}\sin\theta\right) \right] \right\}\Bigg|_{max}. \tag{5.60}$$

Thus, the optimal dimension of the mSPA can be easily determined from (5.58). Consider an mSPA with $\varepsilon_r = 2.2$, $\delta = 0.001$, $l = 26$ mm, $w_1 = 17$ mm, $l_g \times w_g \times h = 150 \times 150 \times 3$ mm^3, $x_1 = w/2$, $y_1 = 5.5$ mm, $p = 0.9$ mm, and $g = 1$ mm. By applying (5.53) and (5.54), the calculated H-plane radiation components of the mSPA on an infinite ground plane as a function of χ is depicted in Figure 5.17. As can be seen, there are a static zero at $\theta = 0°$ and a pair of dynamic zeros. The dynamic zeros are controlled by χ. When $\chi = 0.48$, the dynamic zeros are at $\theta = \pm 60°$, the overall cross-polarization component is the lowest. It can be easily understood that the XPL of the mSPA will be degraded to that of the basic SPA when χ approaches 0 or 1.

Similarly, the co-polarization component E_θ and the cross-polarization component E_ϕ can be found for the mDSPA as

$$E_\theta = j\frac{h}{\lambda} E_{01} \frac{e^{-jk_0 r}\left(e^{jk_0 l \sin\theta} + 1\right)}{r} \left(\frac{w - w_2}{2} + \frac{w_1 - g}{2} \right), \tag{5.61}$$

$$E_\phi = 0 \tag{5.62}$$

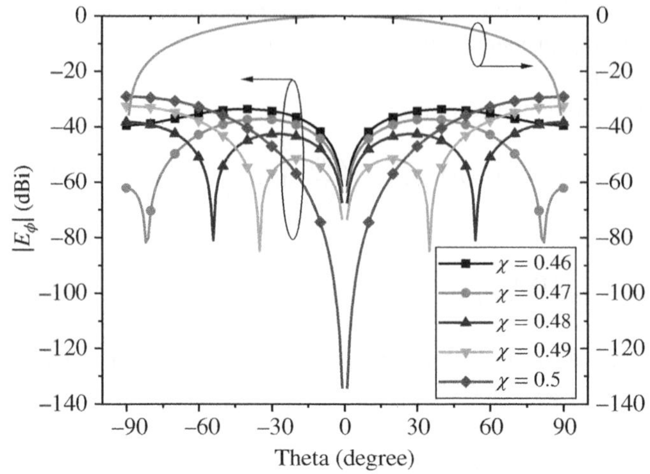

Figure 5.17 Calculated *H*-plane radiation components of the mSPA on an infinite ground plane with different χ in the dual-coupled $TM_{0,1/2}$ mode.

for the *E* plane ($\phi = 90°$) and the co-polarization component E_ϕ and the cross-polarization component E_θ can be found for the mDSPA as

$$E_\phi = j\frac{2h}{\tan\theta}E_0\frac{e^{-jk_0r}}{\pi r}\left[\begin{array}{l}\cos\left(\dfrac{\pi(w_1+g)}{2\lambda}\sin\theta\right)\sin\left(\dfrac{\pi(w_1-g)}{2\lambda}\sin\theta\right)\\ +A\cos\left(\dfrac{\pi(w+w_2)}{2\lambda}\sin\theta\right)\sin\left(\dfrac{\pi(w-w_2)}{2\lambda}\sin\theta\right)\end{array}\right], \tag{5.63}$$

$$E_\theta = j\frac{8lh}{\lambda_0}E_0\frac{e^{-jk_0r}}{\pi r}\left[\begin{array}{l}\sin\left(\dfrac{\pi(w_1+g)}{2\lambda}\sin\theta\right)\sin\left(\dfrac{\pi(w_1-g)}{2\lambda}\sin\theta\right)\\ -A\sin\left(\dfrac{\pi(w+w_2)}{2\lambda}\sin\theta\right)\sin\left(\dfrac{\pi(w-w_2)}{2\lambda}\sin\theta\right)\end{array}\right]. \tag{5.64}$$

For a small gap width of *s*, the coupling coefficient *A* can approximately be 1. Hence, (5.63) and (5.64) can be rewritten as

$$E_\phi = j\frac{2h}{\tan\theta}E_0\frac{e^{-jk_0r}}{\pi r}\left[\begin{array}{l}\cos\left(\dfrac{\pi(w+g)}{2\lambda}\sin\theta\right)\sin\left(\dfrac{\pi(w-g)}{2\lambda}\sin\theta\right)\\ +\cos\left(\dfrac{\pi(w_1+w_2)}{2\lambda}\sin\theta\right)\sin\left(\dfrac{\pi s}{\lambda}\sin\theta\right)\end{array}\right], \tag{5.65}$$

$$E_\theta = j\frac{8lh}{\lambda_0}E_0\frac{e^{-jk_0r}}{\pi r}\begin{bmatrix}\cos\left(\frac{\pi(w+g)}{2\lambda}\sin\theta\right)\cos\left(\frac{\pi(w-g)}{2\lambda}\sin\theta\right)\\ -\cos\left(\frac{\pi(w_1+w_2)}{2\lambda}\sin\theta\right)\cos\left(\frac{\pi s}{\lambda}\sin\theta\right)\end{bmatrix}. \tag{5.66}$$

When $s \ll \lambda/2\pi$, (5.66) can be further simplified as

$$E_\theta = j\frac{8lh}{\lambda}E_0\frac{e^{-jk_0r}}{\pi r}\begin{bmatrix}\cos\left(\frac{\pi(w+g)}{2\lambda}\sin\theta\right)\cos\left(\frac{\pi(w-g)}{2\lambda}\sin\theta\right)\\ -\cos\left(\frac{\pi(w_1+w_2)}{2\lambda}\sin\theta\right)\end{bmatrix}. \tag{5.67}$$

From (5.67), it can be found that the cross-polarization component E_θ can always be canceled out by choosing proper w_1 and w_2. Hence, one can rewrite (5.65) and (5.67) as

$$E_\phi = j\frac{h}{\tan\theta}E_0\frac{e^{-jk_0r}}{\pi r}\begin{bmatrix}\sin\left(\frac{\pi w}{\lambda}\sin\theta\right)-\sin\left(\frac{\pi g}{\lambda}\sin\theta\right)\\ +2\cos\left(\frac{\pi(w_1+w_2)}{2\lambda}\sin\theta\right)\sin\left(\frac{\pi s}{\lambda}\sin\theta\right)\end{bmatrix}, \tag{5.68}$$

$$E_\theta = j\frac{4lh}{\lambda}E_0\frac{e^{-jk_0r}}{\pi r}\begin{bmatrix}\cos\left(\frac{\pi w}{\lambda}\sin\theta\right)+\cos\left(\frac{\pi g}{\lambda}\sin\theta\right)\\ -2\cos\left(\frac{\pi(w_1+w_2)}{2\lambda}\sin\theta\right)\end{bmatrix}. \tag{5.69}$$

When $g \ll \lambda/2\pi$, (5.69) can be further simplified as

$$E_\theta = j\frac{4lh}{\lambda}E_0\frac{e^{-jk_0r}}{\pi r}\left[1+\cos\left(\frac{\pi w}{\lambda}\sin\theta\right)-2\cos\left(\frac{\pi(w_1+w_2)}{2\lambda}\sin\theta\right)\right]. \tag{5.70}$$

Suppose $w_1 + w_2 = 2\chi w$ or $w_1 + s = \chi w$ $(0 < \chi < 1)$, (5.70) can be rewritten as

$$E_\theta = \frac{4lh}{\lambda}E_0\frac{e^{-jk_0r}}{\pi r}\left[1+\cos\left(\frac{\pi w}{\lambda}\sin\theta\right)-2\cos\left(\frac{\pi\chi w}{\lambda}\sin\theta\right)\right]. \tag{5.71}$$

Therefore, for an infinite ground plane, the ratio of the cross-polarization component of the mDSPA at any elevation angle of the H-plane pattern to the co-polarization component can be expressed as

$$\left|\frac{E_\theta}{E_\phi}\right| = \frac{\frac{4l\tan\theta}{\lambda}\left[1+\cos\left(\frac{\pi w}{\lambda}\sin\theta\right)-2\cos\left(\frac{\pi\chi w}{\lambda}\sin\theta\right)\right]}{\sin\left(\frac{\pi w}{\lambda}\sin\theta\right)-\sin\left(\frac{\pi g}{\lambda}\sin\theta\right)+\sin\left(\frac{\pi w_1}{\lambda}\sin\theta\right)-\sin\left(\frac{\pi w_2}{\lambda}\sin\theta\right)}. \tag{5.72}$$

As the co-polarization component E_ϕ of the H plane reaches the maximum in the broadside direction ($\theta = 0°$), the H-plane XPL of the mDSPA can be expressed as

$$XPL = \left|\frac{E_\theta|_{max}}{E_\phi|_{\theta=0°}}\right| = \frac{4l}{\pi(w-g-2s)}\left[1+\cos\left(\frac{\pi w}{\lambda}\sin\theta\right)-2\cos\left(\frac{\pi\chi w}{\lambda}\sin\theta\right)\right]\Big|_{max}. \tag{5.73}$$

Thus, the optimal dimension of the mDSPA can be easily determined from

$$1+\cos\left(\frac{\pi w}{\lambda}\sin\theta\right)=2\cos\left(\frac{\pi\chi w}{\lambda}\sin\theta\right). \tag{5.74}$$

Consider an mDSPA with $\varepsilon_r = 2.2$, $\delta = 0.001$, $l = 26$ mm, $w_1 = 29$ mm, $l_g \times w_g \times h = 150 \times 150 \times 3$ mm^3, $x_1 = w - 21.75$ mm, $y_1 = 5.5$ mm, $x_2 = w - 6.75$ mm, $y_1 = 21.5$ mm, and $g = s = 1.0$ mm. By applying (4.64) and (4.65), the calculated H-plane radiation components of the mDSPA as a function of χ is depicted in Figure 5.18. As can be seen, there is a static zero at $\theta = 0°$. A pair of dynamic zeros occurs when $0.7 < \chi < 0.71$. It can be easily understood that the XPL of the mDSPA will be degraded to that of the basic DSPA when χ approaches 0 or 1.

5.3.2 Design and Experiment

The gap affects the resonant frequency of the mSPA. Figure 5.19 shows the simulated effect of gap width on $|S_{11}|$. As can be seen, the gap width indeed affects the resonant frequency but has little effect on the impedance matching. The wider the

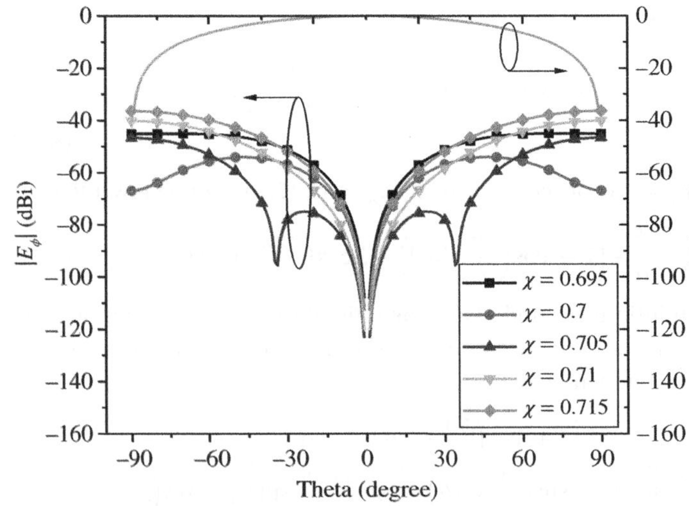

Figure 5.18 Calculated H-plane radiation components of the mDSPA on an infinite ground plane with different χ in the dual-coupled $TM_{0,1/2}$ mode.

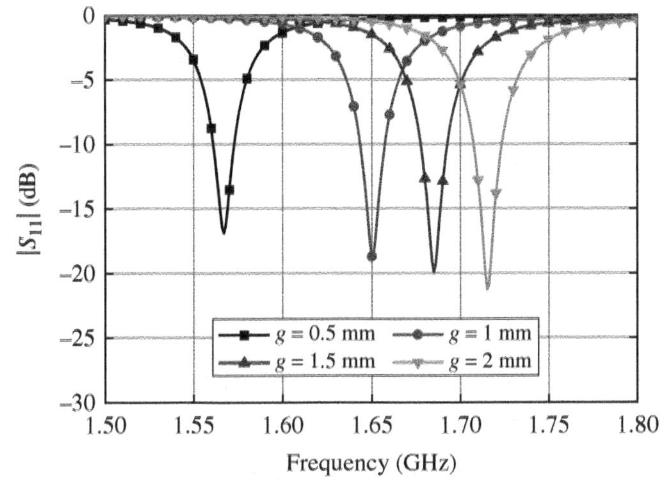

Figure 5.19 Simulated $|S_{11}|$ of the mSPA with different gap width g.

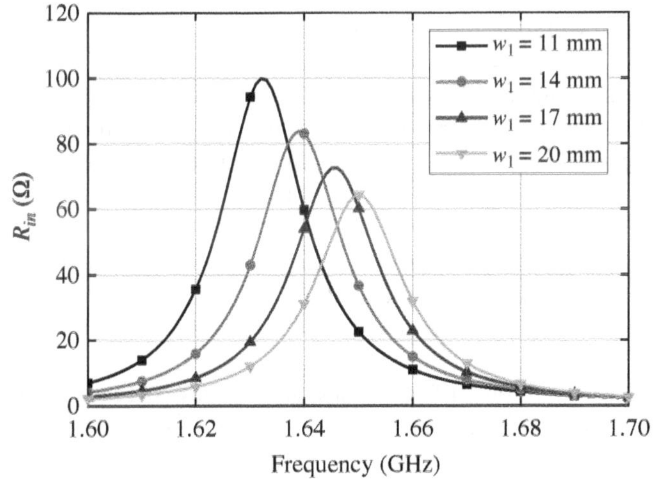

Figure 5.20 Simulated input resistance of the mSPA with different center-patch width w_1.

gap, the higher is the resonant frequency, which can be attributed to the fact that a larger gap width results in less gap capacitance. Hence, one can trim the resonant frequency by controlling the gap width with either a conductive paste or a laser. Figure 5.20 shows the simulated input resistance of the mSPA with different center-patch width w_1. It is evident that the input resistance decreases as w_1 increases. Besides, it can be found that w_1 has little effect on the resonant frequency. Thus, the feeding distance y_1 should be adjusted for a different center-patch width w_1.

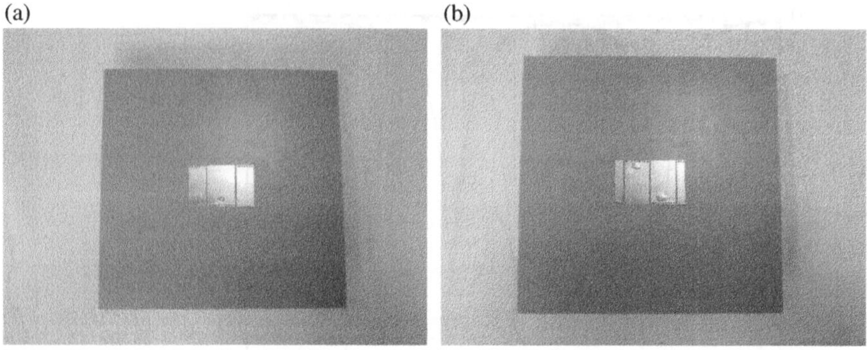

Figure 5.21 Photo of fabricated samples on the F4B substrate. (a) mSPA and (b) mDSPA.

Figure 5.21 shows the photo of the fabricated samples of the mSPA and mDSPA on F4B substrates of $l_g \times w_g \times h = 50 \times 50 \times 1.575\ \text{mm}^3$ with $\varepsilon_r = 2.2$ and $\delta = 0.001$ for $l = 26$ mm. The gap width for mSPA is 0.2 mm. The feed probe has a radius of 0.45 mm. The shorting pins have a radius of 0.2 mm, and they are spaced 0.8 mm apart. For the mSPAs, the feeding distance y_1 was chosen as 2.0, 2.9, and 4.5 mm for the patch ratio $l/w = 2$, 1, and 1/2, respectively. For the mDSPAs, the corresponding y_2 was chosen as 24.4, 17.9, and 16.8 mm, respectively.

Figure 5.22 shows the simulated and measured $|S_{11}|$ and $|S_{d11}|$ as a function of frequency. Note that the mSPA and mDSPA achieve excellent matching. Measured resonant frequencies are slightly higher than those of the simulated frequencies, due to fabrication tolerance.

Figure 5.23 shows the simulated and measured radiation patterns of the mSPA and mDSPA at their resonant frequencies. Note that the measured patterns match well with the simulated patterns. Both mSPA and mDSPA achieve XPL as large as 30 dB in both E and H planes.

Table 5.3 lists the measured resonant frequencies, gain values, XPL, and half-power beamwidth of SPAs. It can be concluded that mSPA and mDSPA have the highest gain and the highest XPL among those SPAs with the most compact size and smallest difference of half-power beamwidth in the E and H planes.

5.4 PIFA and DPIFA

Figure 5.24 shows the geometry and the coordinate system of the PIFA. It is formed with a rectangular patch above a ground plane and a partial shorting wall. The patch has a size of width w and length l and is kept at a height of h above the ground plane of dimensions $w_g \times l_g$. The partial shorting wall has a width of w_s,

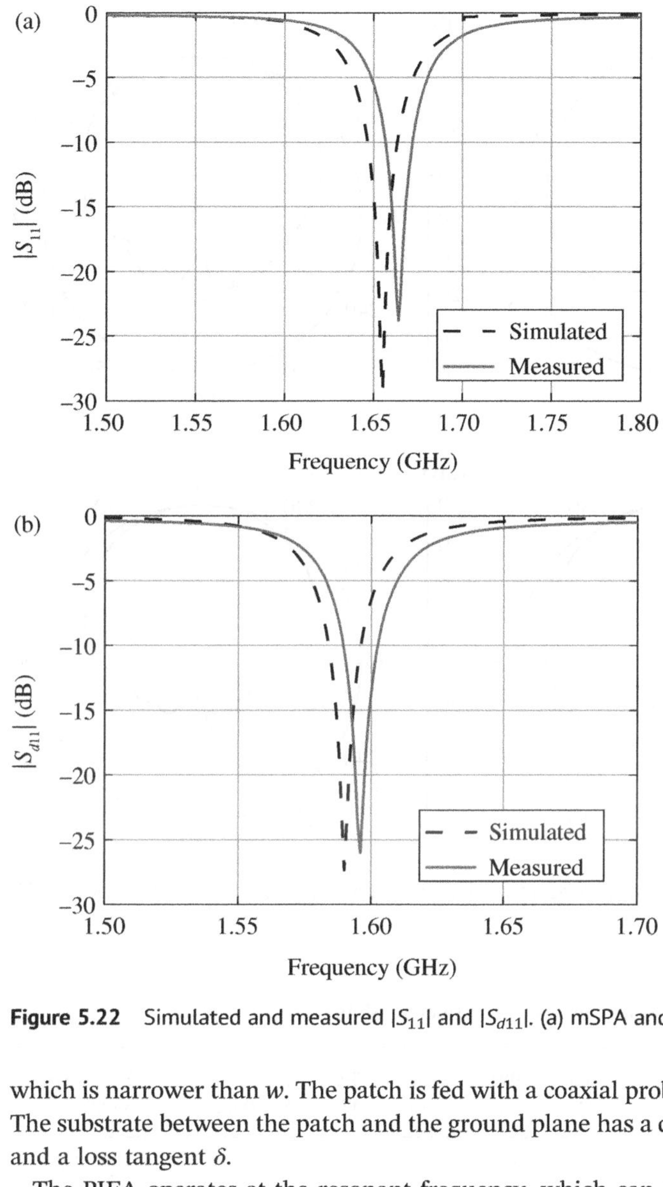

Figure 5.22 Simulated and measured $|S_{11}|$ and $|S_{d11}|$. (a) mSPA and (b) mDSPA.

which is narrower than w. The patch is fed with a coaxial probe located at (x_1, y_1). The substrate between the patch and the ground plane has a dielectric constant ε_r and a loss tangent δ.

The PIFA operates at the resonant frequency, which can be estimated by the effective length of the current flowing on the partial shorting wall and patch. Then, in the case of $w_s = w$, the resonant frequency is expressed by

$$f = \frac{c}{4\sqrt{\varepsilon_r}(l + h)} \tag{5.75}$$

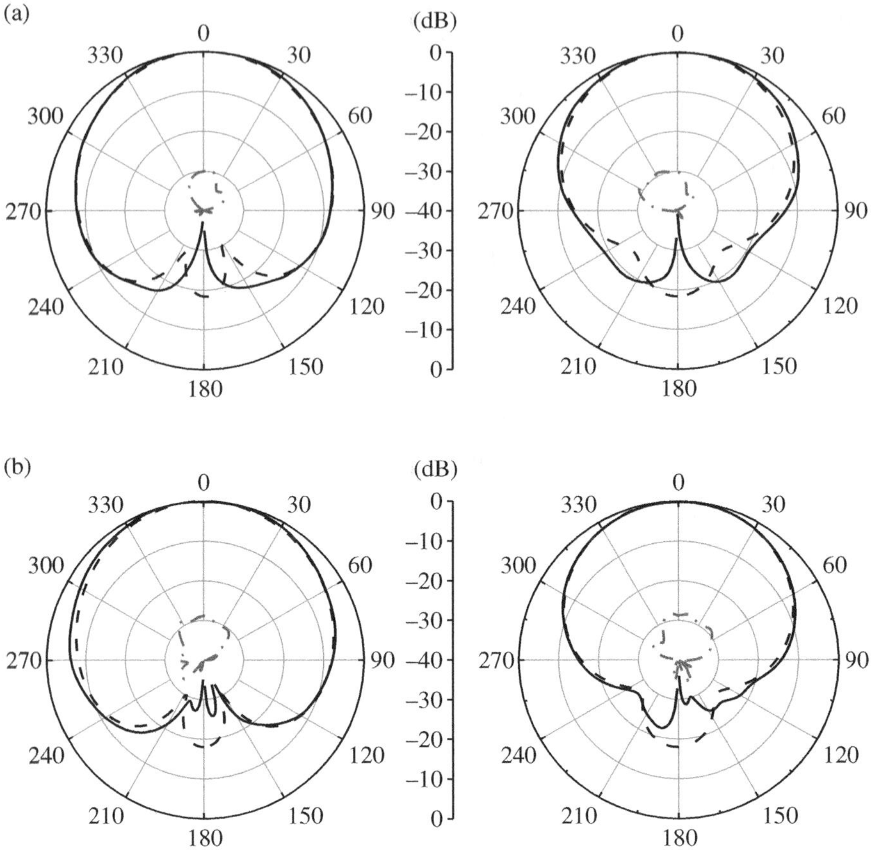

Figure 5.23 Simulated and measured radiation patterns. (a) For mSPA and (b) for mDSPA.

Table 5.3 Measured results of several distinctive SPAs.

Antennas	Frequency (GHz)	Area (λ^2)	BW (%)	Gain (dBi)	XPL (dB)	H-plane $\theta_{3\,dB}$ Sim/Mea	E-plane $\theta_{3\,dB}$ Sim/Mea
SPA	1.89	0.041	1.7	5.0	4	80°/90°	106°/120°
DSPA	1.76	0.036	1.1	6.2	16	78°/86°	84°/88°
mSPA	1.66	0.032	0.9	6.2	29	77°/89°	85°/88°
mDSPA	1.59	0.029	0.9	6.3	29	76°/78°	87°/100°
Ref. [24]	2.63	0.096	0.9	5.0	21.4	80°/	144°/

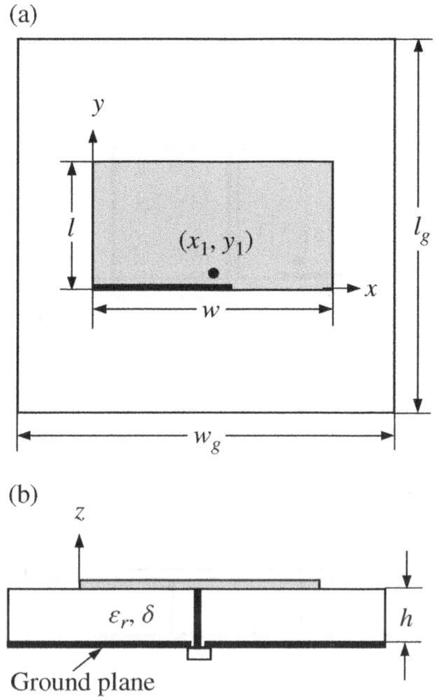

Figure 5.24 Geometry and the coordinate system of the PIFA. (a) Top view and (b) side view.

and, in the case of $w_s \ll w$, it is expressed by

$$f = \frac{c}{4\sqrt{\varepsilon_r}(w + l + h)}. \tag{5.76}$$

When h is sufficiently smaller than the free space wavelength, the resonant frequency calculated by (5.75) or (5.76) agrees well with measured results, with an error of about 3% [8]. The resonant frequency, in the case of $w_s < w$, can be expressed by

$$f = uf_1 + (1 - u)f_2 \tag{5.77}$$

for $w/l \leq 1$ and

$$f = u^{v}f_1 + (1 - u^{v})f_2 \tag{5.78}$$

for $w/l > 1$ where $u = w_s/w$, $v = w/l$, f_1 by (5.75), and f_2 by

$$f_2 = \frac{c}{4\sqrt{\varepsilon_r}(w + l + h - w_s)}. \tag{5.79}$$

The resonant frequency calculated from (5.77) or (5.78) agrees well with the measured results.

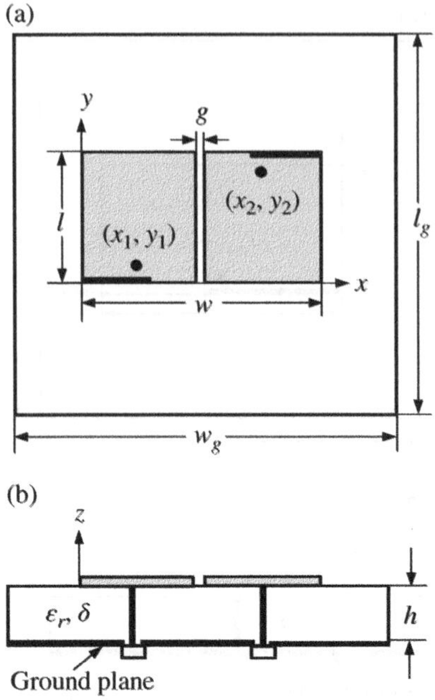

Figure 5.25 Illustration of the differential PIFA and coordinate system. (a) Top view and (b) side view.

Figure 5.25 shows the geometry and dimensions of a basic DPIFA modified from the PIFA by adding another feed and row of pins symmetrically at the diagonal corner and cutting a small gap of g on the patch. The number of shorting pins on each row of the DPIFA needs to be adjusted to make it operate at the same frequency as the PIFA.

The gap in a DPIFA introduces a capacitance and, thus, affects the resonant frequency of the DPIFA. Figure 5.26 shows the cross-section perpendicular to the gap of the DPIFA. The gap effect on the resonant frequency can be modelled as an increase in the width of a half patch of the DPIFA. The half patch with an increased width would function as if it were an isolated PIFA.

It is found that the increased width of the half patch w_e can be calculated as

$$w_e = \frac{(w-g)C_s}{2C_d} \tag{5.80}$$

where C_d is the parallel plate capacitance of the half patch of the DPIFA, and C_s can be expressed as

$$C_s = C_d + C_{ga} + C_{gd} - C_f \tag{5.81}$$

where C_f represents the fringe capacitance of the edge perpendicular to the row of shorting pins of the half patch, and C_{ga} and C_{gd} denote the fringe capacitances

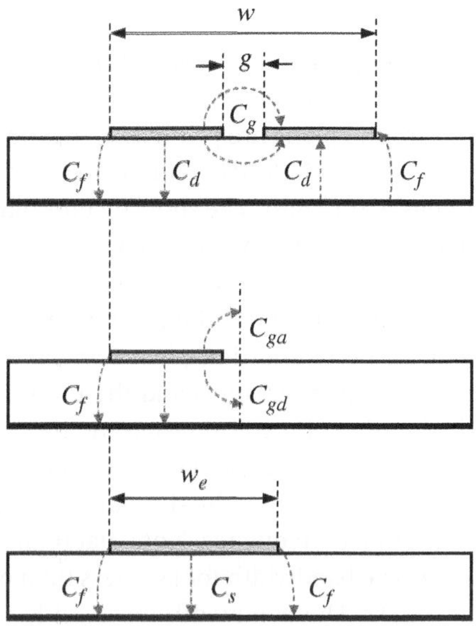

Figure 5.26 Illustration of modelling the gap effect of the DPIFA for predicting its resonant frequency.

across the gap for the air and dielectric regions, respectively. All these fringe capacitances can be calculated with the formulas in [23]. The effective length of the half patch is given by

$$l_e = l - y_p + \Delta l \tag{5.82}$$

where Δl accounts for the length extension caused by the fringe field. Thus, the available formula for the resonant frequency of the PIFA in [10] can be revised to determine the resonant frequency of the DPIFA as follows.

$$f = uf_1 + (1 - u)f_2 \tag{5.83}$$

where

$$u = \frac{(N-1)d + 2p}{w_e}, \tag{5.84}$$

$$f_1 = \frac{c}{4\sqrt{\varepsilon_e}(l_e + h)}, \tag{5.85}$$

$$f_2 = \frac{c}{4\sqrt{\varepsilon_e}(w_e + l_e + h - Nd - d - 2p)}. \tag{5.86}$$

In (5.84) and (5.86), N is the number of shorting pins in a row, p is the radius of a shorting pin, and d is the distance between two adjacent shorting pins. The above equations clearly show that the resonant frequency of the DPIFA can be trimmed by changing the gap width.

5.4.1 Analysis and Design

Assuming that both PIFA and DPIFA are realized on FR4 substrate of dimensions $l_g \times w_g \times h = 50 \times 50 \times 0.8\ \text{mm}^3$ with $\varepsilon_r = 4.4$ and $\delta = 0.02$, the radiating patch and the ground plane are printed on the top and bottom surfaces of the substrate. The patch size for 2.45 GHz is found to be $l \times w = 12.5 \times 12.5\ \text{mm}^2$. The feed probe has a radius of 0.15 mm. The shorting pins have a radius $p = 0.2$ mm and are spaced in a distance $d = 0.6$ mm. For the PIFA, the feed probe is located at $(x_1, y_1) = (1.0\ \text{mm}, 3.0\ \text{mm})$ and the number of shorting pins is 10. For the DPIFA, the gap of width $g = 0.2$ mm separates the patch into two equal halves. Each half has a feed probe and a row of shorting pins. For the left half, the feed probe is located at $(x_1, y_1) = (1.0\ \text{mm}, 2.0\ \text{mm})$ and the number of shorting pins is 4. For the right half, the location of the feed probe and the number of shorting pins can be determined according to the diagonally symmetrical layout.

Figure 5.27 shows the simulated $|S_{11}|$ for the PIFA and of $|S_{d11}|$ for the DPIFA as a function of frequency. Note that both PIFA and DPIFA are well matched at 2.45 GHz. The bandwidth for $|S_{11}|$ less than or equal to − 10 dB is 39.4 MHz or 1.61% at 2.45 GHz. The bandwidth for $|S_{d11}|$ less than or equal to − 10 dB is 38.2 MHz or 1.56% at 2.45 GHz.

Figure 5.28 shows the simulated $|S_{d11}|$ for the DPIFA as a function of frequency with the gap width g as a parameter. Note that the resonant frequency of the DPIFA increases as the gap width g gets wider. This effect is related to the coupling level, which can be used to trim the resonant frequency.

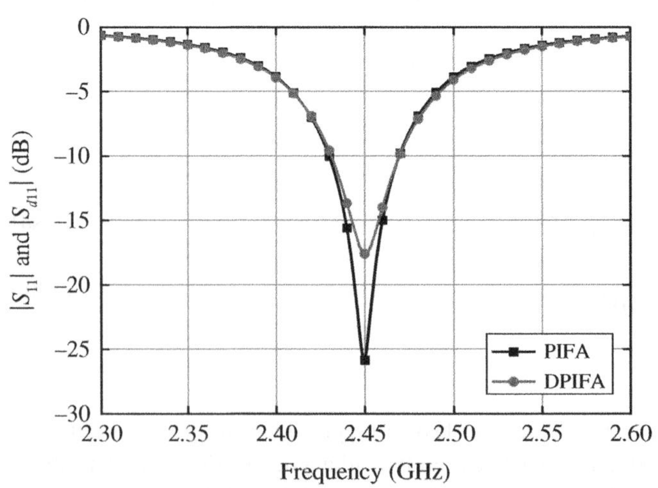

Figure 5.27 Simulated $|S_{11}|$ for the PIFA and $|S_{d11}|$ for the DPIFA.

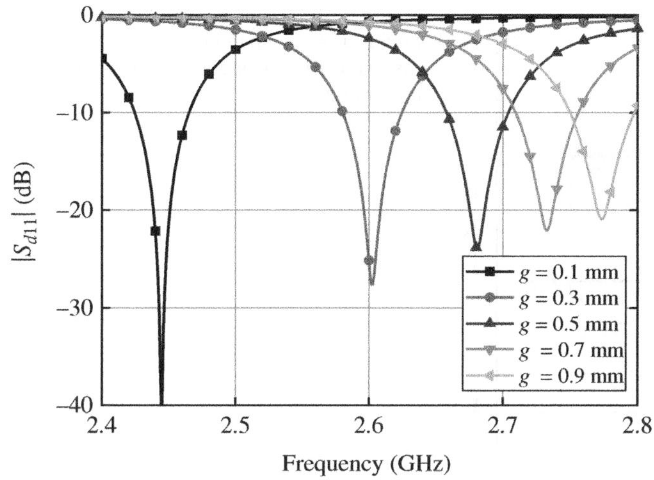

Figure 5.28 Simulated $|S_{d11}|$ for the DPIFA with the gap width as a parameter.

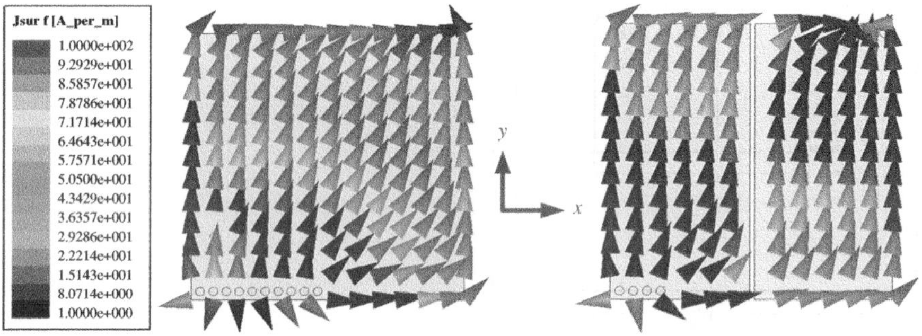

Figure 5.29 Current distributions on the PIFA and DPIFA at 2.45 GHz.

Figure 5.29 shows the simulated current distributions of the PIFA and the DPIFA at 2.45 GHz. Note that the current on the PIFA is asymmetrically distributed, which leads to asymmetrical radiation pattern cuts for the E_θ component. In contrast, the current on the DPIFA is central and symmetrically distributed, which results in symmetrical radiation pattern cuts on the *xoz* and *yoz* planes. In addition, the currents on the probes and shorting pins are inherently out-of-phase for the DPIFA and, thus, reduce the cross-polarization components. The simulated peak directivity values at 2.45 GHz are 5.5 dBi for the PIFA and 5.9 dBi for the DPIFA.

Figure 5.30 shows two gap schemes for the DPIFA. The upper one has the larger separation of the gap from the shorting pins, while the lower one has the smaller

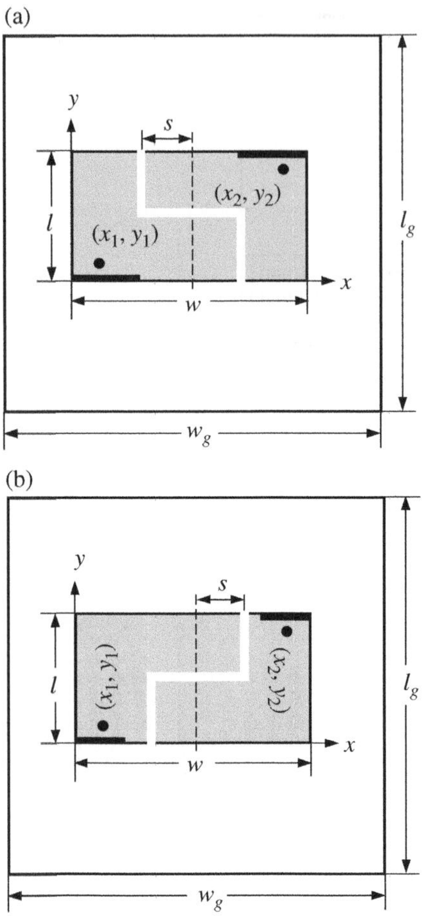

Figure 5.30 Gap schemes for the DPIFA. (a) For a larger changing range of the resonant frequency and (b) for patch miniaturization.

separation of the gap from the shorting pins. It is found that the upper one can accommodate more shorting pins to benefit the changing range of the resonant frequency. More importantly, it is also found that the lower one can be explored to miniaturize the DPIFA. For example, let $g = 0.2$ mm and $s = 3$ mm, the patch size for operation at 2.45 GHz is found to be $w \times l = 10 \times 10\ \text{mm}^2$, which is 36% smaller as compared with the unminiaturized patch size $w \times l = 12.5 \times 12.5\ \text{mm}^2$.

5.4.2 Experiment

Both PIFA and DPIFA were fabricated and measured. Figure 5.31 depicts the photo of the fabricated samples. All these antennas are linearly polarized. The Agilent network analyzer E5071C was employed for measuring S_{11} and S_{d11}.

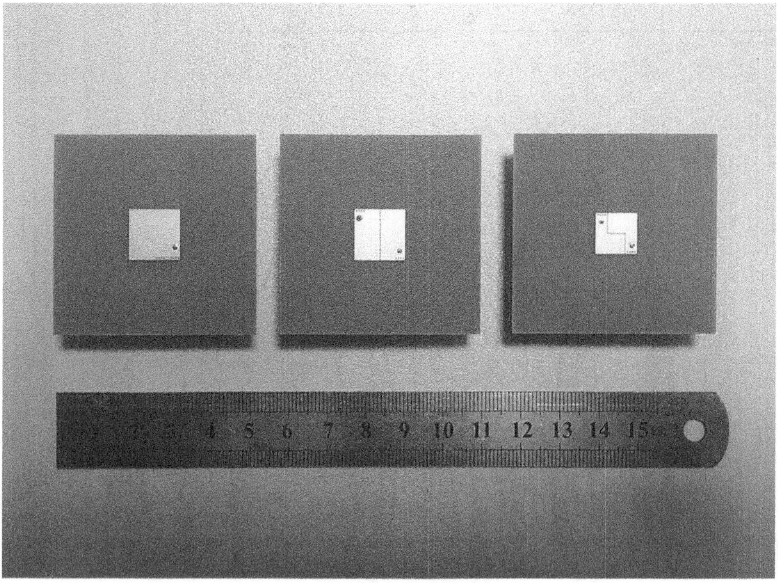

Figure 5.31 Photo of the fabricated PIFA and DPIFA samples.

The two ports of the DPIFA for the differential operation were directly connected to the network analyzer in the measurement.

Figure 5.32 shows the simulated and measured $|S_{11}|$ for the PIFA and $|S_{d11}|$ for the DPIFA as a function of frequency [13]. Note that there are acceptable shifts in the resonant frequencies. The measured resonant frequency shifts down 0.4% for the PIFA and up 0.2% for the DPIFA with respect to the simulated resonant frequency of 2.45 GHz. Both simulated and measured results show that the PIFA and DPIFA achieved good impedance matching. The measured bandwidth for $|S_{11}|$ less than or equal to −10 dB is 36.4 MHz (or 1.49% at 2.44 GHz) for the PIFA. The measured −10 dB $|S_{d11}|$ bandwidth is 36.8 MHz (or 1.50% at 2.455 GHz) for the DPIFA.

Figure 5.33 shows the simulated and measured $|S_{d11}|$ for the miniaturized DPIFA as a function of frequency. The feed probe was located at $(x_1, y_1) = (1$ mm, 1.8 mm). The simulated bandwidths are 37.4 MHz (or 1.53% at 2.44 GHz) and 42.6 MHz (or 1.74% at 2.45 GHz), respectively. They are both slightly wider than those of the unminiaturized one. It should be mentioned that this is remarkably different from the miniaturization of the PIFA, where a size reduction is always accompanied by a bandwidth penalty [10].

A balun was used for measuring the radiation patterns of the DPIFA and the miniaturized DPIFA. Figure 5.34 shows the simulated and measured radiation patterns at 2.45 GHz. The effects of the balun and additional cables are calibrated out to obtain

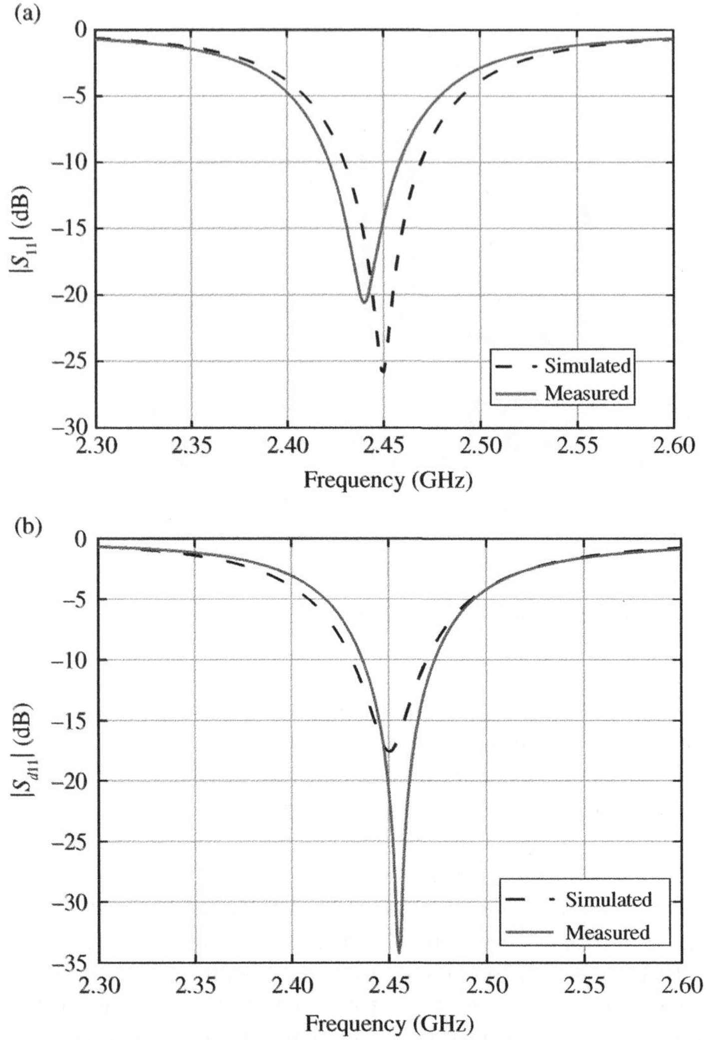

Figure 5.32 Simulated and measured $|S_{11}|$ and $|S_{d11}|$. (a) For PIFA and (b) for DPIFA.

gain values of the DPIFAs. The simulated and measured total gain values are − 4.6 and − 4.7 dBi for the PIFA, − 3.9 and − 4.2 dBi for the DPIFA, and − 6.0 and − 6.8 dBi for the miniaturized DPIFA, respectively. Lower gain values for all antennas are due to larger dielectric loss and electrically thin substrate ($h/\lambda = 0.65\%$). Nevertheless, it is important to note that the gain for the DPIFA is 0.5 and 0.7 dB higher than that for the PIFA in simulation and measurement, respectively. The radiation efficiency was measured in a Wheeler cap and a reverberation chamber, respectively. The simulated

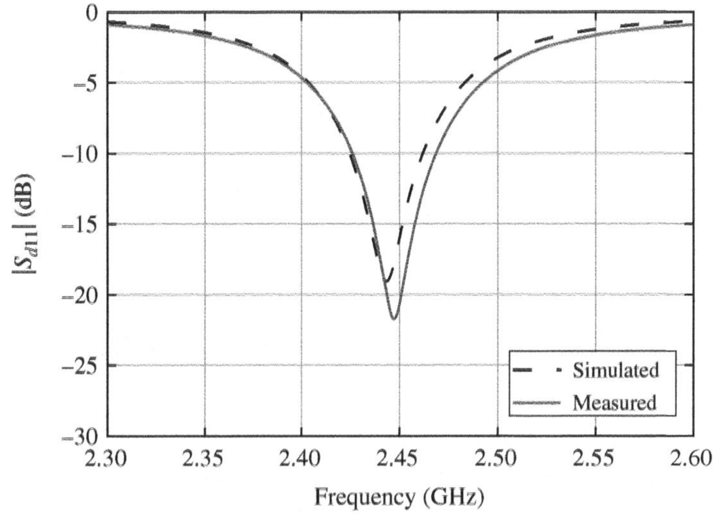

Figure 5.33 Simulated and measured $|S_{d11}|$ for the miniaturized DPIFA.

and measured radiation efficiency values are listed in Table 5.4, which shows that the DPIFA has a higher radiation efficiency than PIFA.

Figure 5.35 shows the calculated, simulated, and measured resonant frequencies for the DPIFA as a function of the gap width g. It is evident from the figure that the simulated and measured resonant frequencies agree very well. There are errors of less than 2% between the calculated and measured resonant frequencies, indicating that the simple formula in (5.83) is rather accurate. They all show that the wider the gap, the higher is the resonant frequency, which can be attributed to the fact that a larger gap width results in a smaller patch size and a lower gap capacitance. Hence, one can trim the resonant frequency by controlling the gap width with either a conductive paste or a laser.

The simulated and measured $|S_{d11}|$ are shown in Figure 5.36 as a function of frequency with the offset s as a parameter. It is interesting to note that the resonant frequency decreases as the offset s increases, due to an elongated current path [13]. Thus, one can miniaturize the size of the DPIFA with a similar bandwidth by controlling the offset s. The miniaturized DPIFA has potential applications in NB-IoT devices.

5.5 ME and DME Dipoles

The ME dipole was proposed by Luk and Wong in 2006 [14]. Figure 5.37 shows the photo of the first ME dipole. As can be seen, a flat dipole is placed in front of a ground plane, with the two inner edges shorted to the ground plane through

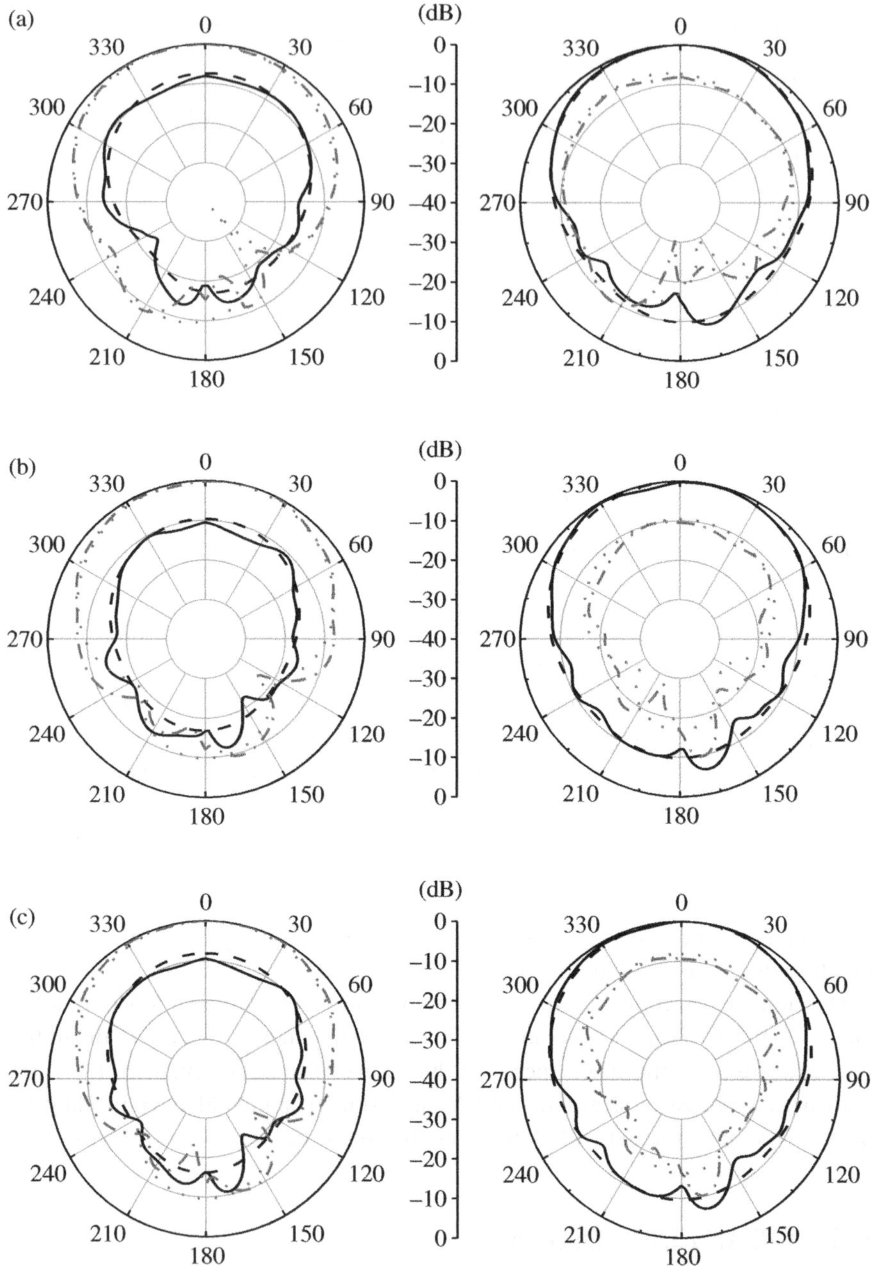

Figure 5.34 Simulated and measured radiation patterns. (a) PIFA, (b) DPIFA, and (c) miniaturized DPIFA.

Table 5.4 Comparison of radiation efficiency.

Efficiency	Simulated	Wheeler cap	Reverberation chamber
PIFA	9.67%	8.12%	9.95%
DPIFA	10.71%	10.34%	10.42%
mDPIFA	6.66%	5.97%	5.81%

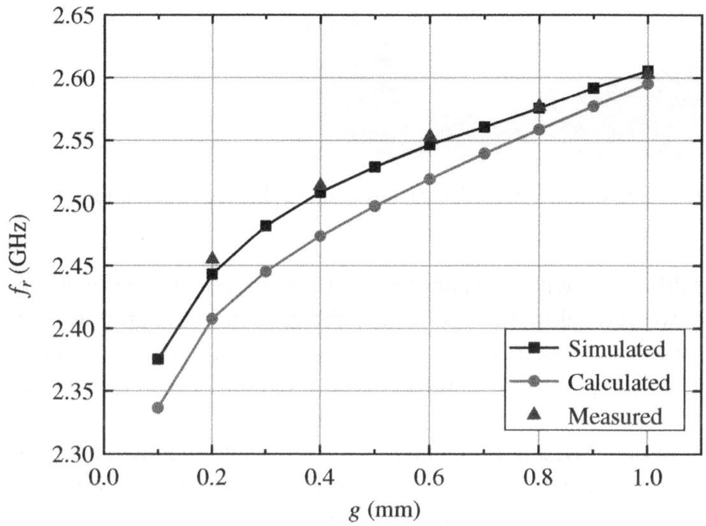

Figure 5.35 Calculated, simulated, and measured resonant frequencies as a function of the gap width g.

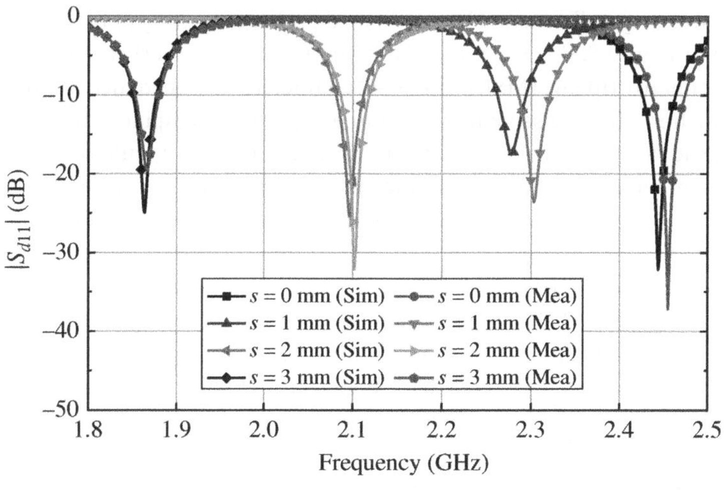

Figure 5.36 Simulated and measured $|S_{d11}|$ as a function of frequency with the offset s as a parameter.

Figure 5.37 Photo of a basic ME dipole. *Source:* Luk and Wu [15]/with permission of IEEE.

two vertical metallic walls with a separation of a few percent of the operating wavelength. The two vertical metallic walls and the portion of the ground plane connecting the two walls together can be interpreted as a vertically oriented SPA, which has a ground plane size identical to the patch size. The flat dipole acts as the electric dipole and the SPA as the magnetic dipole. As a new type of complementary antennas, the ME dipole exhibits a wide impedance bandwidth, a constant gain, and a stable radiation pattern with low cross-polarization and back radiation levels over the operating frequencies [15].

5.5.1 Feeding Techniques

The ME dipole was originally excited by a Γ-shaped probe, consisting of three portions as shown in Figure 5.37. The first portion and the adjacent vertical metallic wall form an air microstrip line (MSL) with 50-Ω characteristic impedance. It links the coaxial launcher with the second portion to excite the ME dipole. The third portion and the adjacent vertical metallic wall form an air open-circuited microstrip stub that compensates for the inductance introduced by the second portion. The third portion can be changed to other shapes or even discarded to facilitate the implementation of standard PCB technology.

The ME dipole, due to its balanced structure, is more suitable for differential feeding [16, 17]. Figure 5.38 shows three simple differential feeding techniques based on the coplanar stripline (CPS), MSL, and metal via, respectively. Both CPS and MSL can be designed together with the electric dipole on the top metal layer of the PCB, and therefore, a single-layered PCB can be used to design the ME antenna or array. However, the CPS introduces asymmetry to the ME dipole which degrades the radiation patterns and the MSL causes spurious radiation. The metal

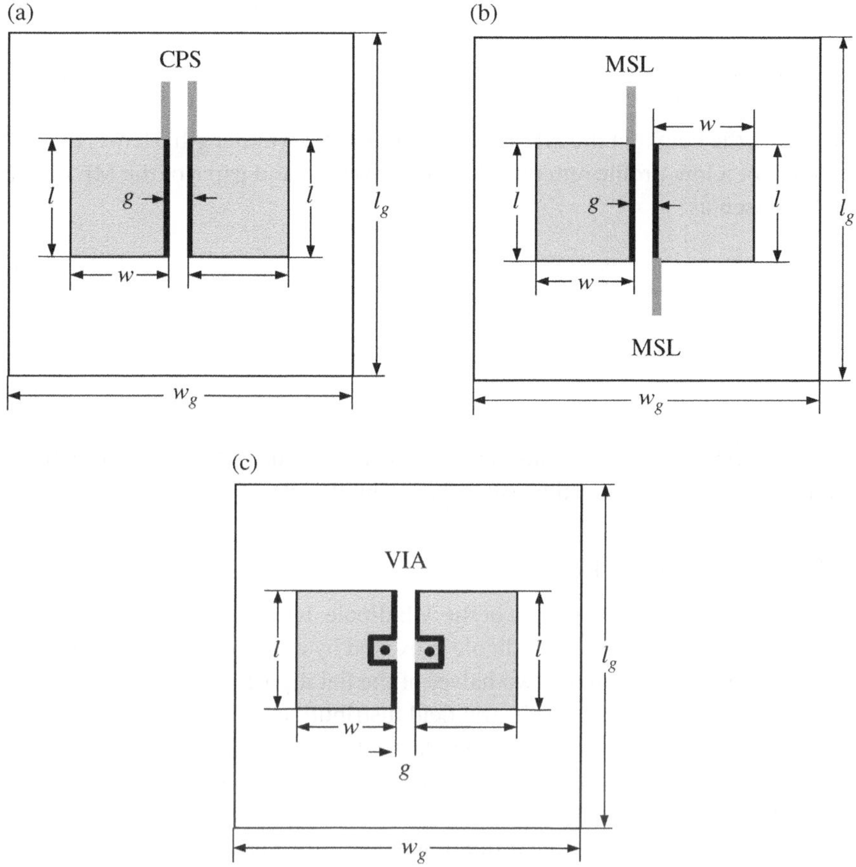

Figure 5.38 Illustrations of three simple differential feeding configurations. (a) CPS, (b) MSL, and (c) VIA.

via with the opening on the ground plane acts as a coaxial probe, which has an undesirable parasitic inductance that limits the bandwidth.

5.5.2 Resonant Frequency

The ME dipole operates at the resonant frequency, which can be estimated by

$$f = \frac{c}{4l\sqrt{\varepsilon_r}} \tag{5.87}$$

for the flat dipole and

$$f = \frac{c}{4h\sqrt{\varepsilon_r}} \tag{5.88}$$

for the SPA. As a result, the dimensions l and h of the ME dipole can be found as

$$l = h = \frac{c}{4f\sqrt{\varepsilon_r}} = \frac{\lambda}{4\sqrt{\varepsilon_r}} = \frac{\lambda_g}{4}. \tag{5.89}$$

Note that the height of the ME dipole is one-quarter wavelength. Hence, the ME dipole is not a low-profile antenna. The dimension w and gap g of the ME dipole can be chosen as

$$w = \frac{\lambda_g}{2} \tag{5.90}$$

and

$$g \leq \frac{\lambda_g}{10} \tag{5.91}$$

for better performance. The dimension values obtained using (5.89)–(5.91) will not be very accurate. However, they can serve as guidelines.

5.5.3 Operating Principle

To explain the operating principle of the ME dipole, let us simulate the ME dipole in HFSS. Assuming that the ME dipole is excited by a lumped port applied at the middle of the gap between the two halves of the flat dipole. Figure 5.39 shows the simulated current and surface electric field distributions on the ME dipole at the central operating frequency of 28 GHz. At $t = 0$, the current, on the flat dipole, achieves the maximum intensity, which means that the electric dipole is strongly excited. Thus, the electric field, generated by the electric dipole, is in the xoz plane. At $t = T/4$, the currents, on the ground plane and the vertical walls, achieve the maximum intensity. It implies that the SPA or magnetic dipole is strongly excited. The electric field can be viewed equivalently as magnetic currents. Thus, the electric field generated by the magnetic dipole is also in the xoz plane. At $t = T/2$, the current of the electric dipole reaches the maximum intensity again but, in a direction, opposite to that occurs at $t = 0$. At $t = 3T/4$, the current of the magnetic dipole reaches the maximum intensity again but, in a direction, opposite to that occurs at $t = T/4$. Hence, the electric and the magnetic dipoles are alternately excited. Besides, electric field, generated by the electric and the magnetic dipoles, is in the same direction with a $T/4$ delay, corresponding to a 90° phase difference.

5.5.4 Design of DME Dipole and Array

Dual-polarized antennas are most popular diversity antennas today, owing to their higher deployment efficiency. A challenging issue in designing dual-polarized antennas is to achieve very high port isolation. The DME dipole can be easily

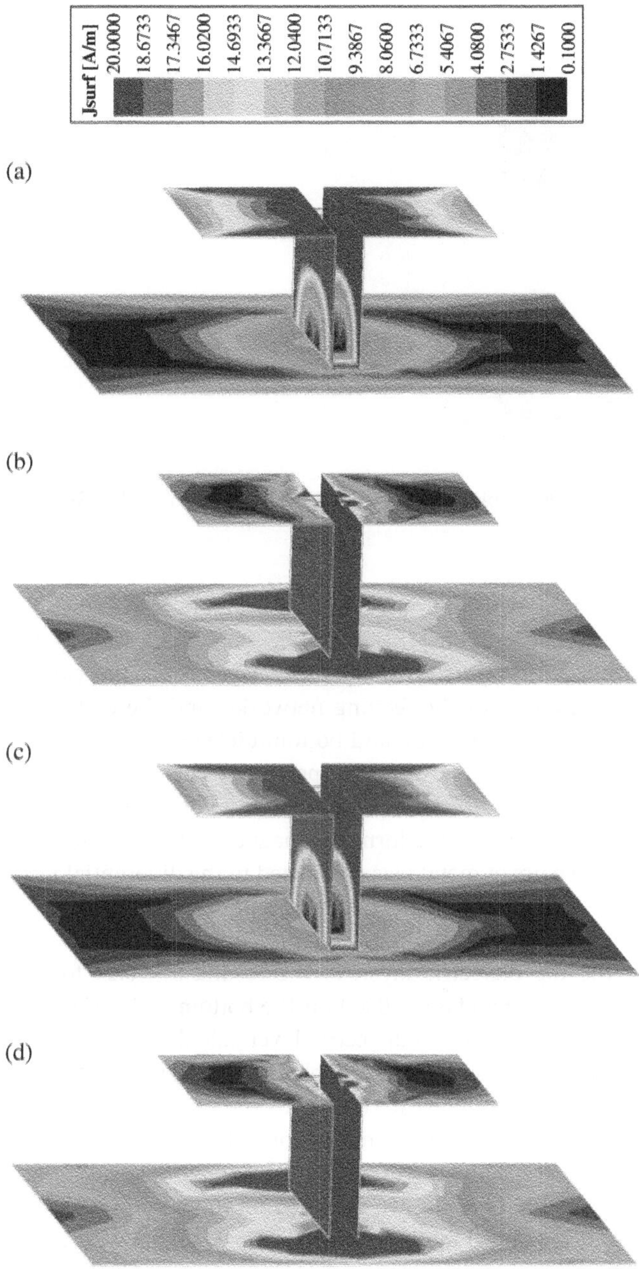

Figure 5.39 Current distributions on the basic ME dipole. (a) $t = 0$, (b) $t = T/4$, (c) $t = T/2$, (d) $t = 3T/4$.

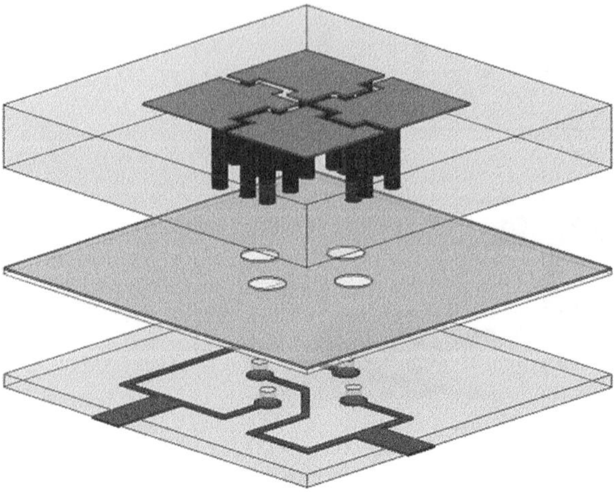

Figure 5.40 The structure of a dual-polarized DME dipole. *Source:* Zhao et al. [18] (© 2022 IEEE).

modified for dual polarizations with high port isolation. Figure 5.40 shows the structure of a dual-polarized DME dipole [18]. As can be seen, it consists of three metal layers and three dielectric layers. The top metal layer is for the electric dipoles, the bottom metal layer is for the feeding networks, and the embedded metal layer is for the ground plane. The top and bottom dielectric layers are the substrates for the dipoles and feeding networks, respectively. The embedded dielectric layer is an adhesive film for bonding. The gaps between the electric dipoles, the blind metal vias, and the ground plane form the magnetic dipoles. The dual-polarized DME is fed by two pairs of metal vias connected to the differential feeding networks through two pairs of openings on the ground plane.

Consider that the dual-polarized DME dipole is designed at 27 GHz with $\varepsilon_r = 3.55$ and $\delta = 0.004$ for the top dielectric layer, $\varepsilon_r = 2.76$ and $\delta = 0.04$ for the embedded dielectric layer, $\varepsilon_r = 2.2$ and $\delta = 0.0009$ for the bottom dielectric layer, respectively. First, the thickness of the top dielectric layer calculated by formula (5.89) is 1.48 mm. Then, the thickness of the available dielectric layer that closes to 1.48 mm is chosen, which is 1.25 mm. Next, the length and width of the electric dipole as well as the gap width can be calculated by formulae (5.89), (5.90), and (5.91) as 1.8, 1.8, and 0.2 mm, respectively. After that, the differential feeding networks formed with 50-Ω and 100-Ω MSLs are designed on the bottom dielectric layer with a thickness of 0.254 mm. Finally, the whole DME dipole with a size of $l_g \times w_g \times h = 8 \times 8 \times 1.604\ \text{mm}^3$ is simulated and optimized in HFSS.

Figure 5.41 shows the simulated *S*-parameters of the dual-polarized DME dipole as a function of frequency. The simulated impedance bandwidths ($|S_{11}| < -10$ dB)

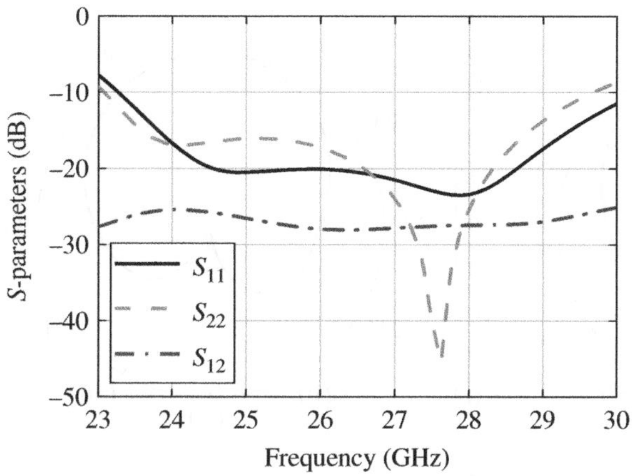

Figure 5.41 Simulated S-parameters of the dual-polarized DME dipole.

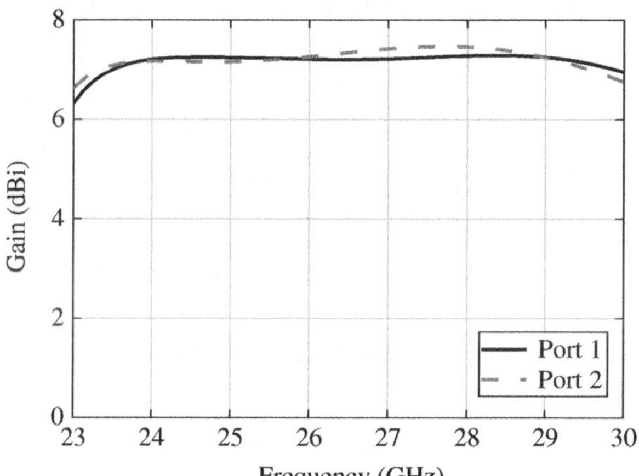

Figure 5.42 Simulated gain of the dual-polarized DME dipole.

are 23% (23.3–29.4 GHz) for the horizontal polarization and 22.2% (52.6–70.6 GHz) for the vertical polarization. The simulated isolation between the ports is better than 25 dB over the impedance bandwidth.

Figure 5.42 shows the simulated gain values of the dual-polarized DME dipole as a function of frequency. It is evident from the figure that the gain values are almost constant about 7.4 dBi over the impedance bandwidth for both polarizations.

Figure 5.43 shows the simulated radiation patterns at 27 GHz. Since the DME dipole is symmetrical about the E and H planes, co-polarized radiation patterns

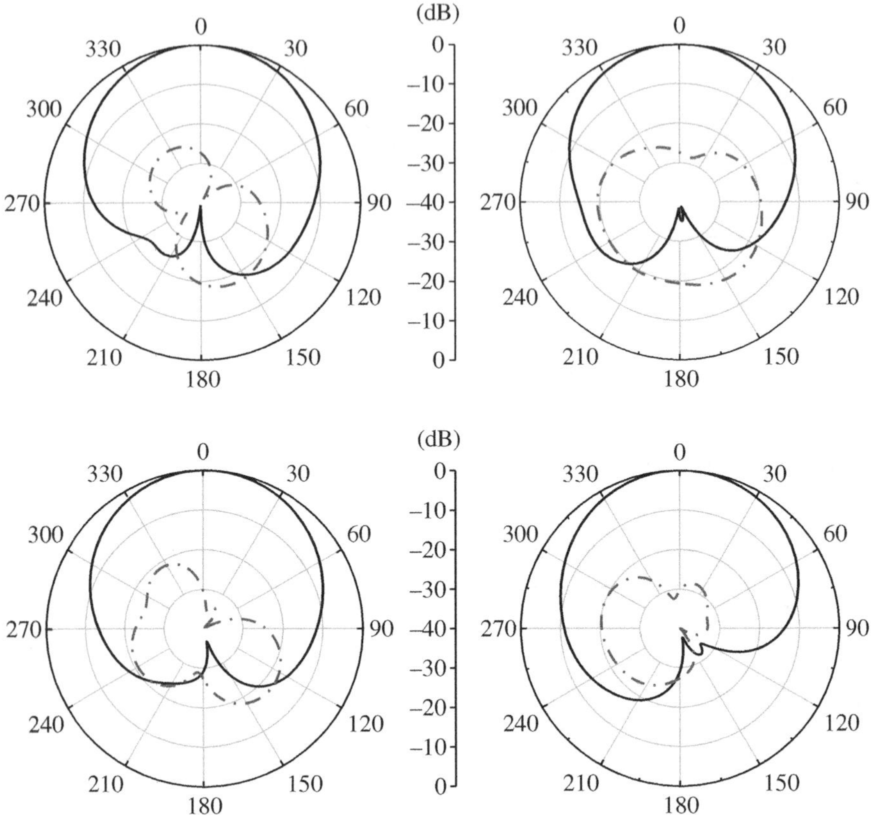

Figure 5.43 Simulated radiation patterns of the dual-polarized DME dipole.

are symmetrical in the *E* and *H* planes and simulated cross-polarization level at the broadside is better than −20 dB. The simulated back radiations are better than −15 dB, and thus, radiation patterns are unidirectional.

The dual-polarized DME dipole has been used as an element in an array. Figure 5.44 shows the top and bottom layouts of a dual-polarized DME dipole array. Note that four dual-polarized DME dipole elements are placed as a 2 × 2 array with a spacing of 7.2 mm (0.624λ at 26 GHz). The elements are connected by two groups of multiple MSLs to the two CPW ports in the two corporate-feed network arrangements.

Figure 5.45 shows the photo of a fabricated sample. The fabricated 2 × 2 dual-polarized DME dipole array has a size of 38 × 25 × 1.604 mm^3. Figure 5.46 shows its simulated and measured *S*-parameters as a function of frequency. The array can cover the impedance bandwidth of 23% (23.3–29.4 GHz) for the dual

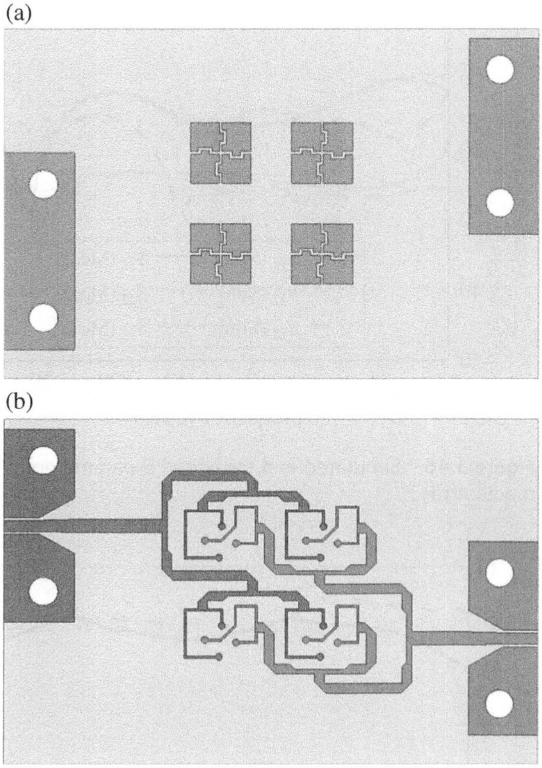

Figure 5.44 Top (a) and bottom (b) layouts of a 2 × 2 dual-polarized DME dipole array. *Source:* Zhao et al. [18] (© 2022 IEEE).

Figure 5.45 Photo of a fabricated 2 × 2 dual-polarized DME dipole array. *Source:* Zhao et al. [18]/with permission of IEEE.

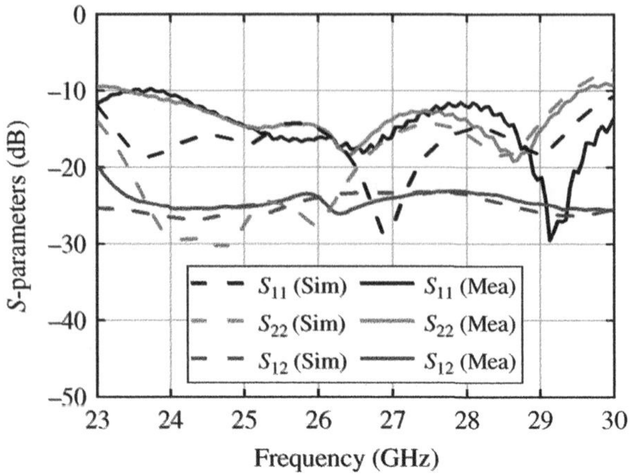

Figure 5.46 Simulated and measured S-parameters of the 2 × 2 dual-polarized DME dipole array.

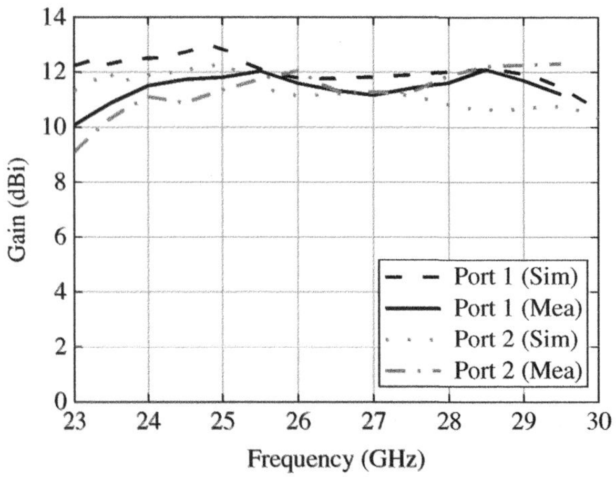

Figure 5.47 Simulated and measured gain values of the 2 × 2 dual-polarized DME dipole array.

polarizations. The measured isolation between the dual ports is better than 23 dB over the impedance bandwidth.

Figure 5.47 shows the simulated and measured gain values of the 2×2 dual-polarized DME dipole array as a function of frequency. The peak gain values are 12.4 and 12.3 dBi, respectively for the horizontal and vertical polarizations, respectively. The gain values vary less than 3 dB over the impedance bandwidth for both polarizations.

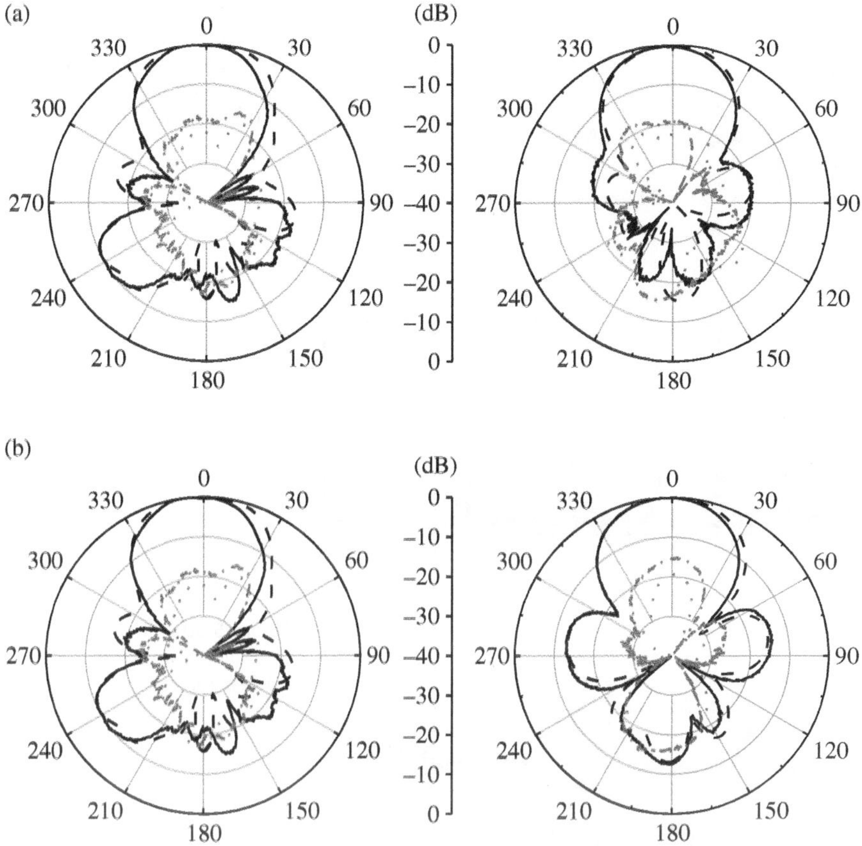

Figure 5.48 Simulated and measured radiation patterns of the 2 × 2 dual-polarized DME dipole array at 27 GHz. (a) Horizontal polarization and (b) vertical polarization.

Figure 5.48 shows the simulated and measured normalized radiation patterns at 27 GHz. The sidelobe is 10 dB lower than the main lobe and the cross-polarization is 15 dB lower than the co-polarization at 27 GHz.

5.5.5 ME Dipole Array Applications

The ME dipole array has found some applications in millimeter-wave 5G mobile communication systems [18–22]. Figure 5.49 shows the photo of an AiP module, which was implemented in the 14-layer (6-2-6) organic build-up package [19]. It has a size of 42.5 mm × 42.5 mm × 1.65 mm and contains 8 × 8 dual-polarized ME dipole elements. The impedance matching is acceptable over the frequency band

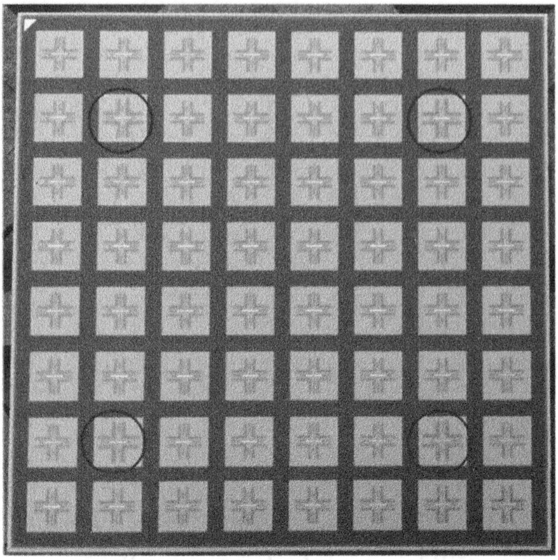

Figure 5.49 Photo of the AiP with ME dipole elements for millimeter-wave 5G new radio. *Source:* Gu et al. [19]/with permission of IEEE.

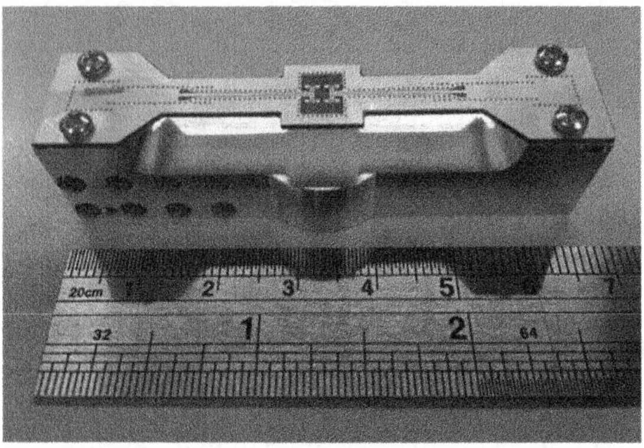

Figure 5.50 Photo of the DME dipole array for 60-GHz radio.

from 24.5 to 29.5 GHz for both the horizontal and vertical polarizations. The element patterns maintain a broad shape suitable for wide beam-steering angles. The patterns are also stable and consistent in the target frequency band. The gain peaks around 5 dBi and remains above 0 dBi across the band with less than 5 dB variation.

Figure 5.50 shows the photo of a 2 × 2 DME dipole array integrated with the feed network for performance validation [20]. The DME dipole is driven by a CPS. A CPW is introduced to form the DME dipole array. One end of the CPW is connected to an input port, while the other end of the CPW is divided into two CPSs. Thus, a 1 × 2 DME dipole array is formed. However, the CPW generates asymmetries in the E plane, which degrades the co-polarization radiation pattern in the E plane and increases the cross-polarization radiation in the H plane. To overcome this problem, the 1×2 DME dipole array is mirrored in structure to realize a differential 2×2 DME dipole array. To further improve radiation characteristics, a rectangular cavity, realized by a wall of metalized vias, is employed. The differential 2×2 DME dipole array was fabricated on Rogers RT/Duroid 5880 substrate with a size of $10 \text{ mm} \times 10 \text{ mm} \times 0.787 \text{ mm}$. The measured impedance bandwidth is from 52.6 to 70.6 GHz. Over the impedance bandwidth, the measured gain varies from 10.9 to 13.7 dBi, the measured co-polarized radiation patterns are symmetrical and unidirectional in the E and H planes, and the measured XPL at the broadside is lower than 20 dB.

5.6 Concluding Remarks

In this chapter, we considered the SPA operating in the TM_{01} mode as a basic antenna. We showed that the SPA is the key to unlock the operation mechanisms of the popular PIFA and ME dipole. However, this chapter focused on the creation, analysis, and design of the DSPA, DPIFA, and DME. It was shown that the DSPA operates in the coupled TM_{01} mode. The electric coupling plays an important role in the DSPA. The DSPA outperforms the SPA. It achieves a symmetrical E-plane radiation pattern and an improved H-plane XPL as compared with the SPA. More importantly, the DSPA has a smaller size than the SPA for the same resonant frequency. For the DPIFA, the formula to predict the resonant frequency of the DPIFA was derived. It was found that the simple formula is rather accurate, and can be used for initial design. For the DME dipole, special emphasis was given to the differential feeding techniques based on CPS, MSLs, and metal vias. Finally, we ended this chapter with application examples of ME and DME dipole arrays in millimeter-wave 5G mobile communication systems.

References

1 C. W. Garvin, R. E. Munson, L. T. Ostwald, K. G. Shroeder, "Missile base mounted microstrip antennas," *IEEE Trans. Antennas Propagat.*, vol. AP-25, no. 5, pp. 604–610, Sep. 1977.

2 E. Penard, J. P. Daniel, "Open and hybrid microstrip antennas," *IEE Proc. Pt. H*, vol. 131, no. 1, pp. 38–44, Feb. 1984.

3 K. F. Lee, Y. X. Guo, J. A. Hawkins, R. Chair, K. M. Luk, "Theory and experiment on microstrip patch antennas with shorting walls," *IEE Proc. Microw. Antennas Propagat.*, vol. 147, no. 6, pp. 521–525, Dec. 2000.

4 Z. Shao, Y. P. Zhang, "Differential shorted patch antennas," *IEEE Trans. Antennas Propagat.*, vol. 67, no. 7, pp. 4438–4444, Jul. 2019.

5 Z. Shao, Y. P. Zhang, "Cross-polarization reduction of shorted patch antenna by coupled TM_{01} mode," *IEEE Trans. Antennas Propagat.*, vol. 67, no. 7, pp. 4438–4444, Jul. 2019.

6 Z. Shao, Y. P. Zhang, "Coupled shorted patch antennas with both broadside and conical radiation patterns," *IEEE Trans. Antennas Propagat.*, vol. 67, no. 7, pp. 4438–4444, Jul. 2019.

7 Z. Shao, Coupled Microstrip Antennas, PhD Thesis, Shanghai Jiao Tong University, 2021.

8 H. Haruki, A. Kobayashi, "The inverted F antenna for portable radio units," *in Convention Rec. IECE Japan*, p. 613, 1982 (in Japanese).

9 T. Taga, K. Tsunekawa, "Performance analysis of a built-in planar inverted F antenna for 800 MHz band portable radio units," *IEEE J. Selected Areas Commun.*, vol. 5, no. 5, pp. 921–929, May 1987.

10 T. Taga, "Analysis of planar inverted-F antennas and antenna design for portable radio equipment," *Analysis, Design and Measurement of Small and Low-Profile Antennas*, K. Hirasawa, M. Haneishi, Eds. Norwood, MA: Artech House, 1992, pp. 161–180.

11 K. Ogawa, H. Iwai, Y. Koyanagi, "Balance-fed planar built-in antenna," *Electron. Lett.*, vol. 37, no. 8, pp. 476–478, Apr. 2001.

12 K. M. Chan, E. Lee, T. Y. Lee, P. Gardner, T. Dodgson, "Aperture coupled, differentially fed planar inverted F antenna," *Electron. Lett.*, vol. 42, no. 11, pp. 608–609, May. 2006.

13 Z. Shao, Y. P. Zhang, "Miniaturization of differentially-driven microstrip planar inverted F antenna", *IEEE Trans. Antennas Propagat.,* vol. 67, no. 2, pp. 1280–1283, Feb. 2019.

14 K. M. Luk, H. Wong, "A new wideband unidirectional antenna element," *Int. J. Microw. Opt. Technol.*, vol. 1, no. 1, pp. 35–44, Jun. 2006.

15 K. M. Luk, B. Q. Wu, "The magnetoelectric dipole – A wideband antenna for base stations in mobile communications," *Proc. IEEE*, vol. 100, no. 7, pp. 2297–2307, Jul. 2012.

16 M. J. Li, K. M. Luk, "A differentially-fed magneto-electric dipole antenna for UWB applications," *IEEE Trans. Antennas Propagat.*, vol. 61, no. 1, pp. 92–99, Jan. 2013.

17 Q. Xue, S. W. Liao, J. H. Xu, "A differentially-driven dual-polarized magneto-electric dipole antenna," *IEEE Trans. Antennas Propagat.*, vol. 61, no. 1, pp. 425–430, Jan. 2013.

18 Y. X. Zhao, X. W. Zhu, W. L. Song, "A 2×2 dual-polarized magneto-electric dipole antenna array at 28 GHz," in *IEEE AP-S Intl. Symp. Digest*, pp. 371–372, Singapore, 4–10 Dec. 2021.

19 X. Gu, et al., "Antenna-in-package integration for a wideband scalable 5G millimeter-Wave phased-array module," *IEEE Microw. Wirel. Comp. Lett.*, vol. 31, no. 6, pp. 682–684, Jun. 2021.

20 X. X. Ruan, K. B. Ng, C. C. Chan, "A differentially fed transmission-line-excited magnetoelectric dipole antenna array for 5G applications," *IEEE Trans. Antennas Propagat.*, vol. 66, no. 10, pp. 5224–5230, Oct. 2018.

21 B. Yu, et al., "A wideband mmWave antenna in fan-out wafer level packaging with tall vertical interconnects for 5G wireless communication," *IEEE Trans. Antennas Propagat.*, vol. 69, no. 10, pp. 6906–6911, Oct. 2021.

22 D. Liu, Y. P. Zhang, *Antenna-in-Package Technology and Applications*, Wiley-IEEE, 2020.

23 R. Garg, I. J. Bahl, "Characteristics of coupled microstriplines," *IEEE Trans. Microw. Theory Technol.*, vol. 27, no. 7, pp. 700–705, Jul. 1979.

24 N. W. Liu, et al., "Cross-polarization reduction of a shorted patch antenna with broadside radiation using a pair of open-ended stubs," *IEEE Trans. Antennas Propagat.*, vol. 68, no. 1, pp. 13–20, Jan. 2020.

6

Differential Microstrip Slot Antennas

6.1 Introduction

An experimental study of a microstrip slot antenna was published by Yoshimura in 1972 [1], which appears to be the first treatment of microstrip slot antennas. His work ignited interest in this type of antennas. Some of the noteworthy studies include a reciprocity method of analysis for printed slot and slot-coupled microstrip antennas [2] and a full-wave analysis for the equivalent circuit of an inclined slot on a microstrip ground plane [3].

Microstrip slot antennas naturally produce bidirectional radiation patterns, while unidirectional radiation patterns can also be obtained through adding reflectors [1]. Microstrip slot antennas are less sensitive to fabrication tolerances and so can be manufactured at lower cost. In addition, microstrip slot antennas do not add weight and size to the system. Hence, they are suitable for applications where cost, size, and weight are of significance.

Most of the reported microstrip slot antennas are single-ended designs. This chapter presents differential microstrip slot antennas and compares them with single-ended counterparts.

6.2 Resonant Modes

Figure 6.1 shows the geometry of a slot in a metallic ground plane on a substrate with the dielectric constant ε_r and loss tangent δ.

Assume that the slot width w is infinitely small, and the ground plane is an infinite, plane, perfectly conducting, thin sheet. So, only a vertical electric field E_y exists within the slot. In accordance with the boundary condition that the tangential component of an electric field on a perfect conductor surface equals zero,

Differential Antennas: Theory and Practice, First Edition. Yueping Zhang.

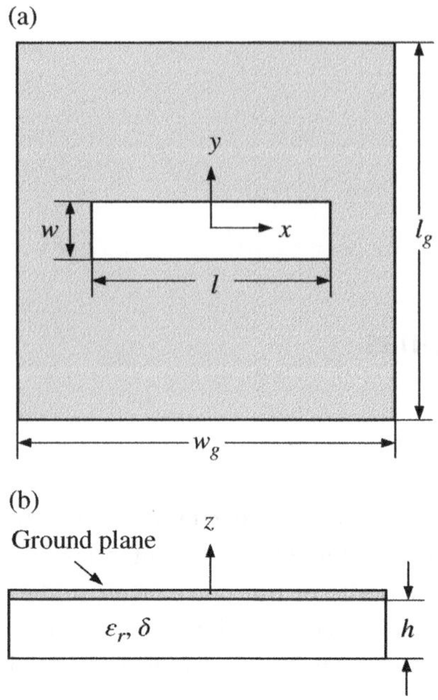

Figure 6.1 Geometry of a rectangular slot. (a) Top view and (b) side view.

the electric field vanishes at the two ends of the slot. The electric field E_y can be written as:

$$E_y = E_0 \sin k\left[\frac{l}{2} \mp x\right] \tag{6.1}$$

where E_0 is the peak electric field strength in the slot, the minus sign is used when $x > 0$, the plus sign when $x < 0$, and

$$k = \frac{2\pi}{\lambda}\sqrt{\varepsilon_e}. \tag{6.2}$$

In (6.2), ε_e is the effective dielectric constant of the uniform medium replacing the two different half spaces on either side of the slot and is given by [4]

$$\varepsilon_e = \frac{2\varepsilon_r}{1+\varepsilon_r}. \tag{6.3}$$

The slot length l can be expressed as

$$l = m\frac{\lambda}{2\sqrt{\varepsilon_e}} \tag{6.4}$$

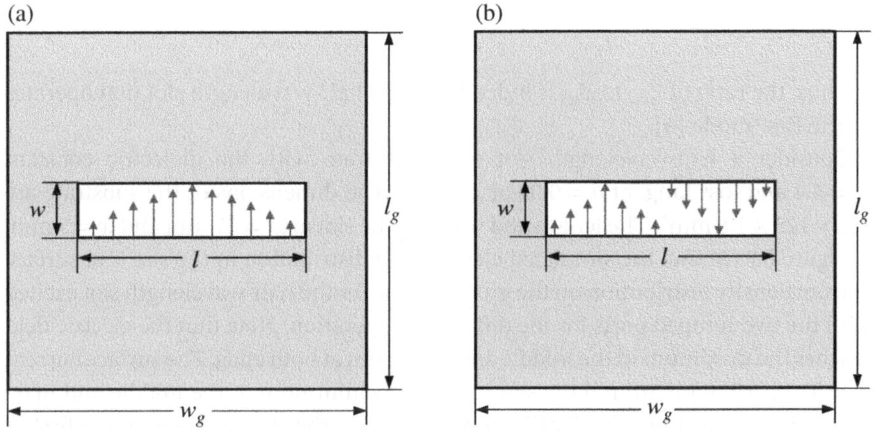

Figure 6.2 Electric field distributions for the first two modes in the half-wavelength slot. (a) $m = 1$ and (b) $m = 2$.

where $m = 1, 2, 3...$ is the mode index in the x direction. Typically, the choice of the slot length is based on the lowest operating frequency of the fundamental mode with a half wavelength [5]. Figure 6.2 sketches the electric field distributions for the first two modes in the half-wavelength slot. It is seen that the electric field for the first mode ($m = 1$) reaches the maximum at the center of the half-wavelength slot but the electric field for the second mode ($m = 2$) goes to the minimum at the center of the half-wavelength slot. It should be mentioned that the second mode of the half-wavelength slot is actually the fundamental mode of the full-wavelength slot.

6.3 Excitation of Resonant Modes

Assume that an ideal current source is applied between the two edges at $x = d$ where $d < l/2$ to excite the slot, the self-impedance Z_{11} is the ratio of the driving-point voltage V_1 to the source current I_1. The voltage V_1 can be found as

$$V_1 = E_y w. \tag{6.5}$$

For the first mode of a half-wavelength slot,

$$V_1 = E_y w = wE_0 \cos kx. \tag{6.6}$$

Imagine that there is a pair of terminals between the two edges at $x = -d$, the terminal voltage V_2 should be equal to V_1 due to the symmetry. The mutual impedance Z_{12} is the ratio of the terminal voltage $-V_2$ to the terminal current I_1, which is

$$Z_{12} = \frac{-V_2}{I_1} = \frac{-V_1}{I_1} = -Z_{11}. \tag{6.7}$$

Thus, the ratio of Z_{dif} to Z_{sig} is indeed 4 for the half-wavelength slot that operates in the first mode [6].

Consider a half-wavelength slot on a substrate with the dielectric constant $\varepsilon_r = 4.4$ and loss tangent $\delta = 0.02$ at 2.4 GHz. The dimensions of the substrate are $125 \times 125 \times 1.6\ \text{mm}^3$. The length and width of the slot are $l = 47$ mm and $w = 3$ mm.

Figure 6.3a shows the simulated electric field distribution in the slot and surface current density distribution on the ground plane for the half-wavelength slot excited with the two lumped ports for the differential operation. Note that the electric field reaches the maximum in the middle and minimum at both ends. The surface current density distribution around the slot reaches the minimum in the middle and maximum near both ends. The distributions reveal that the slot resonates at the first or fundamental mode of the half-wavelength slot at 2.4 GHz. Figure 6.3b shows the simulated electric field distribution in the slot and surface current density distribution on the ground plane for the half-wavelength slot excited with the one lumped

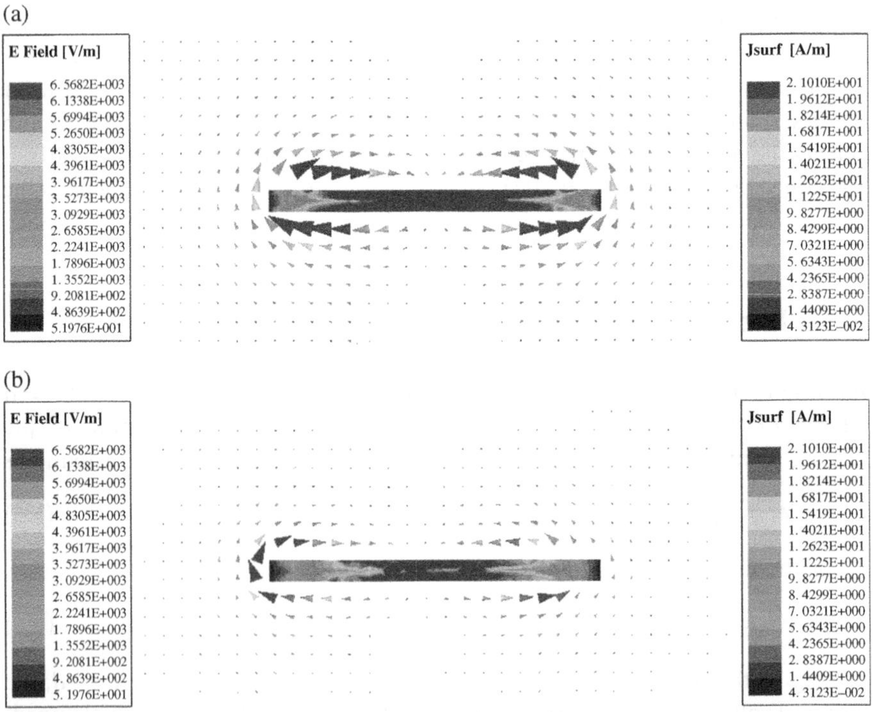

Figure 6.3 Simulated electric field distributions in the slot and surface current density distributions on the ground planes. (a) Differential and (b) single-ended excitations.

port for the single-ended operation. Note that the slot still resonates at the first or fundamental mode of the half-wavelength slot at 2.4 GHz, but the symmetry of the electric field and the surface current density distributions are slightly distorted. The surface current density distribution is obviously stronger at the left end than at the right end because the lumped port is near the left end.

Figure 6.4a shows the simulated input impedances at the frequency of 2.4 GHz for the half-wavelength slot under the differential excitation. The simulated input impedance is generally capacitive. The input resistance value is quite high and decreases as the separation between the two lumped ports increases. The resonance

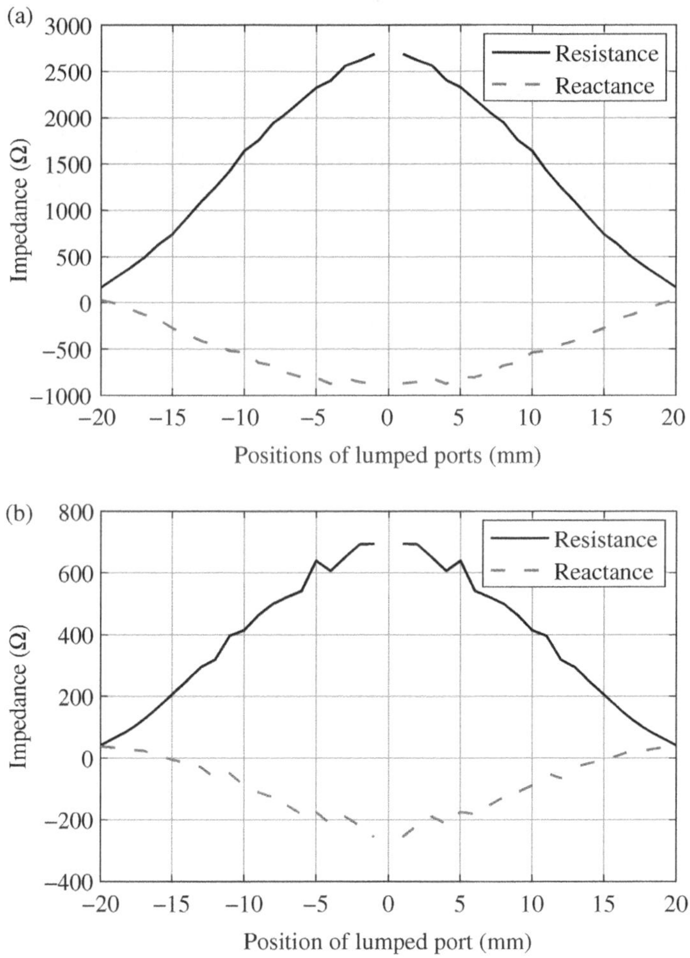

Figure 6.4 Simulated input impedances of the half-wave slot at 2.4 GHz. (a) Differential and (b) single-ended excitations.

occurs when the two lumped ports are 5.5 mm to the respective ends of the slot. Figure 6.4b shows the simulated input impedances at the frequency of 2.4 GHz for the half-wavelength slot under the single-ended excitation. Note that the single-ended input impedance has almost the same profile as the differential input impedance. As expected, the single-ended input impedance is about 4 times smaller than the differential input impedance. The resonance occurs when the lumped port is 8.5 mm to the near end of the slot.

Similarly, for the second mode of the half-wavelength slot,

$$V_1 = E_y w = wE_0 \sin kx, \tag{6.8}$$

which implies that the terminal voltage V_2 is equal to $-V_1$ due to the anti-symmetry. The mutual impedance Z_{12} is still given by (6.7). As a result, the ratio of Z_{dif} to Z_{sig} is also 4 for the half-wavelength slot that operates in the second mode [6].

As mentioned previously, the second mode of the half-wavelength slot is actually the fundamental mode of the full-wavelength slot. Now let us consider a full-wavelength slot on a substrate with the dielectric constant $\varepsilon_r = 4.4$ and loss tangent $\delta = 0.02$ at 2.4 GHz. The dimensions of the substrate are $125 \times 125 \times 1.6\ \text{mm}^3$. The length and width of the slot are $l = 97$ mm and $w = 3$ mm.

Figure 6.5a shows the simulated electric field distribution in the slot and surface current density distribution on the ground plane for the full-wavelength slot excited with the two lumped ports for the differential operation. Note that the

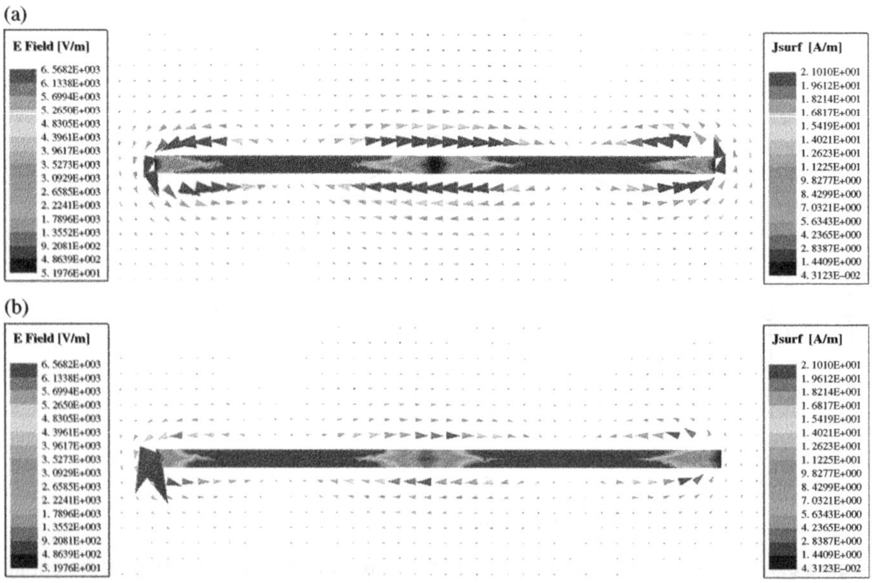

Figure 6.5 Simulated electric field distributions in the slot and surface current density distributions on the ground planes. (a) Differential and (b) single-ended excitations.

electric field reaches the minimum in the middle and at both ends. Its maximum appears between the two minima. The surface current density distribution around the slot reaches the maximum in the middle and near both ends. The distributions reveal that the slot resonates at the fundamental mode of a full-wavelength slot. Figure 6.5b shows the simulated electric field distribution in the slot and surface current density distribution on the ground plane for the full-wavelength slot antenna excited with the one lumped port for the single-ended operation. Note that the slot still resonates at the fundamental mode of the full-wavelength slot at 2.4 GHz, but the symmetry of the electric field and the surface current density distributions are slightly distorted. The surface current density distribution is obviously stronger at the left end than at the right end because the lumped port is near the left end.

Figure 6.6a shows the simulated input impedance at the frequency of 2.4 GHz for the full-wavelength slot under the differential excitation. It is seen that the

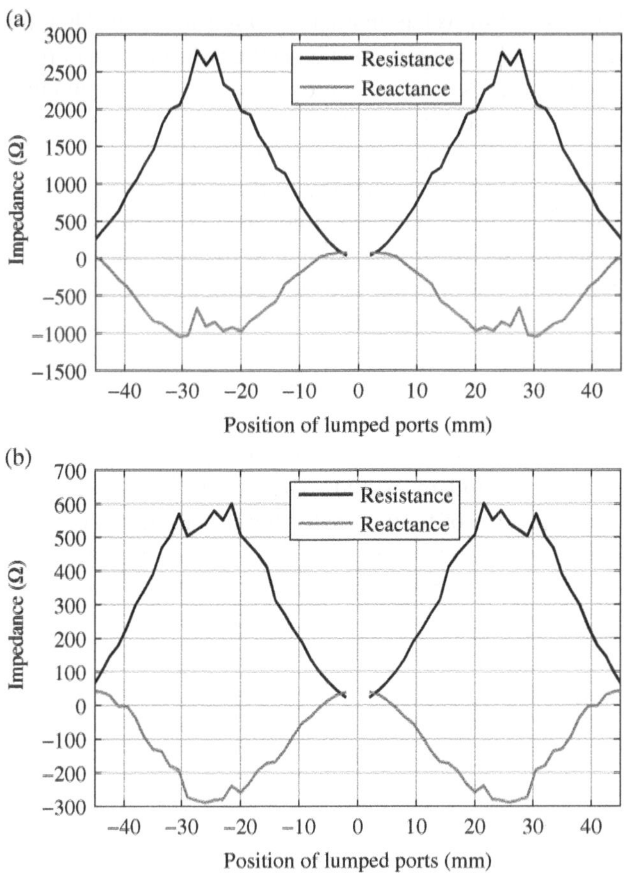

Figure 6.6 Simulated input impedances of the half-wavelength slot at 2.4 GHz. (a) Differential and (b) single-ended excitations.

input resistance first increases and then decreases, while the input reactance first decreases and then increases as the separation between the two lumped ports increases. For the smaller separation, the input impedance is slightly inductive. As the separation increases, the input impedance first becomes more capacitive and then less capacitive. The input impedance becomes slightly inductive when the two lumped ports near the slot ends. Figure 6.6b shows the simulated input impedances at the frequency of 2.4 GHz for the full-wavelength rectangular slot under the single-ended excitation. Note that the single-ended input impedance has almost the same profile as the differential input impedance. As expected, the single-ended input impedance is about four times smaller than the differential input impedance.

6.4 Half-Wavelength Microstrip Slot Antenna

There are two configurations for half-wavelength microstrip slot antennas. They are shown in Figure 6.7. Note that both configurations have the twin microstrip lines oriented in opposite directions, which is required to properly excite the fundamental mode of the half-wavelength slot for the differential operation. In configuration one, the microstrip lines are short-circuited through the dielectric substrate with the slot's longer sides, which are located farther from the feed inputs, respectively. In configuration two, the microstrip lines are open-circuited and extended across the slot to form virtual short circuits for strong coupling to the slot.

The half-wavelength microstrip slot antenna with the opened termination was fabricated on a substrate with the dielectric constant $\varepsilon_r = 4.4$ and loss tangent $\delta = 0.02$. The dimensions of the substrate are $w_g \times l_g \times h\ \mathrm{mm}^3 = 125 \times 125 \times 1.6\ \mathrm{mm}^3$. The dimensions of the half-wavelength slot are $l = 47$ mm and $w = 3$ mm. The width of the microstrip line is $a = 5.1$ mm. The microstrip line is located at $d = 26.2$ mm and $b = 27.5$ mm. Figure 6.8 shows the photos of the fabricated sample.

Figure 6.9 shows the simulated and measured $|S_{d11}|$ of the half-wavelength microstrip slot antenna as a function of frequency for the differential operation. The simulated and measured resonant frequencies are at 2.38 and 2.43 GHz, respectively, which are slightly lower than 2.4 GHz. The simulated impedance bandwidth is wider than the measured one. They are 300 MHz (12.7%) and 271 MHz (11.3%), respectively. The agreement between the simulated and measured results is, in general, acceptable. The discrepancy is likely due to soldering and warpage and the experimental error of our measuring system.

Figure 6.10 shows the simulated and measured radiation patterns of the half-wavelength microstrip slot antenna for the differential operation at 2.4 GHz

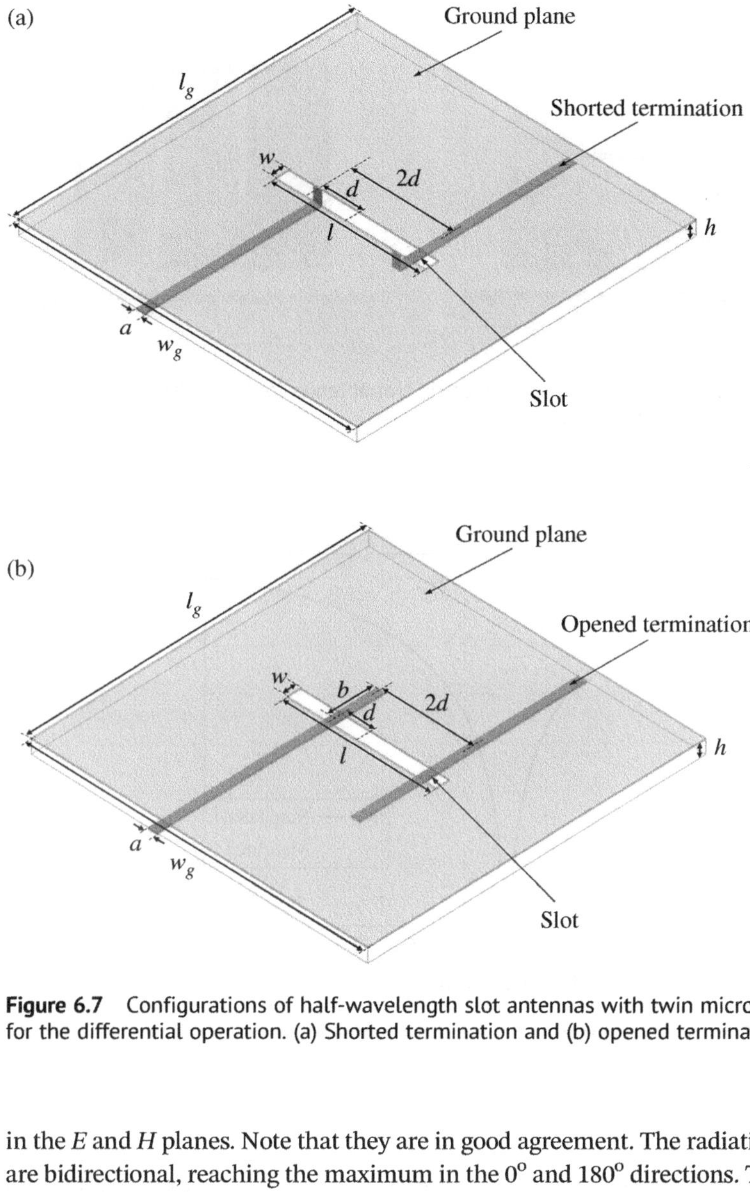

Figure 6.7 Configurations of half-wavelength slot antennas with twin microstrip lines for the differential operation. (a) Shorted termination and (b) opened termination.

in the E and H planes. Note that they are in good agreement. The radiation patterns are bidirectional, reaching the maximum in the 0° and 180° directions. The E-plane radiation patterns with small ripples are considered to have come from the interference of the radiations from the slot and the edges of the ground planes [7]. The H-plane radiation patterns are smoother. The distortions of the radiation

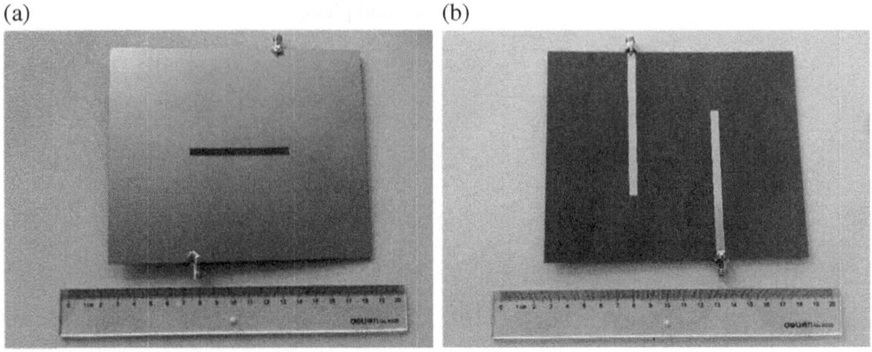

Figure 6.8 Photos of the fabricated microstrip slot antenna for the differential operation. (a) Top view and (b) bottom view.

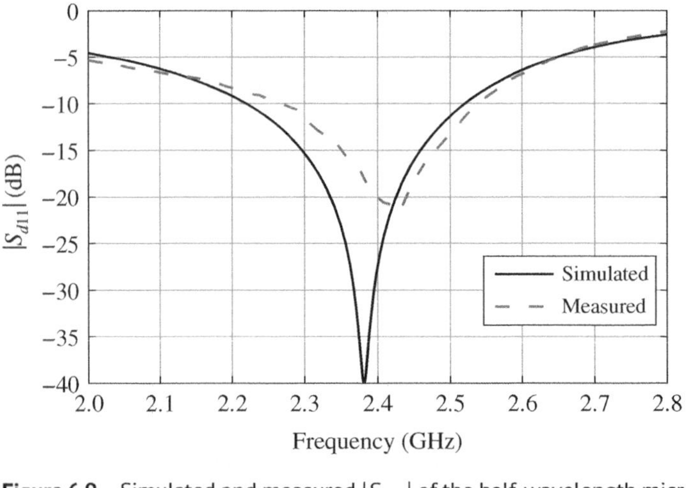

Figure 6.9 Simulated and measured $|S_{d11}|$ of the half-wavelength microstrip slot antenna.

patterns due to the microstrip lines are hardly noticed since the width of the microstrip lines is very small compared with the length of the slot, and the field disturbances due to the existence of the microstrip lines near the slot can hardly cause the far-field distortion. The simulated and measured peak gain values are 5.9 and 5.8 dBi at 2.4 GHz for the half-wavelength microstrip slot antenna for the differential operation, respectively.

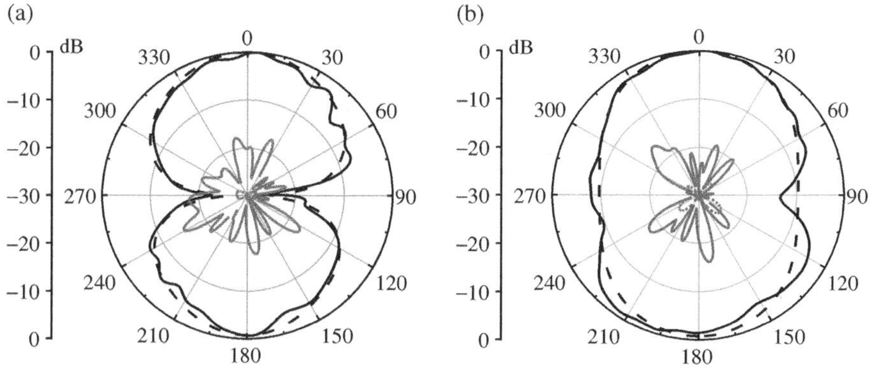

Figure 6.10 Simulated and measured radiation patterns of the half-wavelength microstrip slot antenna for the differential operation at 2.4 GHz. (a) The *E* plane and (b) the *H* plane.

6.5 Full-Wavelength Microstrip Slot Antenna

There are two configurations for full-wavelength microstrip slot antennas. They are shown in Figure 6.11. Note that both configurations have the twin microstrip lines oriented in the same direction, which is required to properly excite the second or fundamental mode of the full-wavelength slot for the differential operation. In configuration one, the microstrip lines are short-circuited through the dielectric substrate with the slot longer side, which is located farther from the feed inputs. In configuration two, the microstrip lines are open-circuited and extended across the slot to form virtual short circuits for strong coupling to the slot.

The full-wavelength microstrip slot antenna with the opened termination was fabricated on a substrate with the dielectric constant $\varepsilon_r = 4.4$ and loss tangent $\delta = 0.02$. The dimensions of the substrate are $w_g \times l_g \times h$ mm^3 = $125 \times 125 \times 1.6$ mm^3. The dimensions of the half-wave slot are $l = 97$ mm and $w = 3$ mm. The width of the microstrip line is $a = 5.1$ mm. The microstrip line is located at $d = 55.5$ mm and $b = 27.5$ mm. Figure 6.12 shows the photos of the fabricated sample.

Figure 6.13 shows the simulated and measured $|S_{d11}|$ of the full-wavelength microstrip slot antenna as a function of frequency for the differential operation. The simulated and measured resonant frequencies are at 2.39 and 2.38 GHz, respectively, which are slightly lower than 2.4 GHz. The simulated impedance bandwidth is wider than the measured one. They are 141 MHz (5.9%) and 120 MHz (5.1%), respectively. The agreement between the simulated and measured results is, in general, acceptable. The discrepancy is likely due to soldering and warpage and the experimental error of our measuring system.

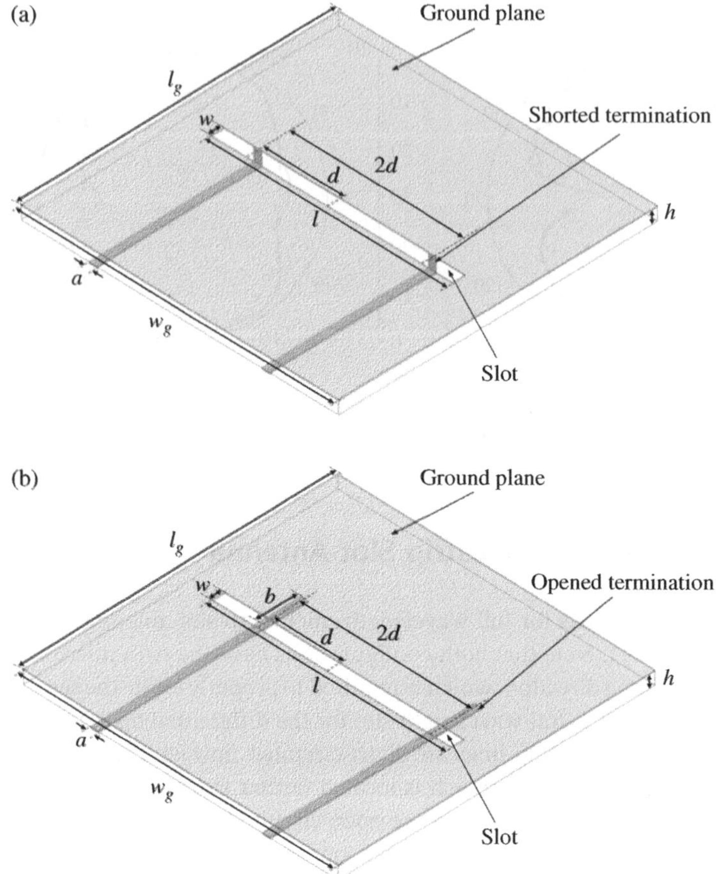

Figure 6.11 Configurations of full-wavelength slot antennas with twin microstrip lines for the differential operation. (a) Shorted termination and (b) opened termination.

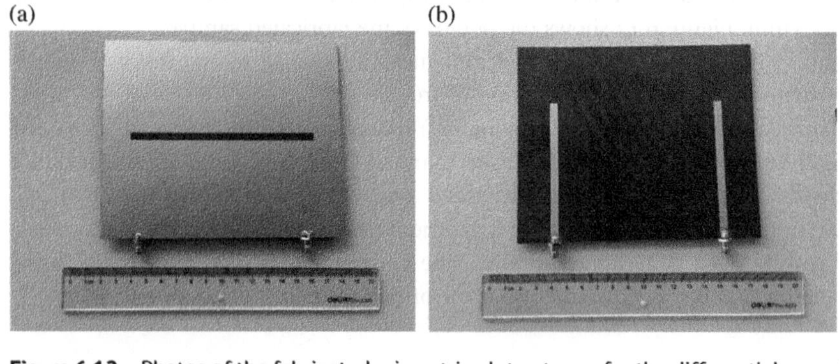

Figure 6.12 Photos of the fabricated microstrip slot antenna for the differential operation. (a) Top view and (b) bottom view.

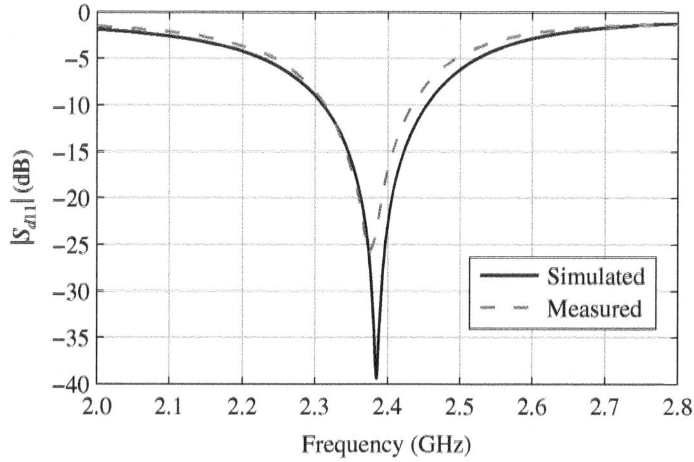

Figure 6.13 Simulated and measured $|S_{d11}|$ of the full-wavelength slot.

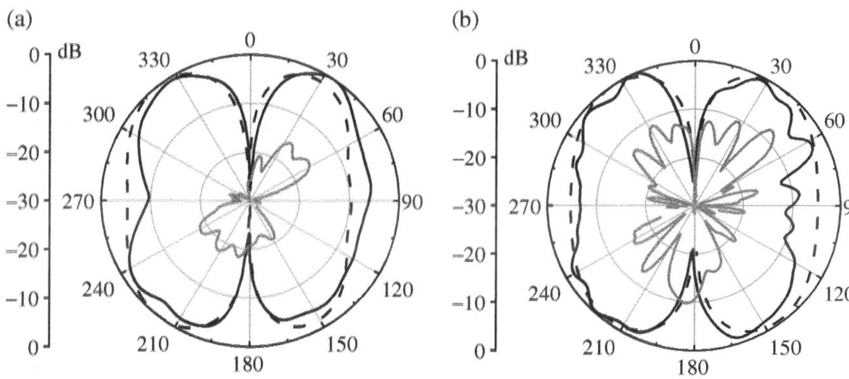

Figure 6.14 Simulated and measured radiation patterns of the full-wavelength slot antenna for the differential operation at 2.4 GHz. (a) The *E* plane and (b) the *H* plane.

Figure 6.14 shows the simulated and measured radiation patterns of the full-wavelength slot antenna for the differential operation at 2.4 GHz in the *E* and *H* planes. Note that they are in acceptable agreement. The radiation patterns are bi-directional, reaching the minimum in the 0° and 180° directions. The distortions of the radiation patterns due to the microstrip lines are hardly noticed since the width of the microstrip lines is very small compared with the length of the slot, and the field disturbances due to the existence of the microstrip lines near the slot can hardly cause the far-field distortion. The simulated and measured peak gain values are 3.9 and 4.8 dBi at 2.4 GHz for the full-wavelength microstrip slot antenna for the differential operation, respectively.

6.6 Concluding Remarks

This chapter first presented the resonant modes in a printed slot. It was shown that the first mode is the fundamental mode of a half-wavelength slot, and the second mode is the fundamental mode of a full-wavelength slot. Then, the chapter examined the impedance relations of the slot excited with the lumped ports for the differential and single-ended operations and confirmed that the differential input impedance is about four times the single-ended input impedance.

Next, the chapter described the two configurations for microstrip slot antennas with one featuring the short-circuited and the other the open-circuited microstrip lines. It was highlighted that the microstrip lines should be made with proper orientation and connection with respect to the slot for the excitation of the desired fundamental modes. Finally, both half-wavelength and full-wavelength differential microstrip slot antennas were fabricated and measured. It was shown that the half-wavelength differential microstrip rectangular slot antenna has wider impedance bandwidth and higher gain as compared with those of the full-wavelength differential microstrip slot antenna.

References

1 Y. Yoshimura, "A microstrip slot antenna," *IEEE Trans. Microwave Theory Technol.*, vol. 20, pp. 760–762, Nov. 1972.

2 D. M. Pozar, "A reciprocity method of analysis for printed slot and slot-coupled microstrip antennas," *IEEE Trans. Antennas Propagat.*, vol. 34, pp. 1439–1446, Dec. 1986.

3 A. K. Bhattacharyya, Y. M. Antar, A. Ittipiboon, "Full wave analysis for the equivalent circuit of an inclined slot on a microstrip ground plane," *IEE Proc. H (Microwaves, Antennas Propagat.)*, vol. 139, no. 3, pp. 245–250, Jun. 1992.

4 B. N. Das, K. K. Joshi, "Impedance of a radiating slot in the ground plane of a microstrip line," *IEEE Trans. Antennas Propagat.*, vol. 30, pp. 922–926, May 1982.

5 H. Y. Wang, D. W. Zhou, L. Xue, S. Gao, H. Xu, "Modal analysis and excitation of wideband slot antennas," *IET Microw, Antennas Propagat.*, vol. 17, no. 13, pp. 1889–1893, Sep. 2017.

6 Y. P. Zhang, "Impedance relations for differential antennas and single-ended counterparts," *IEEE Trans. Antennas Propagat.*, vol. 70, no. 2, pp. 953–959, 2022.

7 W. R. Deal, V. Radisic, Y. X. Qian, T. Itoh, "Integrated-antenna push-pull power amplifiers," *IEEE Trans. Microw. Theory Technol.*, vol. 47, no. 8, pp. 1418–1425, Aug. 1999.

7

Differential Microstrip Grid Array Antennas

7.1 Introduction

The grid array antenna was invented by Kraus in 1962 as a traveling-wave (nonresonant) antenna with the main lobe of radiation in a backward angle-fire direction [1, 2]. The grid array antenna was implemented in microstrip technology by Conti et al. in 1981 as a standing-wave (resonant) antenna with the main beam of radiation in the broadside direction [3]. The grid array antenna was analyzed by Hildebrand and McNamara in 1995 with the method of moments [4], miniaturized by Nakano et al. in 1998 [5], and optimized by Chen et al. in 2010 using a parallel genetic algorithm [6]. They showed that the grid array antenna in microstrip technology has all the usual benefits of conventional patch-type radiators plus broad bandwidth, high gain, and adequate cross-polarization control [3]. However, the grid array antenna had not found any commercial application until Sun and Zhang first applied it to 60-GHz radios in 2008 [7], followed by Bauer et al. to 79-GHz [8], Chen et al. to 94 GHz [9], Beer et al. to 122-GHz radars [10], and Zhang et al. to 140-GHz systems [11, 12], Nariman et al. to 60-GHz wireless power transfer (WPT) [13], and Hu et al. to 2.45-GHz energy harvester [14].

This chapter deals with the analysis and design of differential microstrip grid array antennas. Section 7.2 describes the configuration of a basic microstrip grid array antenna. Section 7.3 explains the principle of operation through careful examination of current distribution and radiation patterns of the microstrip grid array antenna. Section 7.4 reviews the fundamental characteristics of microstrip grid array antennas. Section 7.5 compares the novel with the usual differential excitation technique. Section 7.6 presents the design procedure, formulas, and examples. Section 7.7 discloses an impedance-matching method for large microstrip grid array antennas. Section 7.8 devises the slow-wave structure for miniaturization of microstrip grid array antennas. Section 7.9 considers the microstrip grid array antenna that also acts as a meshed microstrip patch antenna to operate concurrently in dual current modes at high and low frequencies.

Differential Antennas: Theory and Practice, First Edition. Yueping Zhang.

Section 7.10 discusses the use of the empty areas of the loops of a microstrip grid array antenna to house other types of antenna elements to realize co-aperture antenna arrays. Section 7.11 demonstrates the applications of differential microstrip grid array antennas in 60-GHz radios, 79-GHz radars, 24-GHz full duplex and MIMO systems, and 2.45-GHz energy harvester. Finally, Section 7.12 provides concluding remarks and suggestions for further research.

7.2 Basic Configuration

Figure 7.1 shows the geometry and coordinate system of a basic microstrip grid array antenna. The basic microstrip grid array antenna consists of two rectangular loops whose long sides are l long and w_l wide and short sides are s long and w_s wide. It is printed on the top surface of a grounded substrate of dielectric constant ε_r, loss tangent δ, length l_g, width w_g, and thickness h. It is fed by a coaxial probe, having the outer conductor connected to the ground plane and the inner conductor connected to the T junction of one short side and two long sides. The basic configuration is the simplest microstrip grid array antenna that features the necessary bends and T-junctions for radiation [15]. Apparently, more loops can

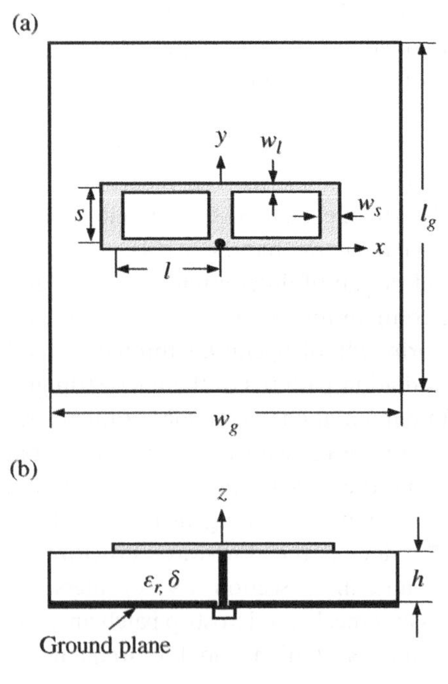

Figure 7.1 Geometry and coordinate system of a basic microstrip grid array antenna. (a) Top view and (b) side view.

be added to the basic configuration in real design to achieve high gain. The microstrip grid array antenna has a simple structure and is relatively easy in construction.

7.3 Principle of Operation

The microstrip grid array antenna usually operates as a resonant antenna [3]. The loops are thus sized such that the current on each short side is essentially in phase, while each long side supports a full wavelength current at the center frequency of operation. As a result, the radiation of the microstrip grid array antenna is largely from the short sides with the long sides acting mainly as guiding or transmission lines. If the short sides (or radiating elements) are arranged to have the same number in the E and H planes, the microstrip grid array antenna radiates the pencil beam [16]. Otherwise, it radiates the fan beam [17].

Let us consider a microstrip grid array antenna printed on a substrate of $\varepsilon_r = 4.5$, $\delta = 0.002$, $w_g = 25$ mm, $l_g = 25$ mm, and $h = 0.254$ mm. The microstrip grid array antenna has nine loops with loop dimensions of $l = 2.84$ mm, $w_l = 0.1$ mm, $s = 1.42$ mm, and $w_s = 0.1$ mm to resonate at 60 GHz. Two more short sides are added to make an equal number of radiating elements in both the E and H planes. Figure 7.2 shows the simulated current distribution at resonance. It is interesting to note that the current on each short side is in phase, while the current on each long side changed the phase at its middle to form a full wavelength current. Consequently, it is the current on the short side that contributes to the far-field radiation more dominantly than that on the long side. Figure 7.3 shows the simulated far-field radiation patterns in both the E and H planes at 60 GHz. As expected, the

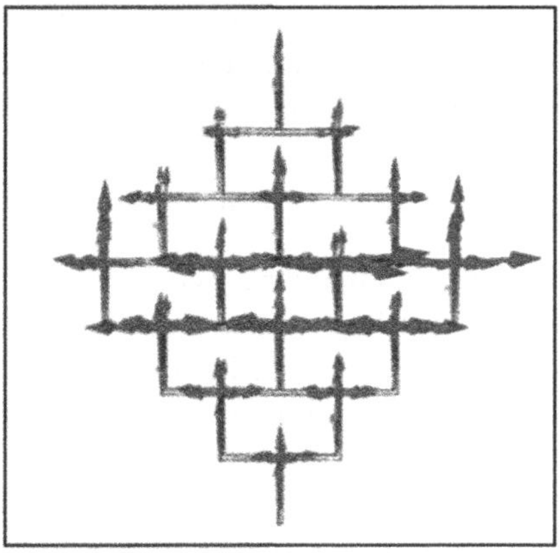

Figure 7.2 The simulated current distribution at the resonant frequency of 60 GHz.

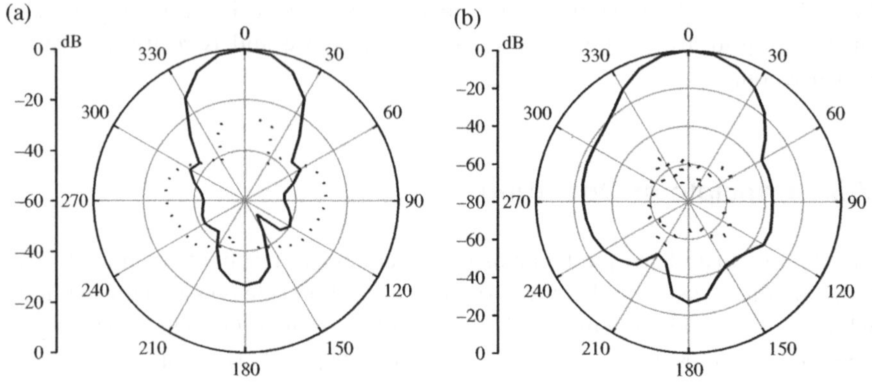

Figure 7.3 The simulated radiation patterns at the resonant frequency of 60 GHz. (a) The *E* plane and (b) the *H* plane.

main beam of radiation is in the broadside direction. The sidelobe level is quite low. The radiation is linearly polarized. The stronger co-polarized component is due to the existence of the same phase of the currents on all the short sides, while the weaker cross-polarized component is due to the presence of antiphase current couplets on all the long sides. The phase transition of the currents on the long sides accounts for the nulls on the cross-polarized patterns.

7.4 Fundamental Characteristics

The radiation properties of the microstrip grid array antenna are uniquely determined by grid geometry. The individual loop dimensions set the center frequency of operation. Loop impedance, interconnections, and feed point determine the amplitude distribution on the array, while the resonant behavior of the multiple loop array determines the overall frequency response [3].

7.4.1 Resonant Frequency

The resonant frequency is the frequency at which the parallel resonance occurs for the microstrip grid array antenna. The resonant frequency for the three different thicknesses as a function of relative permittivity, with the number of radiating elements as a parameter, was examined [18]. It was noted that the resonant frequency changed slightly with the substrate thickness or with the number of radiating elements. For instance, the resonant frequency changed within 5% for most cases with a fixed relative permittivity as the substrate thickness varied from

0.254, 0.381, and 0.508 mm, or the number of radiating elements, n, varied among 9, 16, and 25. It was further noted that the resonant frequency generally decreased with an increase in relative permittivity.

7.4.2 Impedance Bandwidth

The impedance bandwidth of a microstrip grid array antenna is typically from 6% to 12% for $|S_{11}| \leq -10$ dB [3]. The impedance bandwidth for the three different thicknesses as a function of the relative permittivity, with the number of radiating elements as a parameter was evaluated in [18]. As with all microstrip structures, the substrate thickness and relative permittivity have a strong influence on the impedance bandwidth. Generally, a thicker substrate with a lower relative permittivity offers a microstrip grid array antenna with a wider impedance bandwidth. Taking the microstrip grid array antenna with nine radiating elements on the substrate of $\varepsilon_r = 2.2$ and $h = 0.508$ mm (0.1016λ at 60 GHz) as an example, it achieved a 36% impedance bandwidth. Like other resonant array antennas, the impedance bandwidth of the microstrip grid array antenna depends on the number of radiating elements. Simply stated, for a given substrate size and controlled by the gain–bandwidth relationship, the more radiating elements there are, the narrower is the impedance bandwidth.

7.4.3 Half-Power Beamwidth

The half-power beamwidth of a microstrip grid array antenna can be wider or narrower, depending mainly on the number of radiating elements. The half-power beamwidth across the main radiation lobe in the E plane for the three different thicknesses as a function of the relative permittivity, with the number of radiating elements as a parameter was considered [18]. The half-power beamwidth in the H plane was very similar. As expected, the half-power beamwidth decreased with the number of radiating elements. The half-power beamwidth is sensitive to the substrate's relative permittivity but is not sensitive to the substrate's thickness. The higher the substrate's relative permittivity, the wider is the half-power beamwidth, due to the excitation of a stronger surface wave.

7.4.4 Gain

A microstrip grid array antenna is a high-gain antenna [16]. The gain in the broadside direction for the three different thicknesses as a function of relative permittivity, with the number of radiating elements as a parameter, was explored [18]. It was noted that the gain decreased with relative permittivity or thickness but increased with the number of radiating elements. The gain increment with the number of radiating elements will be limited by the attenuation of the microstrip lines that form the array. The gain decrement due to a higher permittivity or a thicker substrate can mainly be attributed to the excitation of a stronger surface wave. The larger attenuation of the microstrip lines due to a higher permittivity and more feeding loss for a thicker substrate also reduces the gain. Furthermore, control over the phase synchronization for more radiating elements on a thicker substrate of a higher permittivity becomes more difficult for a higher gain.

7.4.5 Gain Bandwidth

In addition to the impedance bandwidth, a few bandwidths can be defined under the radiation-pattern characteristics (i.e., pattern shape, sidelobe level, gain, cross-polarization, etc.). It was found that the most stringent bandwidth criterion is seen to be the gain bandwidth [3]. The gain bandwidth for the three different thicknesses as a function of relative permittivity, with the number of radiating elements as a parameter was investigated [18]. It was noted that the gain bandwidth of the microstrip grid array antenna decreased with the number of the radiating elements, which was similar to other resonant array antennas. The gain bandwidth is sensitive to the substrate's thickness. The thicker the substrate, the wider is the gain bandwidth.

7.5 Differential Excitation

Kraus proposed a differential feeding technique [1], which is illustrated by the microstrip grid array antenna with four loops in Figure 7.4a. It uses two wires passing through one opening in the ground plane to connect the source with the opened center of the radiating element. Zhang and his associate proposed another differential feeding technique [19], as shown in Figure 7.4b. It uses two coaxial probes, having their outer conductors connected to the ground plane and their inner conductors connected to the two ends of the radiating element. The novel feeding technique avoids opening the radiating element and results in a large physical separation between the two feeding points that eases the testing. The separation can be half-guided wavelength or farther apart, for example, one and half-guided wavelengths.

Figure 7.5 displays the simulated current distributions on the microstrip grid array antenna with 32 loops at the resonance. The current distribution by Kraus's differential excitation concentrates on the radiating elements near the feeding points, indicating a low aperture efficiency. The current distribution is improved by Zhang's differential excitation [19]. As a result, the impedance matching, gain, and gain bandwidth can be enhanced. The enhancement is mainly attributed to the novel placement of dual coaxial probes, which help maintain current phase

Figure 7.4 The differential feeding techniques proposed by Kraus using twin wires (a) and Zhang using two coaxial probes (b).

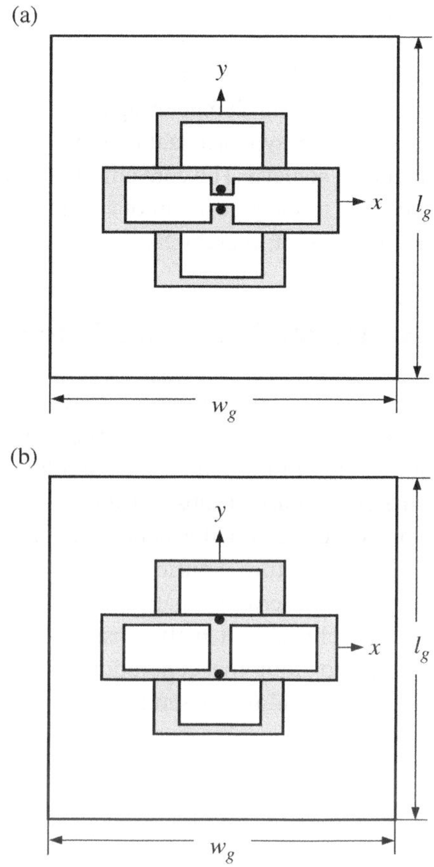

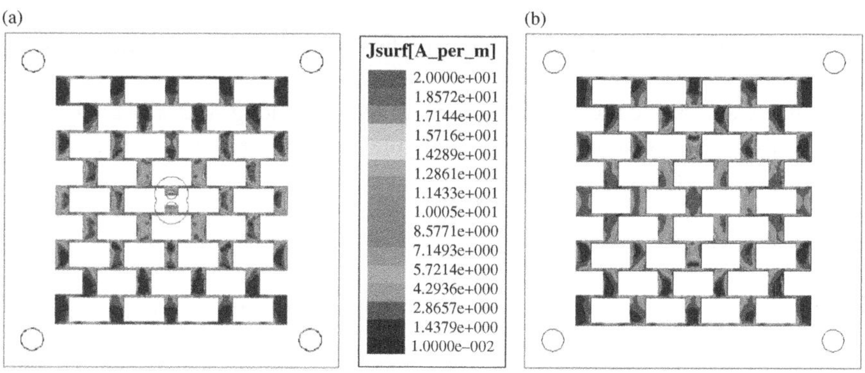

Figure 7.5 Simulated current distributions. (a) Kraus's and (b) Zhang's differential excitation.

synchronization on radiating elements. It is found that the currents are only truly out-of-phase on the long sides and in phase on the short sides of the loops near the feeding point because, at such a high frequency, a slight loop dimension change would cause a big change in signal phase over transmission, thus making the control over phase synchronization of the far loops from the feeding point much more difficult to achieve. The conventional placement of the twin wires in [1] could be thought of as the placement of a single feed, which cannot improve current phase synchronization, thereby having poorer performance.

7.6 Design Procedure, Formulas, and Examples

The design of a microstrip grid array antenna involves the choice of substrate type and size, the determination of the number of loops and feed location, and the optimization of the dimensions of a loop.

First, the choice of substrate for the array has, as with all microstrip structures, a strong influence on the bandwidth. As the feed is an inseparable part of the radiating structure, spurious feed radiation, which often limits the height of single-layer microstrip patch arrays, is not a design limitation for this array, although choosing a high substrate will however mean using a low dielectric constant to limit the generation of surface waves [15]. Hence, the height of substrate should be chosen as

$$h \leq \frac{\lambda}{4\pi\sqrt{\varepsilon_r}}. \tag{7.1}$$

We implement the square-shaped substrate for the pencil-beam radiation with the 3-dB beam width of Φ, and choose the dimensions to satisfy the angular resolution as:

$$w_g = l_g = 60^\circ \frac{\lambda}{\Phi}. \tag{7.2}$$

Similarly, we can implement the rectangle-shaped substrate for the fan-beam radiation.

Then, we determine the length l and width w_l for the long sides and the length s and width w_s for the short sides of rectangular loops. The resonance of the loop requires the length as

$$l = \frac{\lambda}{\sqrt{\varepsilon_e}} \tag{7.3}$$

where ε_e is the effective dielectric constant that can be estimated as

$$\varepsilon_e = \frac{\varepsilon_r + 1}{2} + \frac{\varepsilon_r - 1}{2}\left\{\left[1 + 12\frac{h}{w_l}\right]^{-\frac{1}{2}} + 0.02(\varepsilon_r - 1)\left[1 - \frac{w_l}{h}\right]^2\right\} \tag{7.4}$$

for $w_l/h < 1$ and the third term in (7.4) is zero for $w_l/h > 1$. It is known that the width w_l controls the loss of transmission lines and the radiation of cross-polar components. To lower them, as suggested in [15], we adopt the same width w_l for all long sides for $w_l/h < 1$ with the characteristic impedance as

$$Z_c = \frac{60}{\sqrt{\varepsilon_e}} \ln\left[\frac{8h}{w_l} + \frac{w_l}{4h}\right] \leq 200\,\Omega. \tag{7.5}$$

The resonance of the loop also requires the length s to be half of the length l. It is more involved to determine the widths w_s because the width w_s sets the array radiation coefficients with the first-order approximation of the coefficient being proportional to width [15]. For simplicity, we assign the same radiation coefficient for all radiating elements, leading to the same width w_s for all short sides. To enhance the radiation of co-polar components, the choice of w_s should be

$$w_s \leq \frac{\lambda}{4\sqrt{\varepsilon_e}}. \tag{7.6}$$

Next, we lay out as many rectangular loops as possible on the substrate to form the grid array radiator to achieve directivity. If the grid array radiator has the N number of radiating elements or short sides and the current is uniformly distributed on them, the directivity can be estimated by

$$D = x + 10\log N \tag{7.7}$$

where x in dB accounts for the directivity due to the ground plane and one radiating element.

Finally, we locate the two coaxial probes at the opposite T junctions and optimize the microstrip grid array antenna in the HFSS. The two coaxial probes should be located symmetrically with respect to the broadside direction, taking the differential operation into account.

In the following, we show how to design the microstrip grid array antenna for 24-GHz automotive radars to monitor blind spots. Based on the specifications as the directivity $D \geq 20$ dBi, the angular resolution of the single-shaped beam $\Phi \leq 15^{\circ}$, and the size $w_g \times l_g \leq 100 \times 100\ \text{mm}^2$, we start the design with an RT/Duroid 5880 substrate for its low dielectric constant $\varepsilon_r = 2.2$ and low loss tangent $\delta = 0.0009$. We choose the substrate thickness $h = 0.787$ mm for mechanical strength and other dimensions $w_g \times l_g \geq 4\lambda \times 4\lambda$ from equation (7.2). We decide the substrate size to be smaller than $100 \times 100\ \text{mm}^2$ as $w_g \times l_g = 4.8\lambda \times 4.8\lambda = 60 \times 60\ \text{mm}^2$. Considering realizability from fabrication and reliability for use in automobiles, we prefer $Z_c = 125\,\Omega$. Substituting equation (7.4) into (7.5), we get the width of the long side $w_l \approx 0.4$ mm, which leads to $\varepsilon_e \approx 1.7$ from (7.4) and the length of the long side $l \approx 9.6$ mm from (7.3). The length of the short side s is half of the length l, that is, $s \approx 4.8$ mm. The width of the short side is $w_s \approx 2.4$ mm from equation (7.6). We lay out 32 rectangular loops on the substrate to form the grid array radiator to achieve directivity. There are 41 short sides (or radiating elements). Equation (7.7) indicates that the directivity >20 dBi is achievable. We locate the two coaxial probes symmetrically with a separation of 5s and simulate the performance of the microstrip grid array antenna with the HFSS. In this design, the two SMA coaxial probes have Teflon ($\varepsilon_r = 2.1$) filling, inner diameter of 0.5 mm, and outer diameter of 2 mm. The simulations are conducted mainly to yield the optimized length l and width w_l for the long sides and length s and width w_s for the short sides of rectangular meshes. They are found to be $l = 9.8$ mm, $w_l = 0.4$ mm, $s = 4.9$ mm, and $w_s = 2.5$ mm, which are quite close to our calculated values.

The designed microstrip grid array antenna was fabricated. Figure 7.6 shows the photo of the fabricated microstrip grid array antenna. It has a size of $60 \times 60 \times 0.787\ \text{mm}^3$. There are four holes each with a radius of 2 mm at the four corners of the substrate reserved for stabilizing the radiator to flatten the whole antenna structure. The S-parameter measurements of the fabricated microstrip grid array antenna were made with a Rohde-Schwarz ZVA 67 four-port vector network analyzer, which can be used for differential and single-ended excitations of up to 67 GHz, respectively.

Figure 7.7a shows the simulated and measured $|S_{d11}|$ results for the microstrip grid array antenna under differential excitation at both feeds. The measured 10-dB impedance bandwidth is 1.17 GHz from 23.75 to 24.92 GHz (or 4.85% at 24.15 GHz) and the simulated 10-dB impedance bandwidth is 0.7639 GHz from 23.8894 to 24.6533 GHz (or 3.16% at 24.15 GHz). The difference between the simulated and measured $|S_{d11}|$ results is mainly caused by soldering. The irregular solder joints cannot be accurately modelled and were not included in simulations.

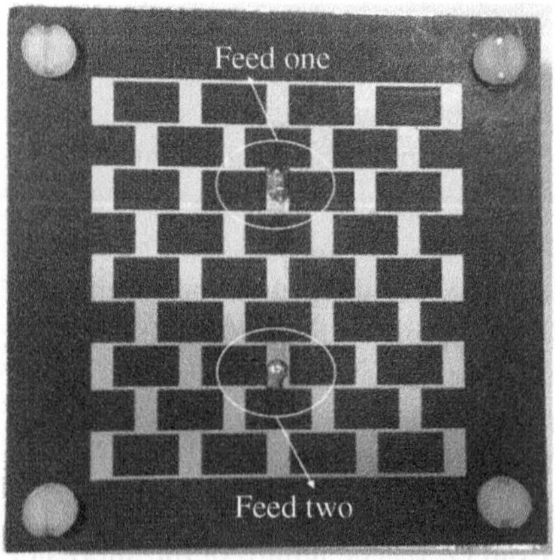

Figure 7.6 Photo of the microstrip grid array antenna.

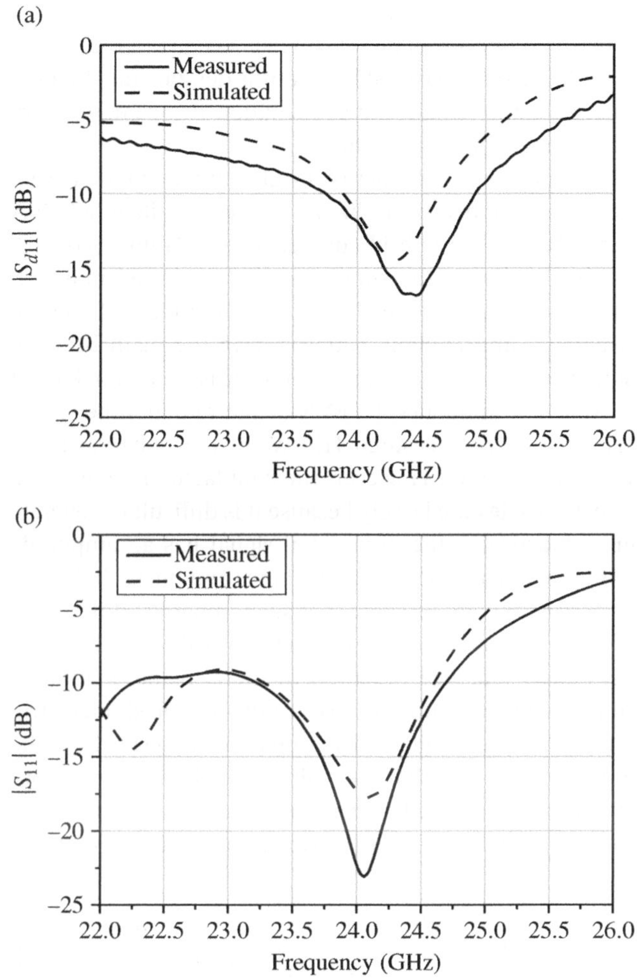

Figure 7.7 Simulated and measured results. (a) $|S_{d11}|$ under differential excitation and (b) $|S_{11}|$ under single-ended excitation.

Nevertheless, both simulated and measured $|S_{d11}|$ results demonstrate that the microstrip grid array antenna under differential excitation has achieved very good matching and enough impedance bandwidth. Figure 7.7b shows the simulated and measured $|S_{11}|$ results for the microstrip grid array antenna excitation at one feed and the other feed opened. The measured 10-dB impedance bandwidth is 1.46 GHz from 23.23 to 24.69 GHz (or 6.05% at 24.15 GHz) for the antenna excited at feed one. The simulated 10-dB impedance bandwidth is 1.3136 GHz from 23.2864 to

24.60 GHz (or 5.44% at 24.15 GHz) for the antenna excited at the feed. The slight difference between the measured and simulated $|S_{11}|$ results is due to the soldering. Nevertheless, both simulated and measured $|S_{11}|$ results demonstrate that the microstrip grid array antenna under single-ended excitation has achieved very good matching and enough impedance bandwidth.

The radiation patterns of the dual-feed microstrip grid array antenna were characterized in an anechoic chamber by utilizing a 2-port Agilent N5230A PNA-L network analyzer, a 180-degree hybrid coupler, and a Satimo SGH-820 ridged horn antenna. Figure 7.8 shows the simulated and measured radiation patterns of the microstrip grid array antenna at 24.15 GHz under differential excitation. As expected, both *E*- and *H*-plane patterns have the main beam of radiation in the broadside direction. The simulated and measured co-polarized radiation patterns agree well. The simulated HPBWs of the radiation beam are both 14° in the *E* and *H* planes. The measured HPBWs of the radiation beam are also both 14° in the *E* and *H* planes. However, the simulated and measured cross-polarized radiation patterns deviate largely because it is difficult to generate an ideal differential signal. Practically, the differential signal has an amplitude imbalance and a phase difference. The amplitude imbalance and phase difference can be up to 1 dB and 10°. Our simulations show that such imperfect differential excitation slightly tilts the main beam of radiation from the broadside direction in the *E* plane and enhances cross-polarized radiation in the *H* plane. Figure 7.9 shows the simulated and measured radiation patterns of the dual-feed microstrip grid array antenna at 24.15 GHz under single-ended excitation at one feed and the other feed opened. It is interesting to note that the main beam of radiation in both simulated and measured *E* planes is deviated 6° from the broadside direction. This is because the phase center of the dual-feed microstrip grid array antenna under

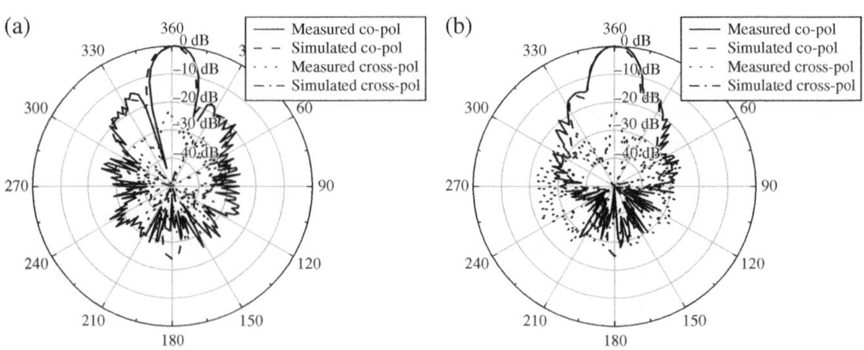

Figure 7.8 Simulated and measured radiation patterns of the dual-feed grid array at 24.15 GHz under differential excitation. (a) The *E* plane and (b) the *H* plane.

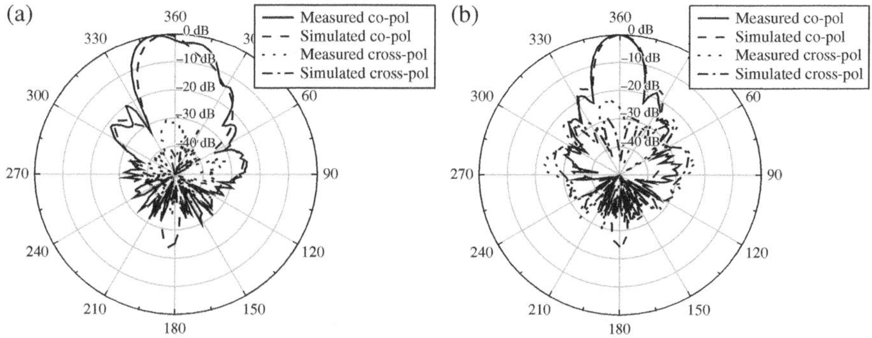

Figure 7.9 Simulated and measured radiation patterns of the dual-feed grid array at 24.15 GHz under single-ended excitation. (a) The *E* plane and (b) the *H* plane.

single-ended excitation is near the feed point rather than the geometrical center of the antenna. The simulated HPBWs of the radiation beam are 17° and 15°, respectively, in the *E* and *H* planes. The measured HPBWs of the radiation beam are 18° and 15°, respectively, in the *E* and *H* planes. Figures 7.8 and 7.9 clearly demonstrate that the dual-feed microstrip antenna radiates pencil-beam patterns under either differential or single-ended excitation.

Figure 7.10 shows the simulated and measured gain values of the microstrip grid array antenna under differential and single-ended excitations. Under differential operation, the measured highest gain is 21.0 dBi at 24.28 GHz and the measured 3-dB gain bandwidth is from 23.09 to 25.27 GHz (or 9.03% at 24.15 GHz), the simulated highest gain is 21.6 dBi at 24.25 GHz and the simulated 3-dB gain bandwidth is from 23.12 to 25.35 GHz (or 9.23% at 24.15 GHz). The simulations also show that the imperfect differential excitation slightly affects the gain. Under single-ended operation, the measured highest gain is 18.4 dBi at 24.48 GHz and the measured gain bandwidth is from 23.39 to 25.26 GHz (or 7.74% at 24.15 GHz), the simulated highest gain is 19.3 dBi at 24.25 GHz, and the simulated 3-dB gain bandwidth is from 23.60 to 25.15 GHz (or 6.42% at 24.15 GHz).

Figure 7.11 displays the simulated current distributions on the microstrip grid array radiator at 24.15 GHz under either differential or single-ended excitation. The higher gain and wider gain bandwidth under differential excitation are mainly attributed to the improved current distribution and phase synchronization on radiating elements. The simulated and measured aperture efficiencies of the dual-feed microstrip grid array antenna are 51% and 44% under differential operation and 32% and 24% under single-ended excitation.

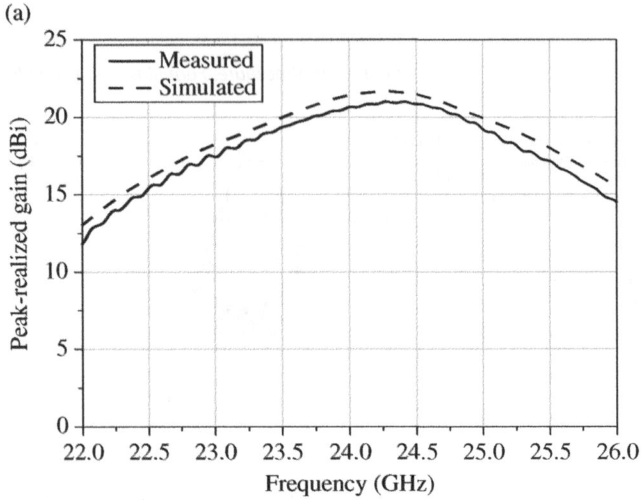

Figure 7.10 Simulated and measured gain values of the microstrip grid array antenna. (a) Under differential and (b) under single-ended excitations.

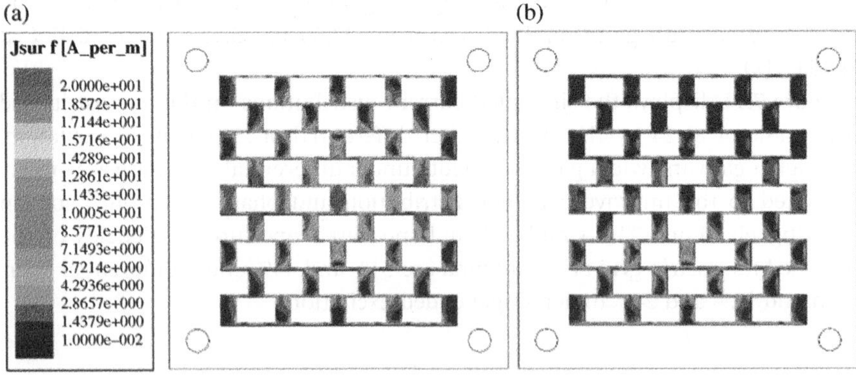

Figure 7.11 Simulated current distributions of the dual-feed microstrip grid array antenna. (a) For differential and (b) for single-ended excitations.

7.7 An Impedance Matching Technique

The microstrip grid array antenna is an interconnected array concept. The input impedance at a feed of the microstrip grid array antenna for the first-order approximation can be expressed as R/n where R is the radiation resistance of a radiating element and n is the number of radiating elements. Obviously, as n increases, the input impedance may decrease to a value much smaller than the source impedance, leading to poor matching. To improve impedance matching, an effective technique based on the stepped-impedance lines as shown in Figure 7.12 is devised as follows [20].

Note that the microstrip grid array antenna is symmetrical about the y-axis. One only needs to consider the half structure. The impedance at the feed node is Z_{in}. The feed node is denoted by a black dot. The impedance at a virtual node is Z_L. The virtual node is denoted by a small circle. Z_L is equal to Z_{in} because the microstrip grid array antenna is resonant; all the node impedances must be essentially identical as they are all a half wavelength apart in the simple transmission line model.

For the half structure, the stepped-impedance line is used to match Z_L to Z_{in}. The impedance Z_{in} should be two times the source impedance of 50 Ω. The stepped-impedance line consists of two sections. For simplicity, we assign the width of the first section w_1 to be the same as the width of the long side of the loop w_l. Hence, its characteristic impedance Z_1 is known. The width of the second section w_2 should be less than $\lambda_g/4$ to suppress higher-order modes. The

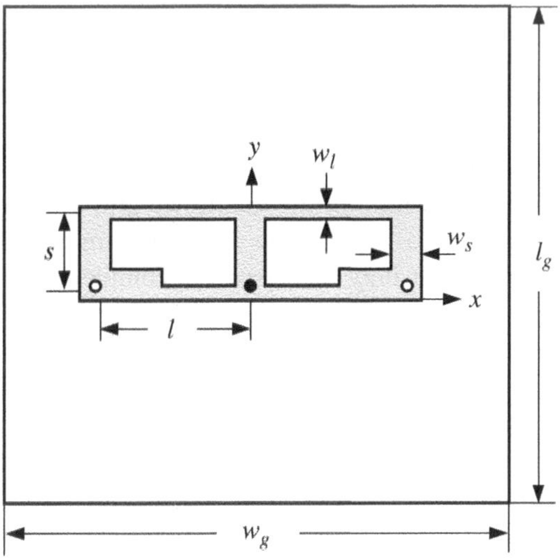

Figure 7.12 A microstrip grid array antenna model with stepped-impedance lines.

characteristic impedance of the second section is Z_2, which should be smaller than Z_1. The lengths of the first and second sections are denoted as l_1 and l_2, respectively, and $l_1 + l_2 = l$.

Based on the stepped impedance line in Figure 7.12, a large microstrip grid array antenna that has 128 loops was simulated [19]. The microstrip grid array antenna was printed on a Rogers RT Duroid 5880 substrate with a size of $w_g \times l_g \times h = 100 \times 100 \times 0.381\ \text{mm}^3$. The loop has $s = 4.48$ mm, $w_s = 1.5$ mm, $l = 9.68$ mm, and $w_l = 0.4$ mm. Figure 7.13a shows the simulated $|S_{11}|$ of the microstrip grid array antenna excited at the feed near the geometrical center versus l_2 with a fixed $w_2 = 1.2$ mm. Note that $l_2 = 2.5$ mm satisfies the impedance matching and bandwidth requirements. Figure 7.13b shows the simulated $|S_{11}|$ of the microstrip grid array antenna excited at the feed versus w_2 with a fixed $l_2 = 2.5$ mm. Note that as the w_2 increases from 0.4 to 1 mm, the $|S_{11}|$ value drops from −7 to −11 dB, and the resonant frequency remains unchanged. Consequently, in the optimization of the stepped impedance line, the resonant frequency can be adjusted with the length of l_2 and the matching is controlled by the width of w_2. Due to the interconnection between ports in the proposed structure, adding a stepped impedance line to one feed affects the input impedance at the other feed. However, it is found that this effect is not critical which can be alleviated by slightly modifying the dimensions of the stepped-impedance line.

7.8 A Slow-Wave Structure

As stated before, the long side of the microstrip grid array antenna acts mainly as a transmission line with one guided wavelength. A meander line was explored to replace a conventional line, resulting in a miniaturized microstrip grid array antenna. The miniaturized microstrip grid array antenna was 38% smaller but at the expense of a gain reduction of 2 dB [5]. A sinusoid line was optimized for the miniaturization of the microstrip grid array antenna. A 47% miniaturization was achieved. However, the addition of an air layer for an enhanced impedance bandwidth not only increased the complexity in fabrication but also decreased the reliability of the microstrip grid array antenna [6]. It should be noted that both meander and sinusoid lines actually increase the physical lengths of the long sides of the loops to miniaturize the microstrip grid array antennas. Recently, a slow-wave structure was investigated, which increases the propagation constant rather than the physical length of the line, for the miniaturization of microstrip grid array antennas [21].

Figure 7.14 shows the geometry of the slow-wave structure. It consists of a main-line and periodic stubs. The main line has a width of w_l and a length of l. The stubs

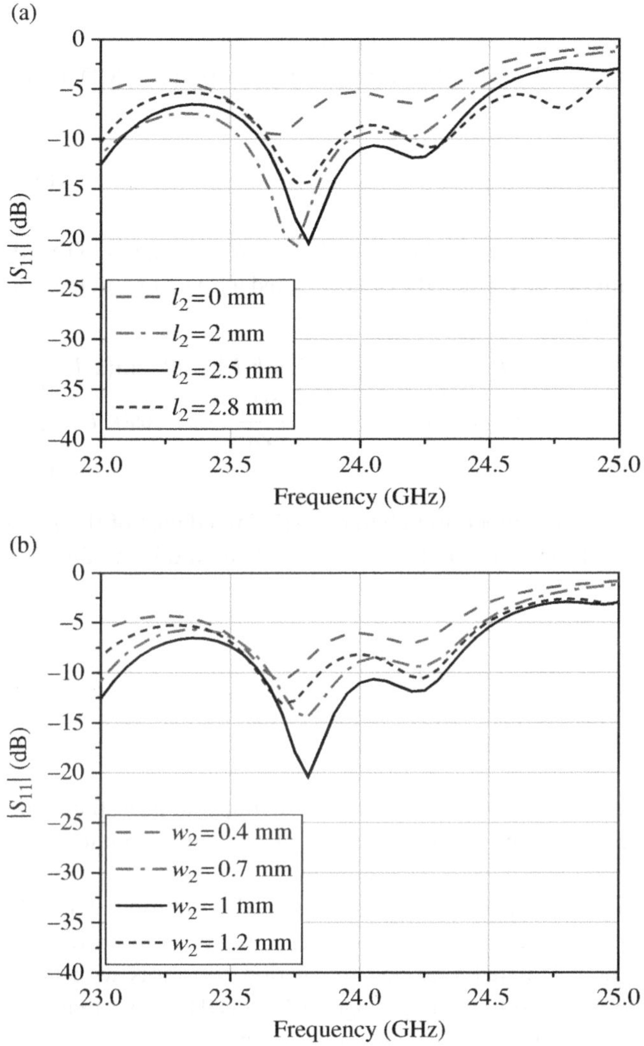

Figure 7.13 Simulated $|S_{11}|$ of the microstrip grid array antenna excited the feed near the geometrical center versus (a) l_2 and (b) w_2.

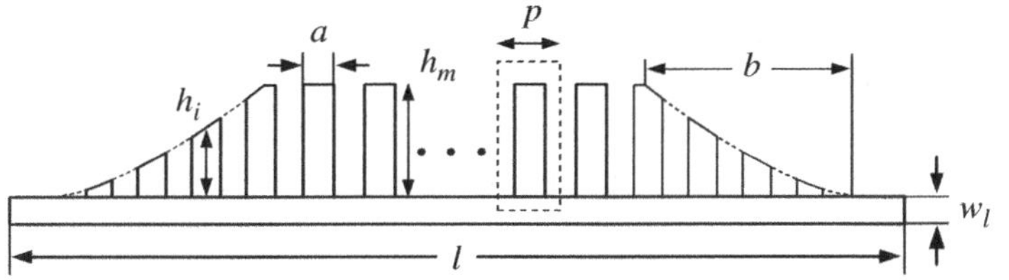

Figure 7.14 Geometry of the slow-wave structure.

at both ends are tapered exponentially, with a gradient length b. The stubs have a period of p, a width of a, and a maximum height of h_m [22].

The slow-wave structure can be treated as a cascaded network. Its ABCD matrix M can be written as

$$M = M_t\left(\prod_{i=1}^{N} M_i\right)M_t \tag{7.8}$$

where M_t is the ABCD matrix of the main line and M_i is the ABCD matrix of the ith unit. Note that the ith unit can be modelled as a line loaded with an open-circuited stub. Its M_i can be expressed as

$$M_i = \begin{bmatrix} \cos(k_1 l_1) & jZ_1\sin(k_1 l_1) \\ j\dfrac{\sin(k_1 l_1)}{Z_1} & \cos(k_1 l_1) \end{bmatrix} \begin{bmatrix} 1 & 0 \\ Y_i & 1 \end{bmatrix} \begin{bmatrix} \cos(k_1 l_1) & jZ_1\sin(k_1 l_1) \\ j\dfrac{\sin(k_1 l_1)}{Z_1} & \cos(k_1 l_1) \end{bmatrix} \tag{7.9}$$

where k_1 and Z_1 are the wavenumber and characteristic impedance of the main line, respectively. Note that the equivalent length of line l_1 should be modified as

$$l_1 = \frac{a}{2} + 0.325(p - a). \tag{7.10}$$

The admittance of the ith stub is given by

$$Y_i = j\frac{\tan(k_2 l_2)}{Z_2} \tag{7.11}$$

where k_2 and Z_2 are the wavenumber and characteristic impedance of the open-circuited branch, respectively, and h_i is the length at the center point of the ith stub.

Consider the slow-wave structure printed on a grounded RT/duriod 5880 substrate with $w = 0.\ 2$ mm, $l = 6.2$ mm, $p = 0.2$ mm, $a = 0.25$ mm, $h_m = 1.2$ mm. Figure 7.15 compares the calculated average propagation constants with those simulated from the HFSS for the frequency range of 20–36 GHz. Note that they are in good agreement. In addition, the average propagation constants for the normal line with the same length l are also provided for reference. Apparently, good slow-wave characteristics are obtained in the whole frequency band for the slow-wave line. It is interesting to note that as the gradient length b decreases from 2.6 to 1.8 mm, the average propagation constant k increases. The larger k determines the smaller dimension at the same frequency, and we can choose an appropriate value of k by varying the gradient length b to suit the need.

Figure 7.16 shows the geometry of the microstrip grid array antenna with five loops using the slow-wave lines. The optimized dimensions at 28 GHz on the RT/Duroid 5880 substrate are given in Table 7.1. Note that the length of the

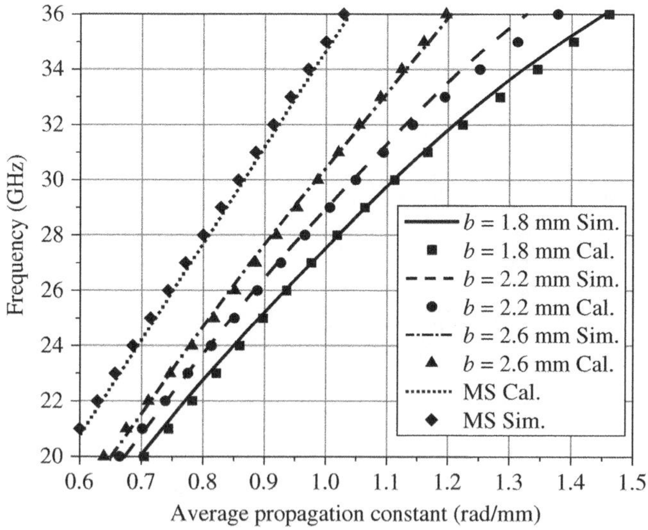

Figure 7.15 Average propagation constant as a function of frequency.

Figure 7.16 Geometry of the microstrip grid array antenna using the slow-wave lines.

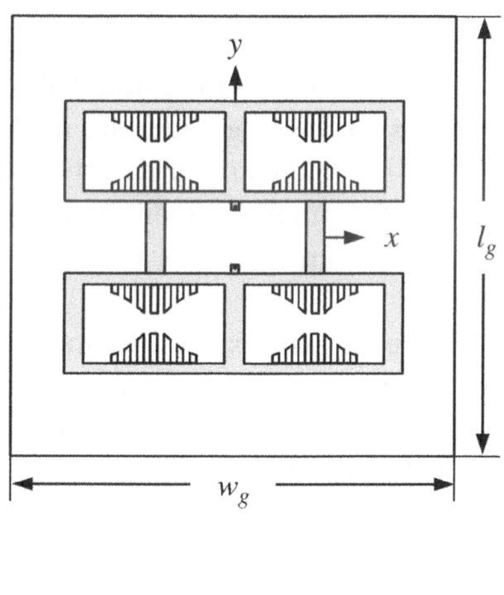

Table 7.1 Dimensions in mm.

w_g	l_g	h	l	w_l	s	w_s	h_s
20	20	0.787	7.14	0.4	4	0.7	1.2

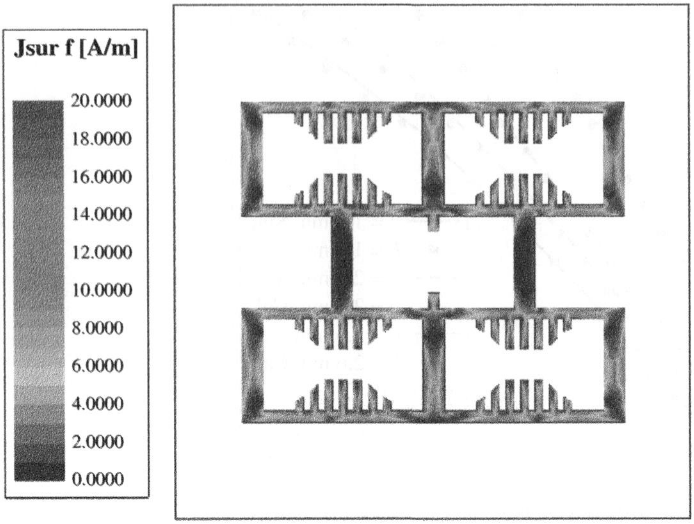

Figure 7.17 Simulated current distribution of the microstrip grid array antenna at 28 GHz.

slow-wave lines is 7.14 mm, which is shorter than the length of the normal lines by 1.86 mm, indicating miniaturization of 21%.

Figure 7.17 shows the simulated current distribution at 28 GHz. As expected, the currents on the short sides are in the same direction, indicating the proper operation of the microstrip grid array antenna. The stubs alter the current distribution, causing the suppression on the short sides and the enhancement on the stubs. The radiation from the stubs will affect the radiation from the short sides to some extent.

Figure 7.18 compares the simulated $|S_{d11}|$ values as a function of frequency for the microstrip grid array antenna using the slow-wave structure (SGAA) and the normal line (NGAA). The fractional impedance bandwidth is 4.6% at 28 GHz, which is much broader than that (2.2%) of the microstrip grid array antenna using the normal line.

Figure 7.19 compares the simulated gain values as a function of frequency for the microstrip grid array antenna using the slow-wave structure and the normal line. The realized gain is 15.0 dBi at 28 GHz, which is slightly lower than that (15.5 dBi) of the microstrip grid array antenna using the normal line. However, the 3-dB gain bandwidth of the microstrip grid array antenna using the slow-wave structure is 10.3%, which is wider than that (5%) of the microstrip grid array antennas using the normal line.

Figure 7.20 compares the simulated radiation patterns at 28 GHz for the microstrip grid array antenna using the slow-wave structure and the normal line.

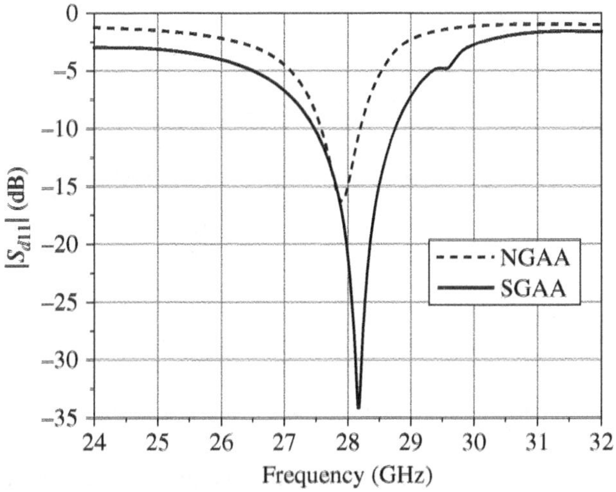

Figure 7.18 Simulated $|S_{d11}|$ as a function of frequency.

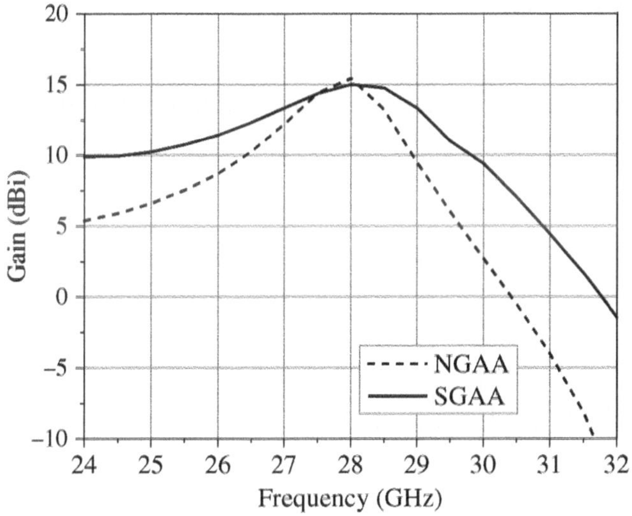

Figure 7.19 Simulated peak realized gain as a function of frequency.

The beamwidth and side lobe level of the microstrip grid array antenna using the slow-wave structure are slightly narrower and higher in the E plane but are slightly wider and lower in the H plane as compared with those of the microstrip grid array antenna using the normal line. These variations are mainly caused by the

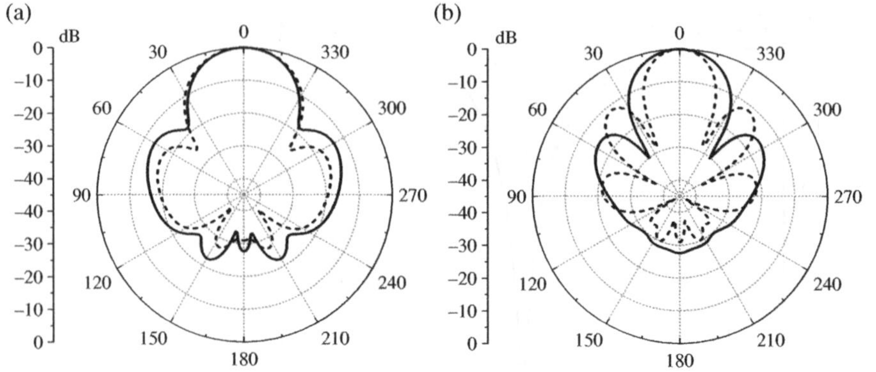

Figure 7.20 Simulated radiation patterns at 28 GHz. (a) The E plane and (b) the H plane.

influence of the loaded stubs, which could be adjusted by changing the length of stubs and gradient transition.

7.9 Dual Current Modes

A microstrip grid array antenna can be designed to operate in dual current modes at high and low frequencies [23, 24]. Figure 7.21 shows the geometry of such a microstrip grid array antenna. Note that the loops are arranged so that the outline

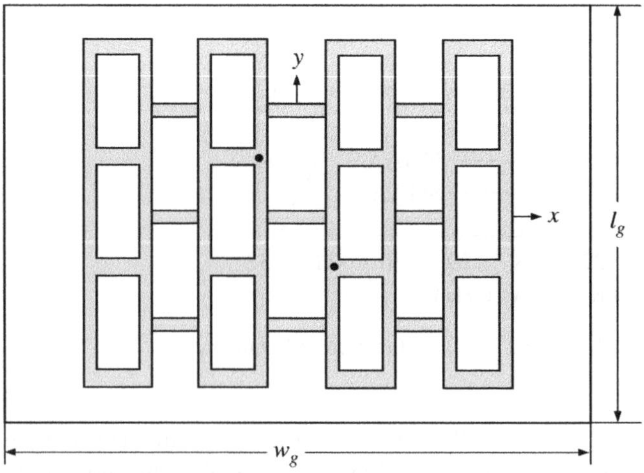

Figure 7.21 Geometry of the microstrip grid array antenna that can operate in the dual current modes.

of the microstrip grid array antenna can be approximated as a rectangle. The locations of the feeds are denoted by the black dots. At the high-frequency f_h, it operates just like a normal microstrip grid array antenna, while at the low-frequency f_l, it operates as a meshed microstrip patch antenna does. Based on the current distributions, the high-frequency f_h and the low-frequency f_l can be calculated as

$$f_h \approx \frac{c}{l\sqrt{\varepsilon_e}} \approx \frac{c}{2s\sqrt{\varepsilon_e}}, \tag{7.12}$$

$$f_l \approx \frac{c}{2L\sqrt{\varepsilon_e}} \tag{7.13}$$

where L is the equivalent patch length, which can be estimated as

$$L \approx 3l + s \tag{7.14}$$

for this specific microstrip grid array antenna. It should be clear that the low resonant frequency f_l decreases obviously as the number of long sides increases, while the high resonant frequency f_h shifts up slightly as the number of long sides increases, which is due to the reduced effect of fringing field. Based on (7.12) and (7.13), the ratio of f_h to f_l can be found as

$$\frac{f_h}{f_l} \approx \frac{L}{s}, \tag{7.15}$$

which is useful to estimate the initial values for f_h and f_l to start the design.

Let us consider the microstrip grid array antenna printed on an RT Duroid 5880 substrate with a size of $w_g \times l_g \times h = 70 \times 70 \times 0.787\ \text{mm}^3$. The loop has $s = 4.4$ mm, $l = 8.2$ mm, and $w_s = w_l = 0.8$ mm. The microstrip grid array antenna is excited for the differential operation at $f_h = 28$ and $f_l = 3.7$ GHz. The simulated impedance bandwidths are 2.52% (27.73–28.44 GHz) and 0.59% (3.674–3.696 GHz), respectively. The impedance bandwidths are narrow at both frequencies because the substrate is electrically thin, especially at 3.7 GHz. The simulated gain values are 20 dBi at 28 GHz and 8 dBi at 3.7 GHz. The gain at 3.7 GHz is almost the same as that of the normal microstrip patch antenna.

7.10 Co-Aperture Dual Bands

The empty areas of the loops of a microstrip grid array antenna can be utilized to house other types of antenna elements to realize co-aperture antenna arrays. Figure 7.22 shows the basic microstrip grid array antenna with each loop containing two microstrip patches. The dimensions of the patch are l_p and w_p. The separation between the two patches in the loop is s_p. The separation between the long

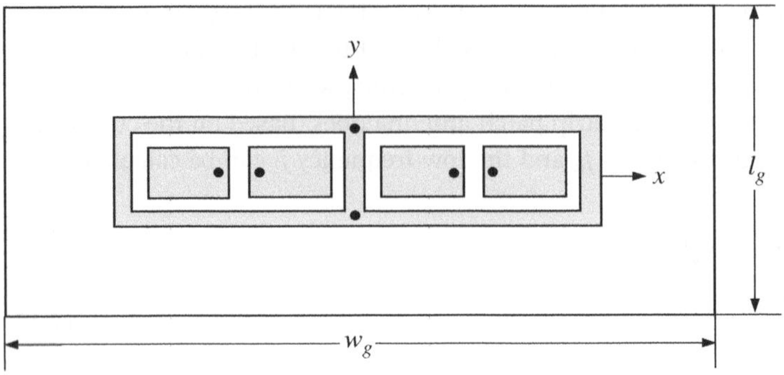

Figure 7.22 Geometry of co-aperture microstrip grid and patch array antennas.

Table 7.2 Dimensions in mm.

w_g	l_g	h	l	w_l	s	w_s
28	32	0.787	9.4	0.8	0.8	0.8
l_p	w_p	s_p	s_x	p_f	g_f	
2.45	2.1	1.75	0.8	0.625	3	

sides of the patch and loop is s_y. The separation between the short sides of the patch and loop is s_x.

Xu et al. reported such a design that embedded 16 microstrip patches in the 8 loops of the microstrip grid array antenna on an RT/Duroid 5880 substrate [25]. Table 7.2 lists the optimized dimensions of the co-aperture microstrip grid and patch array antennas to operate at 28 and 38 GHz, respectively.

Figures 7.23 and 7.24 show the simulated electric field distributions on the co-aperture structure at 28 and 38 GHz, respectively. It can be seen that the field strength of the microstrip patch antenna element at the center of the narrow side is very weak. When the feed point of the microstrip patch antenna element is at the center of the narrow side, the coupling from the microstrip grid array antenna at 28 GHz will be small and the isolation will be improved. On the other hand, when the microstrip patch array antenna is under differential excitation at 38 GHz, it is seen that the electric field at the two feed points of the microstrip grid array antenna is weak. Therefore, the isolation of the microstrip grid and patch array antennas in the co-aperture configuration is guaranteed due to polarization orthogonality.

Figure 7.25 shows the simulated $|S_{d11}|$ of the microstrip grid array antenna with the microstrip patch elements as a function of frequency. Two resonant

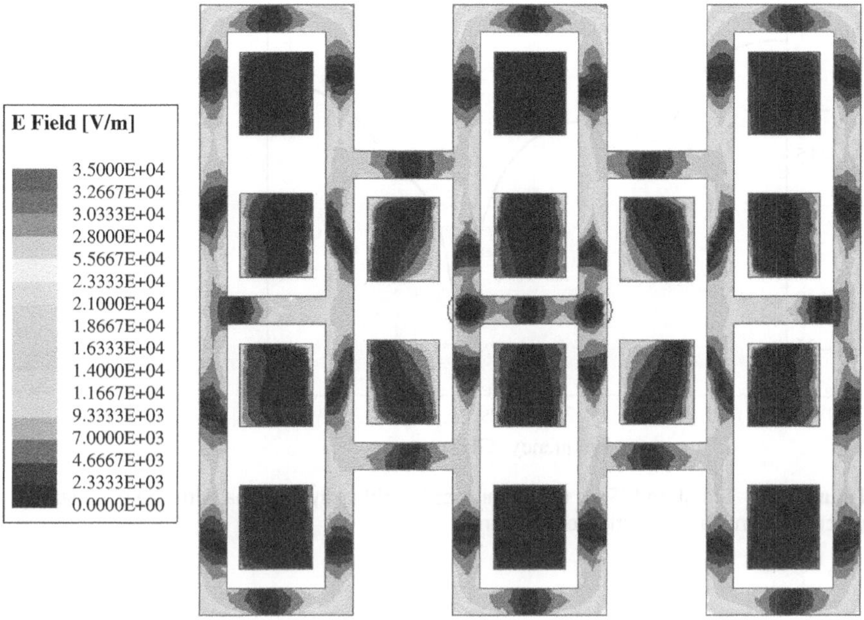

Figure 7.23 Simulated electric field distribution of the microstrip grid array antenna fed by the differential signal at 28 GHz.

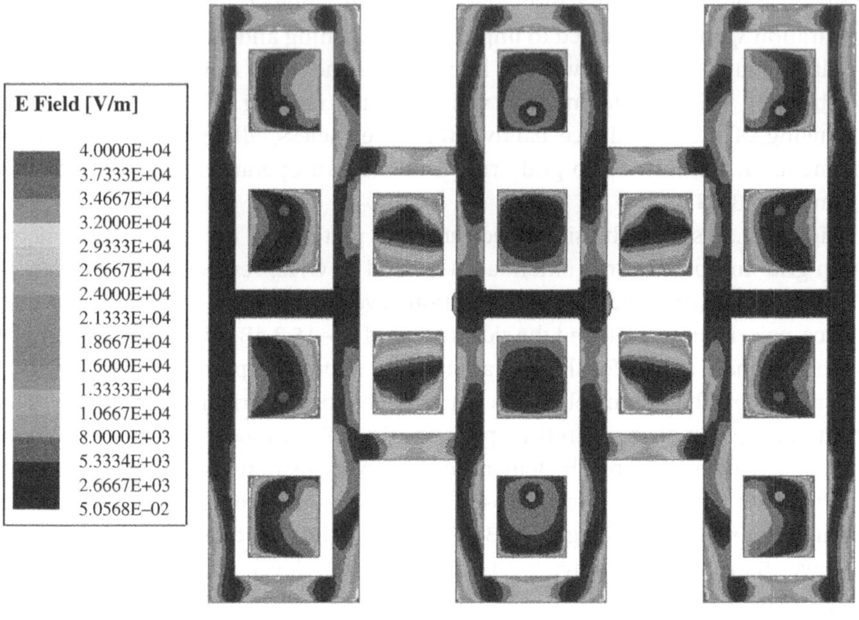

Figure 7.24 Simulated electric field distribution of the patch antenna array fed by the differential signal at 38 GHz.

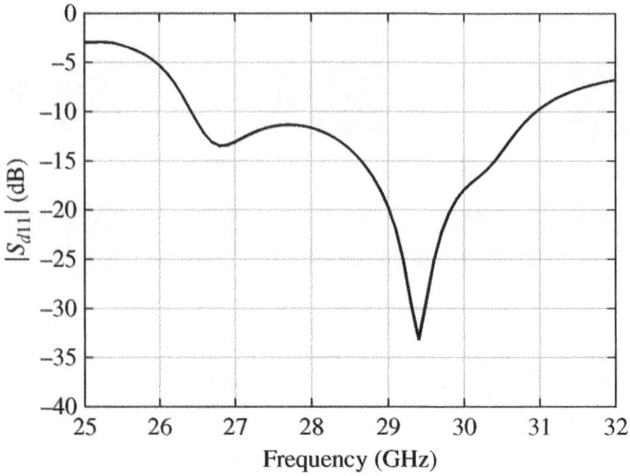

Figure 7.25 Simulated $|S_{d11}|$ of the microstrip grid array antenna with the microstrip patch elements as a function of frequency.

frequencies at 27 and 29.5 GHz can be seen. It is the loop that accounts for the resonance at 27 GHz, while it is the coupling between the loop and patch that creates the resonance at 29.5 GHz. The dual resonances make the impedance bandwidth much wider. It is found that the patch width w_p controls this resonance. When w_p gets larger, the resonance moves to a lower frequency. The patch separation s_p can be adjusted to improve the matching and bandwidth. The patch itself resonates at 38 GHz with an impedance bandwidth from 35.2 to 41.2 GHz. Embedding it in a different loop will cause the shifting of the resonance and changing of the impedance bandwidth. Nevertheless, all the microstrip patch elements in the microstrip grid array antenna can operate over the bandwidth from 37 to 39 GHz.

Figures 7.26 shows the simulated radiation patterns of the co-aperture microstrip grid and patch array antennas at 28 and 38 GHz, respectively. The gain at 28 GHz is 16.4 dBi, the cross-polarization level is less than −50 dB relative to the co-polarization level, and the side lobe level is −15.2 dB. At 38 GHz, the amplitude and phase are set for the excitation of the microstrip patch element. The two microstrip patch antenna elements in the same loop are driven differentially. The microstrip patch element in the upper position of each loop is excited with a 0° phase. The microstrip patch element in the lower position of each loop is excited with a 180° phase. All the microstrip patch antenna elements have the same excitation amplitude. The gain at 38 GHz is 16.9 dBi, the cross-polarization level is lower than −55 dB relative to the co-polarization level, and the side lobe level is −11.3 dB.

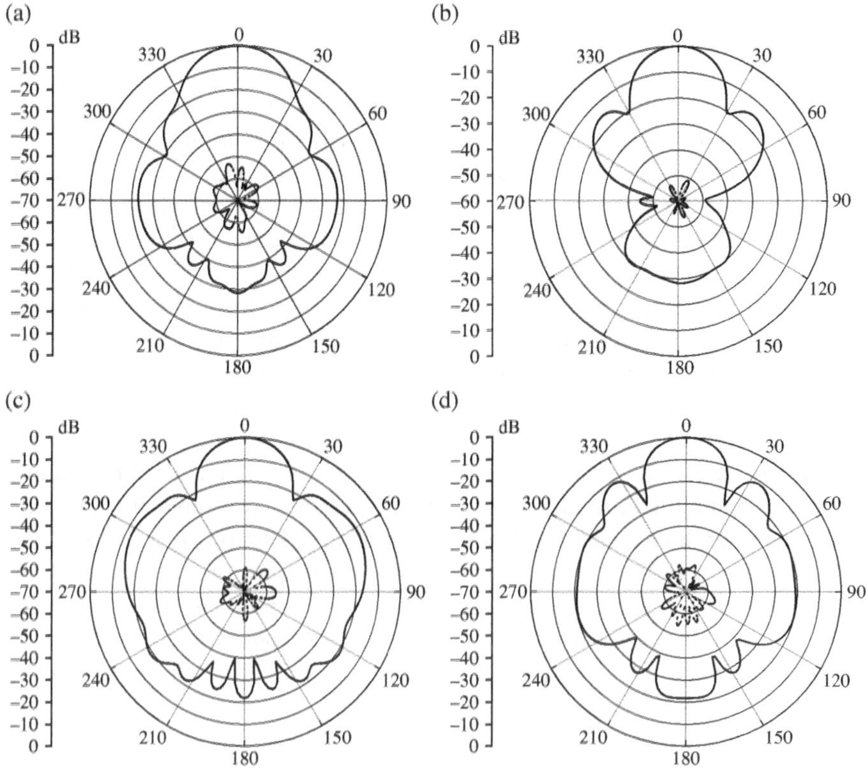

Figure 7.26 Simulated radiation patterns of the co-aperture microstrip grid and patch array antennas. (a) The *E* and (b) *H* planes for the microstrip grid array antenna at 28 GHz, (c) the *E* and (d) *H* planes for the microstrip patch array antenna at 38 GHz.

7.11 Applications

The microstrip grid array antenna provides a low-cost lightweight structure with sufficient design flexibility to be useful in a variety of applications. Adequate gain, bandwidth, cross-polarization quality, and sidelobe performance can be achieved for planar as well as linear-type arrays [3]. In the following, we illustrate the applications of differential microstrip grid array antennas in 60-GHz radios, 79-GHz radars, 24-GHz full-duplex and MIMO systems, and 2.45-GHz energy harvester.

7.11.1 60-GHz Radios

The great Indian scientist Sir Jagadish Chandra Bose was the first man who pioneered 60-GHz technology, using it to ring a bell or to ignite gunpowder remotely. A century later, the spectrum in the 60-GHz band was proposed for use in microcellular network [26]. It was realized in 1995 that unlicensed use could be an appropriate regime for such a spectrum since most of the justifications for radio licensing were not applicable to these frequencies. Japan first issued a 60-GHz regulation for unlicensed utilization in the 60-GHz band in the year 2000. The year 2007 began a new phase for 60-GHz technology, as the IEEE initiated a new standard for the unlicensed 60-GHz band, and many companies became involved in the development of 60-GHz radio chipsets in SiGe or CMOS technologies. It is known that CMOS scaling improves amplifier noise performance and gain but exacerbates the difficulty of generating sufficient output power by the power amplifier (PA) at 60 GHz [12]. Theoretically, a differential PA yields 3 dB more output power than a single-end one does. Hence, the differential antenna is preferred to the single-end antenna in highly integrated 60-GHz radio transmitters [27–33].

Figure 7.27 shows the photo of dual microstrip grid array antennas in a chip package for the IBM first-generation 60-GHz radio chipsets in LTCC technology. There are three cofired laminated ceramic layers for the package. The first ceramic layer is 0.385 mm thick, and the second to third ceramic layers are both 0.095 mm thick. There are four metallic layers for the package. The top layer provides the metallization for the dual microstrip grid array antennas, the first buried layer metallization for the patterned ground plane, the second buried layer the metallization for the antenna feeding traces, and the bottom exposed layer the metallization for the signal traces. The size of the whole package is $13.5 \times 13.5 \times 0.575\ \text{mm}^3$.

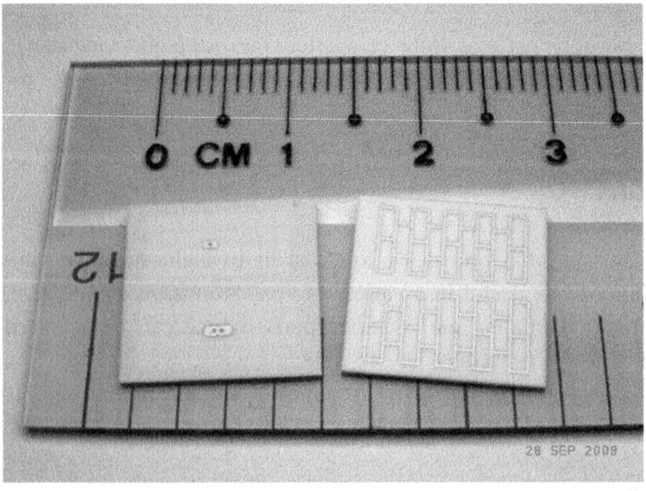

Figure 7.27 Photo of the dual microstrip grid array antennas in the package with one for differential transmitter and the other for a single-ended receiver of the IBM 60-GHz radio chipsets.

The measured results showed that the package achieved good antenna performance in the 60-GHz band with a maximum gain of 13.5 and 14.5 dBi for the single-ended and differential antennas, respectively.

7.11.2 79-GHz Radars

In recent years, with respect to automotive radars, the frequency band between 77 and 81 GHz has been allocated for medium-range and short-range applications, such as collision warning, blind spot detection, and precrash vehicle preparation. The rapid development of highly integrated multichannel transceiver chips SiGe or CMOS technologies enables the realization of compact and low-cost radars, which are a prerequisite for mass-market adoption. A radar sensor based on LTCC has, compared to traditional soft substrate materials, better mechanical stability, and the advantage of higher thermal conductivity. Figure 7.28 shows the photos of a short-range radar sensor built using a four-channel fully differential transceiver chip in SiGe technology and four single-row and four double-row microstrip grid array antennas in LTCC technology [8]. The whole radar module is compact with a size of $23 \times 23 \times 1.416\ mm^3$. The microstrip grid array antennas were fed differentially with laminated waveguides. The measured 3-dB beamwidth at 79 GHz for a single row is 16° in the E plane and 58° in the H plane. The measured 3-dB beamwidth at 79 GHz for a double row is 16.5° in the E plane and 40° in the H plane. The calibrated measured gain taking into account the loss due to the feeding line and transition is 12.4 dBi for the single-row antenna and 14 dBi for the double-row grid antenna. The simulated directivity of both antennas is 15.2 and 17.4 dBi, respectively. The reduced gain was mainly due to high dielectric loss of the ceramic substrate.

(a) (b)

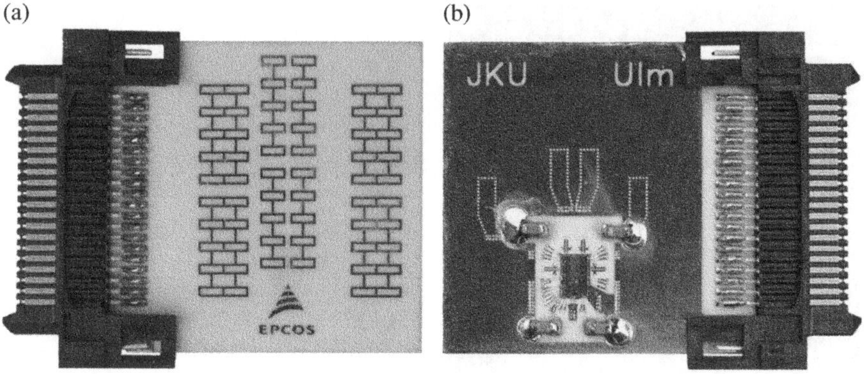

Figure 7.28 Photos of the differential microstrip grid array antennas in LTCC for 79-GHz radars. (a) Top and (b) bottom views. *Source:* Bauer et al. [8]/with permission of IEEE.

7.11.3 Full Duplex Radios and MIMO Systems

Full duplex radios have the potential to double the spectral efficiency of wireless communication systems. A key challenge for full duplex radios is self-interference suppression [34]. Prior full duplex radio designs mainly use multiple antenna techniques [35], which can achieve up to 40 dB of isolation but result in bulky size due to the required physical distances between the antennas. An alternative technique uses a single antenna with a circulator for simultaneous transmission and reception [36]. Unfortunately, the circulator provides a moderate isolation of 20 dB. A third technique uses a multiport antenna, for example, Chen et al. demonstrated that a microstrip grid array structure can be configured into one or multiple antennas for full duplex radios [20]. Figure 7.29 shows the photos of the fabricated multiport microstrip grid array structure with four ports. It has 145 radiating elements on an RT/duroid 5880 substrate with a size of $100 \times 100 \times 0.381\ \text{mm}^3$.

For the structure excited at port 2 as a single-ended antenna, the simulated and measured 10-dB input impedance bandwidths are 0.38 GHz (or 1.57% at 24.15 GHz) and 0.33 GHz (or 1.37% at 24.15 GHz), respectively. The measured maximum gain is 23.26 dBi at 24.18 GHz with the 3-dB gain bandwidth of 0.80 GHz from 23.74 to 24.54 GHz (or 3.31% at 24.15 GHz). The simulated maximum gain is 23.58 dBi at 24.15 GHz with the 3-dB gain bandwidth of 2.52 GHz from 22 to 24.52 GHz (or 10.43% at 24.15 GHz). For the structure excited at ports 2 and 3 as a differential antenna, the simulated and measured 10-dB input impedance bandwidths are 0.33 GHz (or 1.37% at 24.15 GHz) and 0.30 GHz (1.24% at 24.15 GHz), respectively. The measured maximum gain is 26.84 dBi at 24.17 GHz with a 3-dB gain bandwidth of 0.70 GHz from 23.82 to

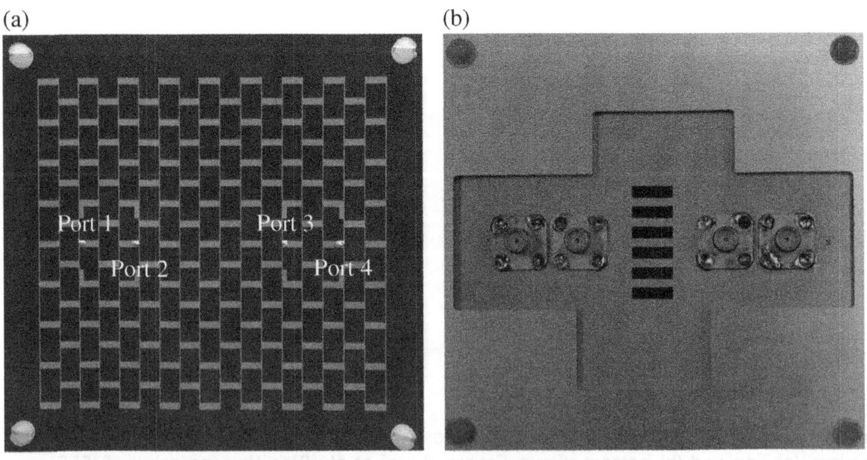

Figure 7.29 Photos of the fabricated multiport microstrip grid array structure with four ports. (a) Top and (b) bottom views.

24.52 GHz (or 2.90% at 24.15 GHz). The simulated maximum gain is 26.39 dBi at 24.15 GHz with the 3-dB gain bandwidth of 0.70 GHz from 23.80 to 24.50 GHz (or 2.90% at 24.15 GHz).

For the structure, the large physical distance between ports 1 and 4 guarantees a high isolation between the two ports, which enables the structure to be used for simultaneous transmission and reception as well as for MIMO application. When the structure was transmitting at port 1 and receiving at port 4, the measured 10-dB impedance bandwidth was 0.24 GHz from 24.17 to 24.41 GHz (or 1% at 24.15 GHz) and 0.37 GHz from 24.12 to 24.49 GHz (or 1.5% at 24.15 GHz) for transmission and reception, respectively. The measured isolation between two ports at 24.15 GHz was 34.24 dB, which is about 10 dB higher than the typical isolation level of a waveguide-based circulator. The simulated and measured peak realized gains at transmitting port 1 and receiving port 4 had small differences in the 24-GHz band from 24.05 to 24.25 GHz. Thus, together with the adaptive leakage cancellation in the receiver, we believe that the structure can function as two single-ended antennas with one for transmission and the other for reception for frequency-modulated continuous wave radar and sensor applications in the band [17, 18].

Popular metrics for MIMO antennas are the ratio of the mean effective gains and the envelope correlation. Usually, the ratio should be smaller than 3 dB and the envelope correlation coefficient should be smaller than 0.5 to ensure good channel characteristics [20]. The simulated and measured ratios of the mean effective gains for the structure at ports 1 and 4 were smaller than 1 dB in the 24-GHz band. The simulated and measured envelope correlation coefficients for the structure at ports 1 and 4 were smaller than −60 dB in the 24-GHz band.

For the structure, ports 1 and 2 can be configured as one differential antenna for transmission and ports 3 and 4 as another differential antenna for reception or as differential antennas for a 2×2 MIMO system. The simulated impedance bandwidths for both transmitting and receiving antennas were 0.40 GHz from 24.15 to 24.55 GHz (or 1.66% at 24.15 GHz). The measured impedance bandwidths for the transmitting and receiving antennas were 0.52 GHz from 24.07 to 24.59 GHz (or 2.15% at 24.15 GHz) and 0.64 GHz from 24.12 to 24.75 GHz (or 2.65% at 24.15 GHz), respectively. The simulated and measured isolations between differential ports 1 and 2 at 24.15 GHz were 32.80 and 34.13 dB, respectively. The simulated and measured gains were different due to the loads connected to the antenna, which were not exactly 50 Ω in the 24-GHz band. For both antennas, the desired broadside radiation patterns and low cross-polarization components were observed.

The ratio of the mean effective gains for the structure at differential ports 1 and 2 was found to be smaller than 0.5 dB in 24-GHz band. The envelope correlation coefficient for the structure at differential ports 1 and 2 was found to be smaller than −50 dB in the 24-GHz band.

7.11.4 RF Energy Harvester and Wireless Power Transfer

RF energy harvesting converts energy from the ambient electromagnetic wave emitted from widespread RF transmitters such as indoor Wi-Fi routers and outdoor base stations into DC voltages. A basic RF energy harvester consists of an antenna and a rectifying circuit. For the antenna, it requires a wide beam and high gain. With one antenna port, the gain and the beamwidth always conflict. Thus, the multiport antenna that uses the same antenna aperture for multiple beams could be a promising solution for RF energy harvesting. The multiport and multibeam antenna could be equivalent to multiple antennas of different high-gain beams. Hu et al. proposed to use a differential microstrip grid array antenna to implement this idea [14].

Figure 7.30 shows the photos of a fabricated RF energy harvester on an FR4 substrate of $w_g \times l_g = 337 \times 226\,\text{mm}^2$. To obtain a stable design, the rectifying circuits and the microstrip grid array antenna are designed on one plane, pins are used for DC connection, and DC lines are separated by a conducting plane. It ensures the antenna performance free from the interference of DC lines. Note from the top view that the microstrip patch antenna has eight loops. The dimensions of the loops are fixed at $l = 137$ mm, $s = 32.6$ mm, and $w_l = w_s = 3$ mm. At 2.45 GHz, s is $\lambda_g/3.5$ and $l = 1.2\lambda_g$. Hence, the microstrip grid array antenna operates in the nonresonant mode. Also note that the two rectifying

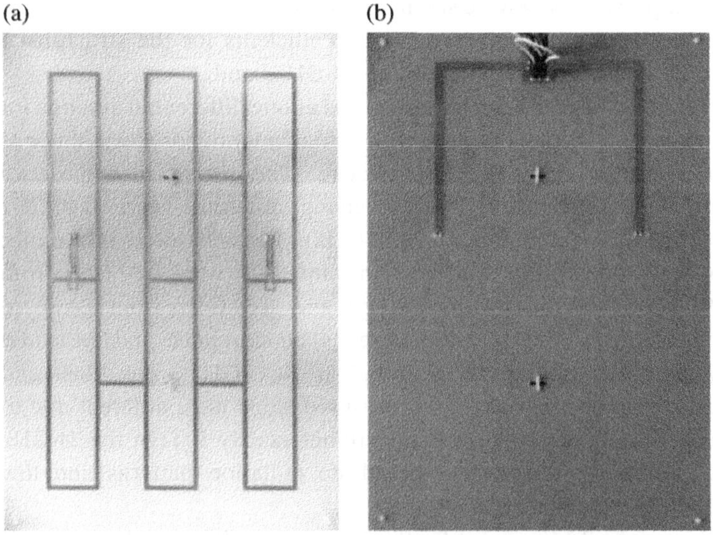

Figure 7.30 Photos of the fabricated multiport microstrip grid array structure with four ports. (a) Top and (b) bottom views. *Source:* Nariman [13]/with permission of IEEE.

circuits are differentially connected to the microstrip grid array antenna through two pairs of CPS lines at the middle of the left and right short sides of the loops. By taking the advantage of interconnecting the circuits with the antenna at the opposite ends, two independent, high-gain, and symmetrically tilted beams are created. Then, effectively harvesting RF energy in a wide-angle space is realized.

The RF energy harvester was measured with a power density of 1 $\mu W/cm^2$ in a frequency band from 2.39 to 2.47 GHz. At 2.45 GHz, the maximum DC output power was more than 200 μW and the maximum RF-to-DC efficiency was calculated to be 45.3%. The angle range reached 96° if the DC power was desired to be more than 10 μW. The angle range dropped to 59° if the DC power was desired to be more than 100 μW. It can be applied for harvesting power in IEEE 802.11 b/g bands to the case that the source location is unclear or multiple sources exist at different places.

Finally, it should be mentioned that the microstrip grid array antennas were used in 60-GHz WPT system [13]. The system was fabricated in a 40-nm digital CMOS process and an inexpensive packaging material. In the rectenna, a microstrip grid array antenna was integrated with a complementary cross-coupled oscillator-like rectifier. At a 4-cm spacing from the transmitter, the rectenna harvested energy at a rate of 1.22 mW with a 32.8% efficiency, which was significantly higher than the prior state of the art. The transmitter was equipped with a quad-core PA that produced a saturated output power of 24.6 dBm. The designed 2×2 microstrip grid array antennas helped the transmitter produce 35.3-dBm peak equivalent isotropically radiated power.

7.12 Concluding Remarks

In this chapter, the analysis and design of microstrip grid array antennas were presented. It was found from the analysis that the substrate has a strong influence on the performance of the microstrip grid array antenna. A thin substrate at millimeter-wave frequencies provides adequate gain, bandwidth, and cross-polarization quality. As the substrate becomes thicker, more surface-wave modes can propagate and can carry more power. They are diffracted from the substrate's edges, possibly causing undesirable effects on sidelobe levels, polarization, or main-beam shape. It was observed that the microstrip grid array antenna follows a gain-bandwidth relationship: the more radiating elements there are, the narrower is the impedance bandwidth, and the higher is the gain. The novel differential excitation using two coaxial probes was compared with the usual one using twin wires. It was shown that the novel feeding technique avoids opening the radiating element and results in a large physical separation that eases the testing and improves the current phase synchronization. A set of formulas was obtained for the initial design and the measured results from the fabricated samples validated these design formulas. It should be pointed out that the design of microstrip grid array antennas for gain less than 20 dBi is easy and for gain larger than 20 but less than 26 dBi is feasible but for gain larger than 26 dBi will be quite challenging if not impossible. The challenge is how to keep the current phase synchronization on radiating elements far away from the feed on a large microstrip grid array antenna. Several techniques for impedance matching, dual mode operations, co-aperture radiation, and miniaturization were presented, which have enriched the design of microstrip grid array antennas. The applications of differential microstrip grid array antennas were illustrated in 60-GHz radios, 79-GHz radars, 24-GHz full-duplex and MIMO systems, and 2.45-GHz energy harvester.

References

1 J. D. Kraus, "Backward angle travelling wave wire mesh antenna array," US patent No. 3290688A (1966).

2 J. D. Kraus, "A backward angle-fire array antenna," *IEEE Trans. Antennas Propagat.*, vol. 12, no. 1 pp. 48–50, Jan. 1964.

3 R. Conti, J. Toth, T. Dowling, J. Weiss, "The wire-grid microstrip antenna," *IEEE Trans. Antennas Propagat.*, vol. 29, no. 1, pp. 157–166, Jan. 1981.

4 L. T. Hildebrand, D. A. McNamara, "Verification of spatial domain integral-equation analysis of wire-grid arrays," *IEE Proc.-Microw. Antennas Propagat.*, vol. 142, no. 4, pp. 314–318, Aug. 1995.

5 H. Nakano, T. Kawano, J. Yamauchi, "Meander-line grid-array antenna," *IEE Proc.-Microw. Antennas Propagat.*, vol. 145, no. 4, pp. 309–312, Aug. 1998.

6 X. Chen, G. Wang, K. Huang, "A novel wideband and compact microstrip grid array antenna," *IEEE Trans. Antennas Propagat.*, vol. 58, no. 2, pp. 596–599, Feb. 2010.

7 M. Sun, Y. P. Zhang, "Design and integration of 60-GHz grid array antenna in chip package," in *Proc. Asia Pacific Microw. Conf.*, Hong Kong, China, 12-16 Dec. 2008.

8 F. Bauer, X. Wang, W. Menzel, A. Stelzer, "A 79-GHz radar sensor in LTCC technology using grid array antennas," *IEEE Trans. Antennas Propagat.*, vol. 61, no. 6, pp. 2514–2521, 2013.

9 Z. Chen, et al., "An LTCC microstrip grid array antenna for 94-GHz applications," *IEEE Antennas Wireless Propagat. Lett.*, vol. 14, pp. 1279–1281, 2015.

10 S. Beer, H. Gulan, C. Rusch, T. Zwick, "Integrated 122-GHz antenna on a flexible polyimide substrate with flip chip interconnect," *IEEE Trans. Antennas Propagat.*, vol. 61, no. 4, pp. 1564–1572, Apr. 2013.

11 B. Zhang, et al., "Integration of a 140 GHz packaged LTCC grid array antenna with an InP detector," *IEEE Trans. Compon. Packag. Manuf. Technol.*, vol. 5, no. 8, pp. 1060–1068, Aug. 2015.

12 B. Zhang, C. Kärnfelt, H. Gulan, T. Zwick, H. Zirath, "A D-band packaged antenna on organic substrate with high fault tolerance for mass production," *IEEE Trans. Compon. Packag. Manuf. Technol.*, vol. 6, no. 3, pp. 359–365, Mar. 2016.

13 M. Nariman, et al., "A compact 60-GHz wireless power transfer system," *IEEE Trans. Microw. Theory Technol.*, vol. 64, no. 8, pp. 2664–2677, Jul. 2016.

14 Y. Y. Hu, S. Sun, H. Xu, H. C. Sun, "Grid-array rectenna with wide angle coverage for effectively harvesting RF energy of low power density," *IEEE Trans. Microw. Theory Technol.*, vol. 67, no. 1, pp. 402–413, 2019.

15 K. D. Palmer, J. H. Cloete, "Synthesis of the microstrip wire grid array," in *Proc. IEE 10th Int. Conf. Antennas Propagat.*, vol. 1, pp. 114–118, 14-17 Apr. 1997.

16 H. Nakano, T. Kawano, Y. Kozono, J. Yamauchi, "A fast MoM calculation technique using sinusoidal basis and testing functions for a wire on a dielectric substrate and its application to meander loop and grid array antennas," *IEEE Trans. Antennas Propagat.*, vol. 53, no. 10, pp. 3300–3307, Oct. 2005.

17 L. Zhang, W. M. Zhang, Y. P. Zhang, "Microstrip grid and comb array antennas," *IEEE Trans. Antennas Propagat.*, vol. 59, no. 11, pp. 4077–4084, Nov. 2011.

18 B. Zhang, Y. P. Zhang, "Analysis and synthesis of millimeter-wave microstrip grid array antennas," *IEEE Antennas Propagat. Mag.*, vol. 53, no. 6, pp. 42–55, Dec. 2011.

19 L. Zhang, Y. P. Zhang, "Differential grid array antenna to radiate pencil beam at 24 GHz for radar and sensor applications," *IET Proc.-Microw. Antennas Propagat.*, vol. 8, no. 10, pp. 765–769, Jul. 2014.

20 Z. H. Chen, L. Zhang, B. Zhang, Y. P. Zhang, "A multi-port microstrip grid array structure," *IEEE Trans. Antennas Propagat.*, vol. 64, no. 11, pp. 4953–4958, Nov. 2016.

21 B. S. Zhao, M. Tang, Z. J. Shao, Y. P. Zhang, J. F. Mao, "Design of broadband compact grid array antennas using gradient slow-wave structures," *IEEE Antennas Wireless Propagat. Lett.*, vol. 21, no. 3, pp. 620–624, Mar. 2022.

22 B. S. Zhao, M. Tang, J. F. Mao, "Miniaturized grid array antenna with gradient slow-wave structures," in *Proc. Int. Conf. Microw. Millimeter Wave Technol.*, Harbin, China, 12–15 Aug. 2022.

23 G. H. Xu, et al., "Frequency band ratio analysis of dual-band microstrip grid array antennas," in *Proc. Asia Pacific Microw. Conf.*, Singapore, 10–13 Dec. 2019.

24 G. H. Xu, et al., "Dual-band differential shifted-feed microstrip grid array antenna with two parasitic patches," *IEEE Trans. Antennas Propagat.*, vol. 68, no. 3, pp. 2434–2439, Mar. 2020.

25 G. H. Xu, et al., "Microstrip grid and patch-based dual-band shared-aperture differentially fed array antenna," *IEEE Antennas Wireless Propagat. Lett.*, vol. 20, no. 6, pp. 1043–1047, Jun. 2021.

26 Y. P. Zhang, D. Liu, "Antenna-on-chip and antenna-in-package solutions to highly-integrated millimeter-wave devices for wireless communications," *IEEE Trans. Antennas Propagat.*, vol. 57, no. 10, pp. 2830–2841, Oct. 2009.

27 Y. P. Zhang, M. Sun, D. Liu, Y. Lu, "Dual grid array antennas in a thin profile package for flip-chip interconnection to highly integrated 60-GHz radios," *IEEE Trans. Antennas Propagat.*, vol. 59, no. 4, pp. 1191–1199, Apr. 2011.

28 M. Sun, Y. P. Zhang, D. Liu, K. M. Chua, L. L. Wai, "A ball grid array package with a microstrip grid array antenna for a single-chip 60-GHz receiver," *IEEE Trans. Antennas Propagat.*, vol. 59, no. 6, pp. 2134–2140, Jun. 2011.

29 B. Zhang, D. Titz, F. Ferrero, C Luxey, Y. P. Zhang, "Integration of quadruple linearly-polarized microstrip grid array antennas for 60-GHz antenna-in-package application," *IEEE Trans. Compon. Packag. Manuf. Technol.*, vol. 3, no. 8, pp. 1293–1300, Aug. 2013.

30 L. Zhang, K. Y. See, B. Zhang, Y. P. Zhang, "Dual-band monopole antenna in PCB and microstrip grid array antenna in package for single-chip tri-band radio," *IEEE Trans. Antennas Propagat.*, vol. 61, no. 1, pp. 439–443, Jan. 2013.

31 B. Zhang, Y. P. Zhang, D. Titz, F. Ferrero, C. Luxey, "A circularly-polarized array antenna using linearly-polarized sub grid arrays for highly-integrated 60-GHz radio," *IEEE Trans. Antennas Propagat.*, vol. 61, no. 1, pp. 436–439, Jan. 2013.

32 B. Zhang, Y. P. Zhang, "Grid array antennas with subarrays and multiple feeds for 60-Gz radios," *IEEE Trans. Antennas Propagat.*, vol. 60, no. 5, pp. 2270–2275, May 2012.

33 M. Sun, Y. P. Zhang, D. Liu, K. M. Chua, L. L. Wai "A ball grid array package with a microstrip grid array antenna for a single-chip 60-GHz receiver," *IEEE Trans. Antennas Propagat.*, vol. 59, no. 6, pp. 1191–1199, June 2011.

34 A. Sabharwa, et al., "In-band full-duplex wireless: Challenges and opportunities," *IEEE J. Sel. Areas Commun.*, vol. 32, no. 9, pp. 1637–1652, Sep. 2014.

35 Y. P. Zhang, "In-band full duplex: How it can work," in *Proc. of Asia Pacific Microw. Conf.*, Nanjing, China, 6–9 Dec. 2015.

36 J. Zhou, et al., "Integrated full duplex radios," *IEEE Commun. Mag.*, vol. 55, no. 4, pp. 142–151, Apr. 2017.

8

Differential Printed Antennas

8.1 Introduction

A printed antenna is an antenna printed on one surface of a dielectric substrate where no ground plane is printed directly underneath the radiator on the other surface of the dielectric substrate. There are many types of printed antennas. We choose only to cover quasi-Yagi, fractal, and spiral antennas in this chapter for endfire radiation, multiband, and wideband operations, respectively. For quasi-Yagi antennas [1–10], we will discuss the excitation and propagation of surface waves as the basis of the operation and describe the design procedure of a differential quasi-Yagi antenna. For fractal antennas [11–19], we will consider the Sierpinski gasket for multiband operation. Special emphasis will be given to the analysis, design, and measurement of a Sierpinski gasket dipole fed by a differential CPS line. For spiral antennas, we will focus on the analysis and design of an equiangular spiral antenna for wideband operation [20–32]. We will show how these printed antennas are configured for real applications, for example, the integration of both single-ended and differential quasi-Yagi antennas in an antenna-in-package (AiP) module for the 60-GHz radio and the implementation of discrete equiangular spiral antennas for ultra-wideband (UWB) radios.

8.2 Quasi-Yagi Antennas

The quasi-Yagi antenna was first reported by Qian et al. in 1998 as a single-ended antenna, although the driver is differential [1]. They integrated a balun to feed the driver, truncated the ground plane as a reflector, and printed a director to enhance the radiation in the forward endfire direction on a substrate of dielectric constant $\varepsilon_r = 10.2$ and thickness $h = 0.635$ mm at X-band frequencies. The antenna total size, including the microstrip-to-coplanar strip transition, was less than 0.5λ in

Differential Antennas: Theory and Practice, First Edition. Yueping Zhang.

width and 0.6λ in length. It achieved a bandwidth of 17%, a gain of 6.5 dBi, a front-to-back ratio (FBR) of 18 dB, and a cross-polarization (X-pol) level lower than 15 dB at 10 GHz [1].

The quasi-Yagi antenna was further optimized by Qian et al. in 1999 for wide bandwidth and good radiation characteristics in terms of beam pattern, FBR, cross-polarization, and low mutual coupling [2]. The antenna obtained a bandwidth of 48% at 10 GHz for a VSWR < 2, a FBR better than 15 dB, and a cross-polarization level lower than 12 dB across the entire frequency band. In addition, a mutual coupling level below 22 dB was measured for the two optimized quasi-Yagi antennas with a half-wavelength separation at 10 GHz [2]. This low mutual coupling level makes the quasi-Yagi antenna a good element candidate for power combining, phased arrays, and active arrays, as well as millimeter-wave imaging arrays [3].

The quasi-Yagi antenna generally includes a balun, which not only makes its size bulky but also affects its radiation pattern. Hence, there have been attempts to get rid of the balun, for instance, using the slot line [4] or monopole driver [5]. A differential quasi-Yagi antenna naturally does not need the balun for differential excitation. The differential quasi-Yagi antenna, like the original single-ended quasi-Yagi antenna, operates in surface waves [6]. The driver is designed to launch the surface wave of TE_0 mode in the dielectric substrate. The director is strongly coupled to the driver because it shares the same field polarization. It guides the electromagnetic field toward the endfire direction while simultaneously acting as an impedance-matching element. The reflector completely reflects the surface wave of TE_0 mode to the director to enhance the endfire radiation at the end edge of the substrate into free space. The total reflection is due to the fact that the surface wave of TE_0 mode is cut off in the grounded dielectric substrate region. The surface wave of TM_0 mode will be also excited. However, it should be excited as weak as possible because it degrades the antenna performance with an increase in cross-polarization radiation and a reduction in gain and FBR.

8.2.1 Surface Wave

A dielectric substrate will guide the surface wave. Two types of dielectric substrates are considered. One has the ground plane while the other does not have the ground plane. The surface wave is classified into TM and TE modes [7].

8.2.1.1 Guided by Grounded Dielectric Substrate

For a grounded dielectric substrate with an infinitely large size, the cut-off frequencies of surface waves in TM and TE modes are given by

$$f_c = \frac{nc}{2h\sqrt{\varepsilon_r - 1}}, n = 0, 1, 2, \ldots \tag{8.1}$$

and

$$f_c = \frac{(2n-1)c}{4h\sqrt{\varepsilon_r - 1}}, n = 1, 2, 3, \ldots \tag{8.2}$$

respectively. It can be seen that the cut-off frequency of the TM_0 mode is zero, which implies that this mode always exists in the substrate. In addition, the lowest high-order mode is TE_1, and then TM_1, TE_2, etc. The cut-off frequency interval of the two adjacent surface wave modes is the same.

To simulate the propagation of the surface wave in the grounded dielectric substrate, we set the side of the substrate in the HFSS model as the radiation boundary to eliminate the influence of the edge reflection of the substrate. We choose the FR4 substrate with a planar size of $8\lambda \times 8\lambda$ and thickness $h = 0.8$ and 4 mm. We print an electric dipole to excite the surface wave at 24 GHz. Figures 8.1 and 8.2 show the HFSS simulated distributions of surface wave fields excited by the electric dipole on the substrates with the thickness of $h = 0.8$ and 4 mm, respectively.

According to equations (8.1) and (8.2), when the substrate thickness is less than 1.69 mm, the grounded FR4 substrate only supports the surface wave of TM_0 mode. Therefore, when $h = 0.8$ mm, the surface wave mainly propagates along the axial direction as shown in Figure 8.1; when $h = 4$ mm, the higher-order mode appears, and the surface wave has both the TM_0 mode along the axial direction and the TE_1 mode along the normal direction as shown in Figure 8.2. It is interesting to note that the electric field of the TM_0 mode is mainly concentrated on the side close to the lower surface, while the electric field of the TE_1 mode is mainly concentrated on the side close to the upper surface. This phenomenon verifies that a magnetic dipole on a grounded dielectric substrate is more likely to excite TM_0 surface wave than an electric dipole [7].

8.2.1.2 Guided by Ungrounded Dielectric Substrate

Unlike the above grounded, an ungrounded dielectric is symmetric. Therefore, the field distributions of TE and TM modes can be represented in symmetric (even/cos) and asymmetric (odd/sin) functions of z. For the ungrounded dielectric substrate with an infinitely large size, the cut-off frequencies of surface waves in TE and TM modes are given by

$$f_c = \frac{nc}{2h\sqrt{\varepsilon_r - 1}}, n = 0, 1, 2, \ldots. \tag{8.3}$$

It can be seen that the cut-off frequencies of the TE_0 and TM_0 modes are zero, which implies that both modes always exist in the substrate. In addition, it was observed by Alexopoulos, et al. that TE modes "turn on" much more quickly than TM modes [6]. The lowest high-order mode is TE_1, and then TM_1, TE_2, etc. The cut-off frequency interval of the two adjacent surface wave modes is the same.

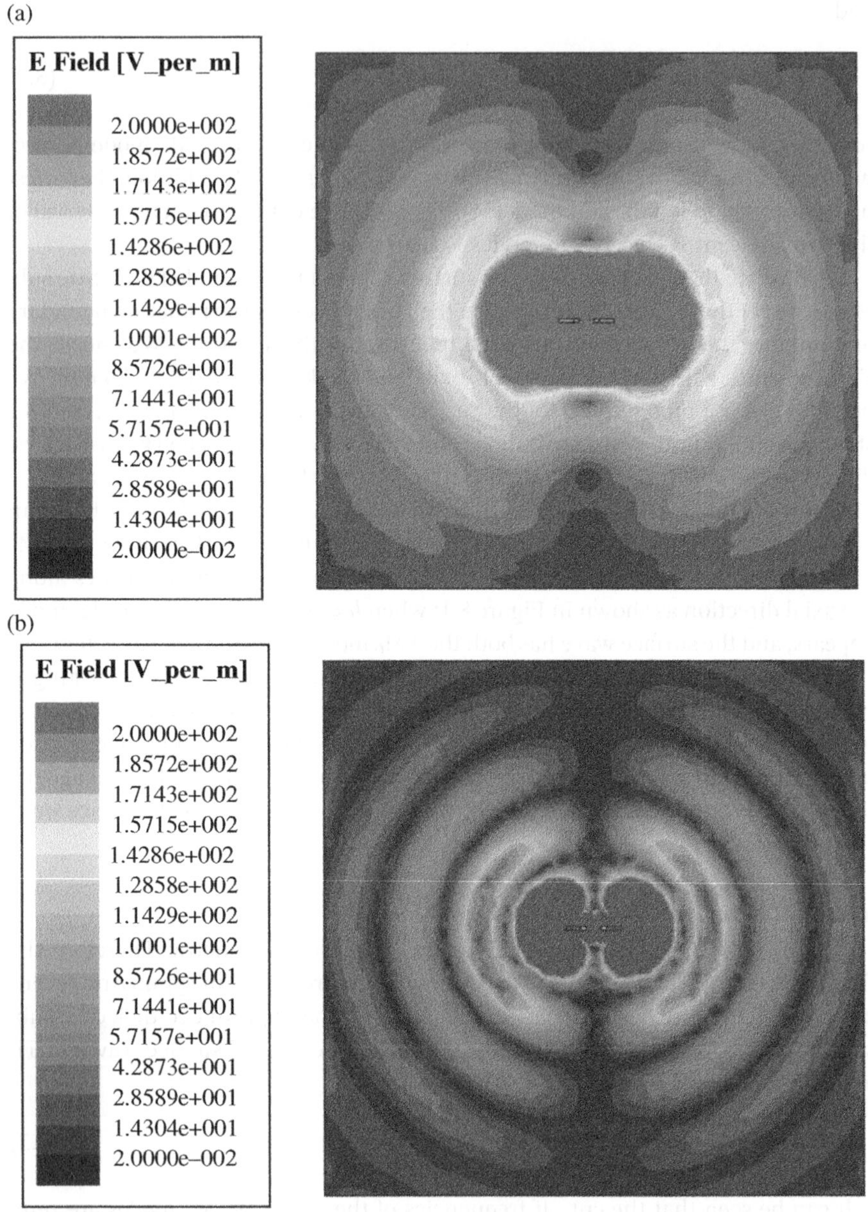

Figure 8.1 HFSS simulated electric field distributions of the surface wave excited by a half-wave electric dipole on the substrate with $h = 0.8$ mm. (a) Upper surface and (b) lower surface.

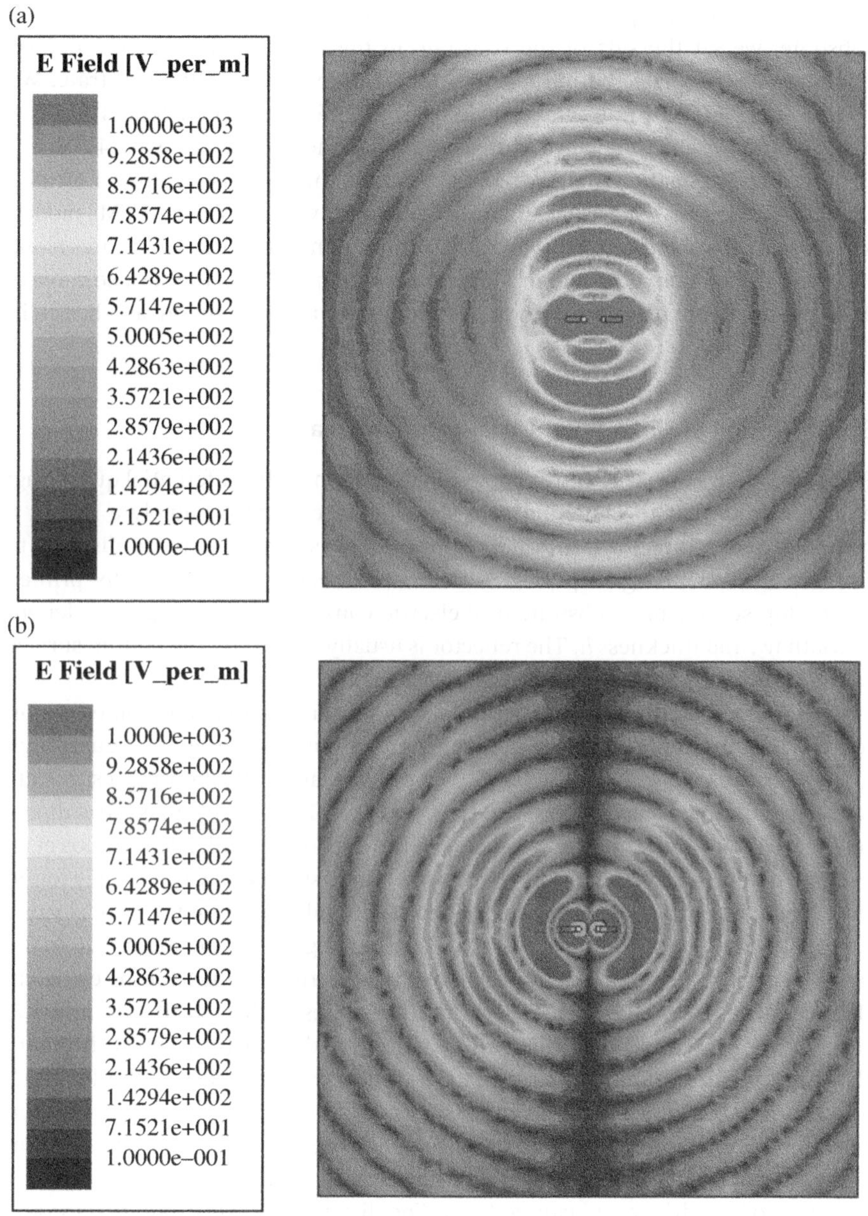

Figure 8.2 HFSS simulated electric field distributions of the surface wave excited by a half-wave electric dipole on the substrate with $h = 4.0$ mm. (a) Upper surface and (b) lower surface.

To simulate the propagation of the surface wave in the ungrounded dielectric substrate, we set the side of the substrate in the HFSS model as the radiation boundary to eliminate the influence of the edge reflection of the substrate. We choose the FR4 substrate with a planar size of $8\lambda \times 8\lambda$ and thickness $h = 0.8$ and 4.0 mm. We print an electric dipole to excite the surface wave at 24 GHz.

According to equation (8.3), when the substrate thickness is less than 1.69 mm, the ungrounded FR4 substrate supports the surface wave of TE_0 and TM_0 modes. When $h = 0.8$ mm, the surface wave of TE_0 mode dominates and propagates along the normal direction as shown in Figure 8.3; when $h = 4$ mm, the surface wave of TM_0 mode dominates and propagates along the axial direction as shown in Figure 8.4.

8.2.2 Design of Differential Quasi-Yagi Antenna

Figure 8.5 shows the structure and coordinate system of the differential quasi-Yagi antenna. Note that the antenna consists of three elements, namely a driver, a reflector, and a director. More commonly, there is more than one director to improve the antenna gain [6]. The driver and director are horizontally printed on the top surface of a substrate of dielectric constant ε_r, loss tangent δ, length l_s, width w_s, and thickness h. The reflector is usually printed on the bottom surface of the substrate, which has another function as the ground plane.

The design procedure for a differential quasi-Yagi antenna is presented as follows. It is found that the design is much simpler than the design of a quasi-Yagi antenna because no balun is involved. The design starts with choosing a substrate for the maximum excitation of the surface wave of TE_0 mode by an electric dipole on the substrate at the central frequency of the operating band. Alexopoulos et al. examined how the substrate affects the excitation of surface waves [7]. They found that the surface wave of TE_0 mode can be maximumly excited if the critical value for the substrate electrical thickness is satisfied. Using the method described in the classical paper [7], Leong and Itoh showed that the critical values for the electrical thickness are 0.03, 0.05, and 0.08 for the substrates with $\varepsilon_r = 10.2$, 4, and 2.2, respectively [8]. Then, the initial values are set for the length l_1 and width w_1 of the director, the distance d_1 between the director and the driver, the length l_2 and width w_2 of the driver, and the distance d_2 from the driver to the reflector. For simplicity, the same width of 0.02λ can be chosen for the driver and director. The length of the driver is about $0.45\lambda_g$. The length of the director should be shorter than that of the driver and can be $0.3\lambda_g$. The distances between the director and driver and between the driver and reflector are about $0.3\lambda_g$ and $0.25\lambda_g$, respectively. Next, the width w_3 and spacing g of the CPS lines can be estimated with the available empirical formula. Finally, the optimum values for the above design parameters can be obtained from HFSS simulations [9].

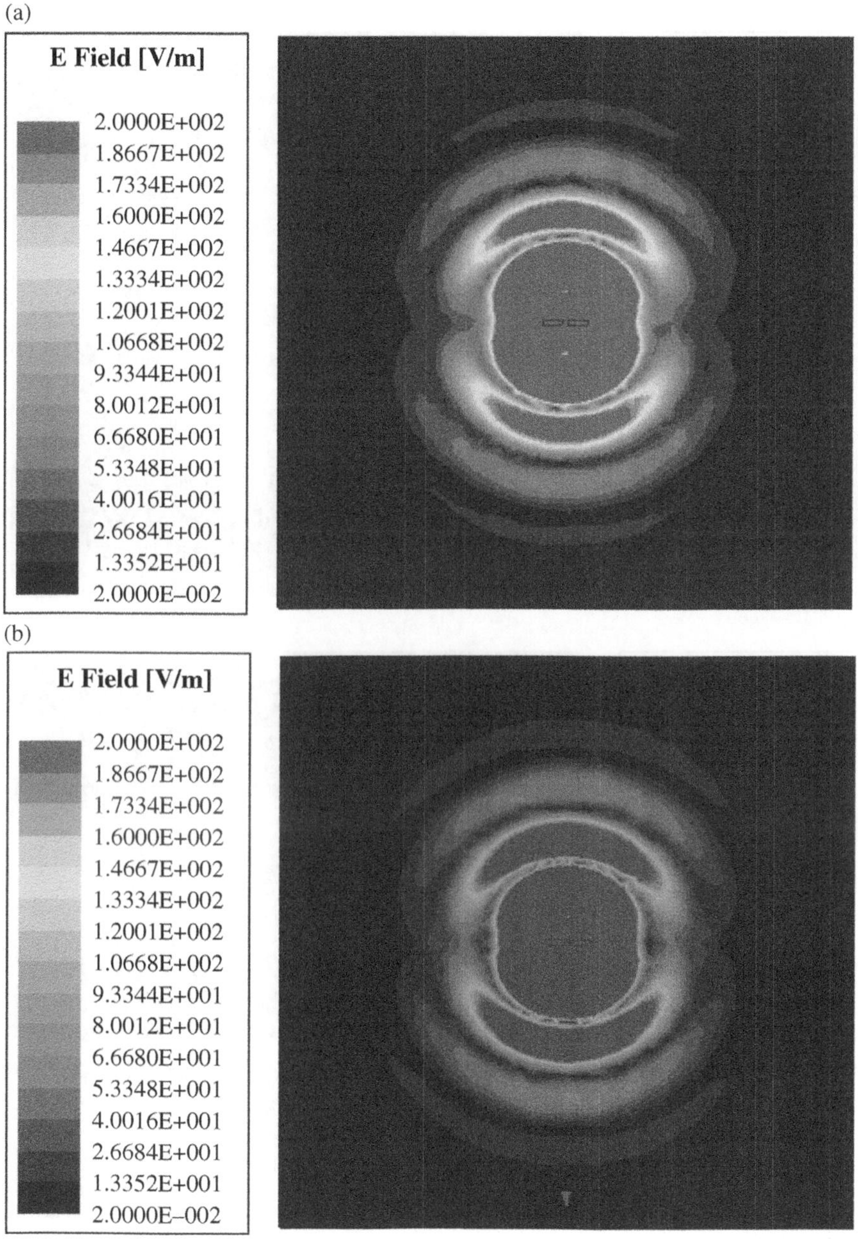

Figure 8.3 HFSS simulated electric field distribution of the surface wave excited by an electric dipole on the substrate with the thickness $h = 0.8$ mm. (a) Upper surface and (b) lower surface.

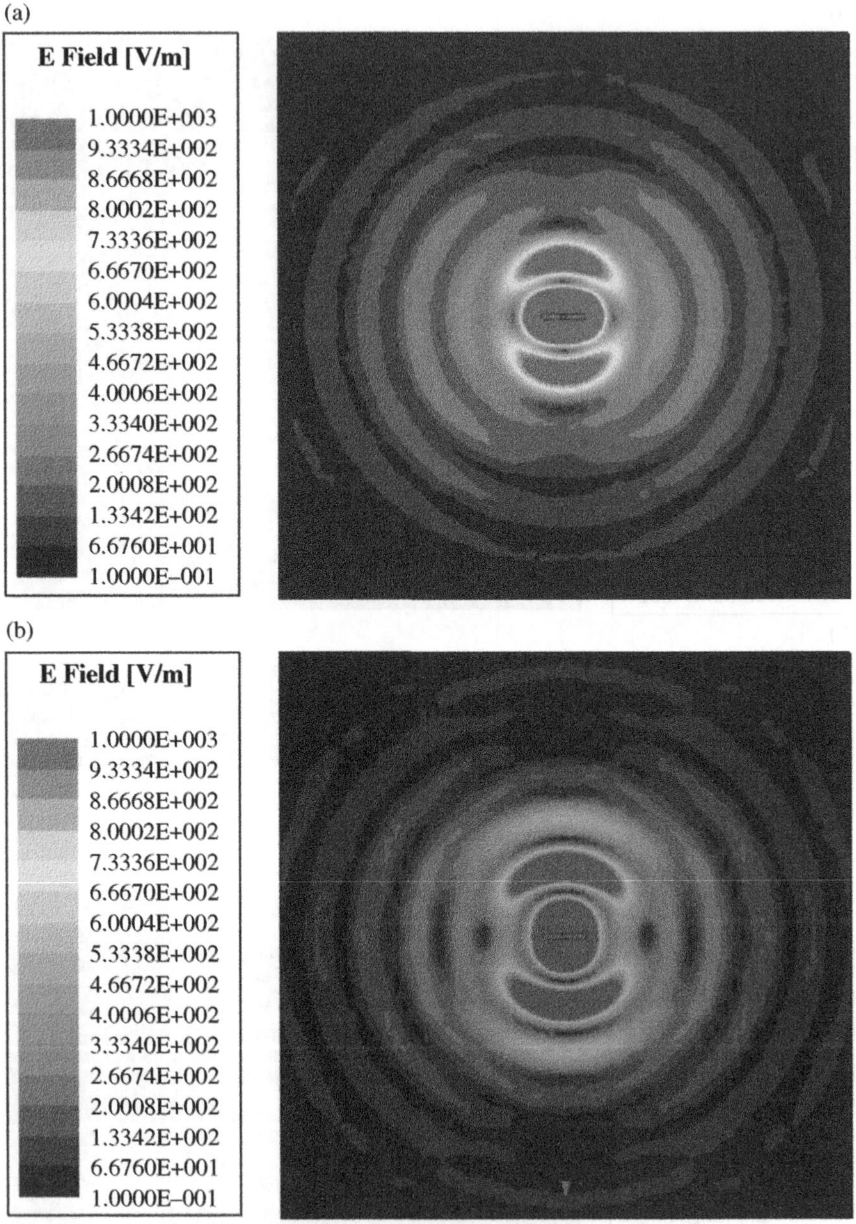

Figure 8.4 HFSS simulated electric field distribution of the surface wave excited by an electric dipole on the substrate with the thickness $h = 4.0$ mm. (a) Upper surface and (b) lower surface.

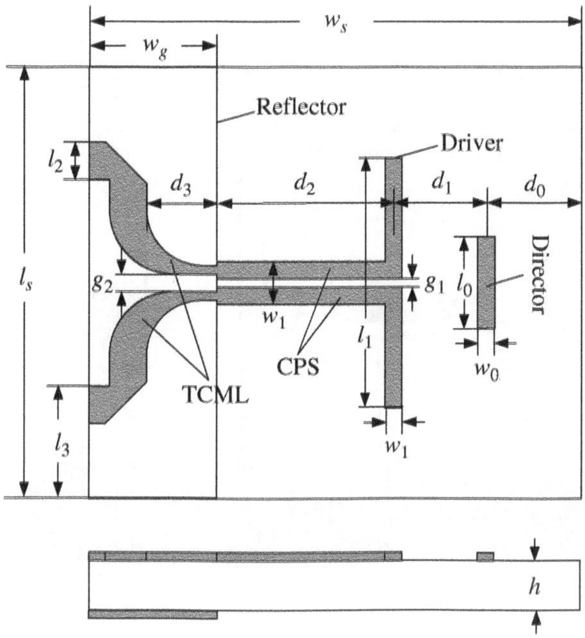

Figure 8.5 Structure and coordinate system of the differential quasi-Yagi antenna.

Figure 8.6 shows the photo of the differential quasi-Yagi antenna designed and fabricated on a substrate of dielectric constant $\varepsilon_r = 10.2$ and thickness $h = 0.635$ mm at X-band frequencies. The fabricated dimensions are $w_1 = w_2 = 0.6$ mm, $l_1 = 5$ mm, $l_2 = 8$ mm, $d_1 = 3$ mm, $d_2 = 2.5$ mm, $w_3 = 1$ mm, and $g = 0.2$ mm.

Figure 8.7 shows the simulated electric field distribution on the top surface of the differential quasi-Yagi antenna at 10 GHz. As expected, the surface wave of TE_0 mode is excited by the driver, coupled to the director, and reflected by the reflector on the top surface of the ungrounded substrate region. The surface wave of TE_0 mode is cut off and the surface wave of TM_0 mode is weakly excited in the grounded dielectric substrate region. Hence, the electric field distribution on the top surface of the grounded dielectric substrate region is unnoticeable.

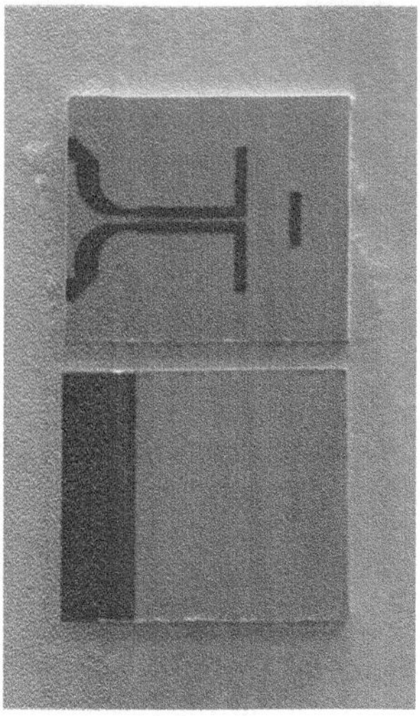

Figure 8.6 Photo of the differential quasi-Yagi antenna.

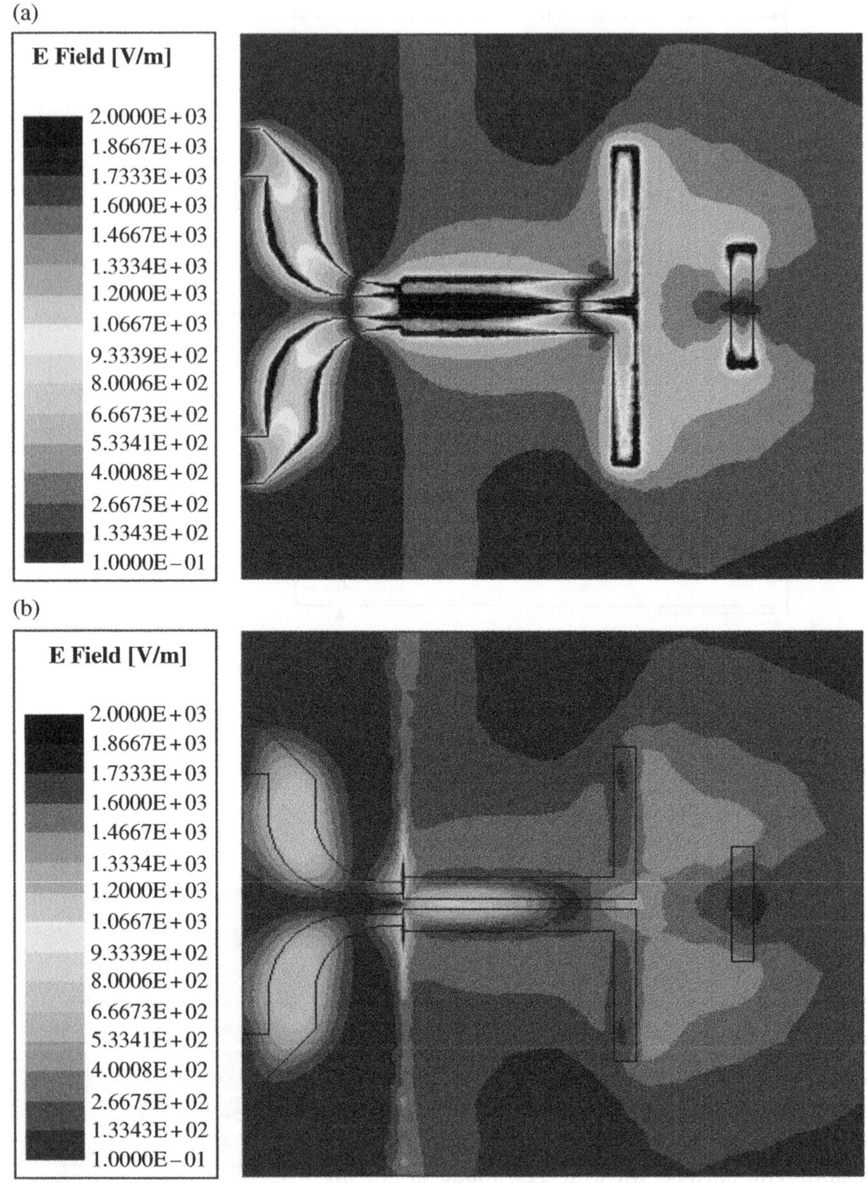

Figure 8.7 HFSS simulated electric field distribution of the differential quasi-Yagi antenna at 10 GHz. (a) Top surface and (b) bottom surface.

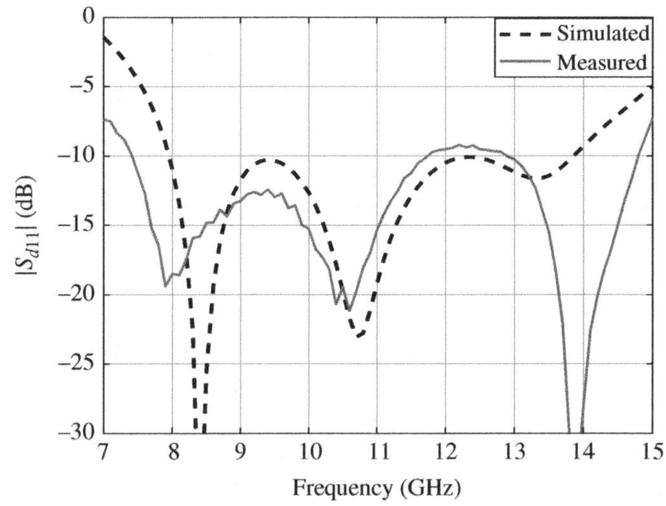

Figure 8.8 Simulated and measured $|S_{d11}|$ of the differential quasi-Yagi antenna.

Figure 8.8 shows the simulated and measured $|S_{d11}|$ of the differential quasi-Yagi antenna as a function of frequency from 7 to 15 GHz. It is evident from the figure that although there are differences between the simulated and measured values, the antenna achieved acceptable matching from 7.5 to 14.8 GHz or a fractional bandwidth of 66% at 11.15 GHz. Figure 8.9 shows the simulated and measured radiation patterns at 8.2, 10.6, and 12.3 GHz. As expected, the antenna radiates an endfire beam. Figure 8.10 shows the simulated and measured gain values. The measured gain values fluctuate between 3.7 and 5.4 dBi from 8 to 12.3 GHz. The simulated radiation efficiency is 94% at 10 GHz.

Table 8.1 lists the bandwidth, gain, efficiency, X-pol, and FBR for the differential and the single-ended quasi-Yagi antennas at 10 GHz. It should be mentioned that these quasi-Yagi antennas have the same size and were fabricated with the same material. Note from the table that the differential quasi-Yagi antenna outperforms the single-ended counterparts.

8.2.3 Quasi-Yagi Antenna Arrays

The application of the Quasi-Yagi antenna as an array element is explored in this section. We limit our effort to E-plane linear arrays because we target their potential use in portable and mobile devices [10]. A key feature of the original single-ended quasi-Yagi antenna is to use a broadband uniplanar microstrip-to-CPS transition to feed the differential driver. The transition is a 3-dB power divider or a

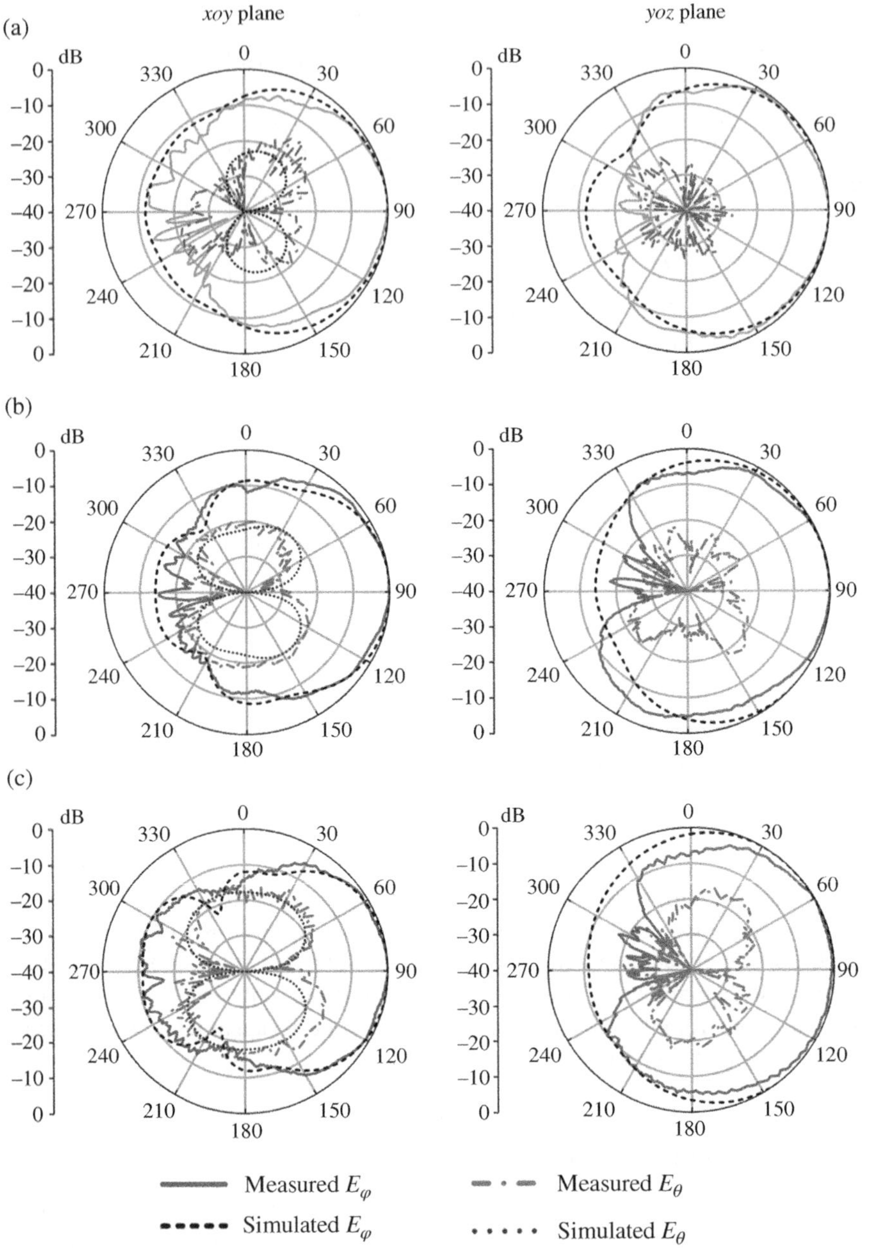

Figure 8.9 Simulated and measured radiation patterns of the differential quasi-Yagi antenna at (a) 8.2 GHz, (b) 10.6 GHz, and (c) 12.3 GHz.

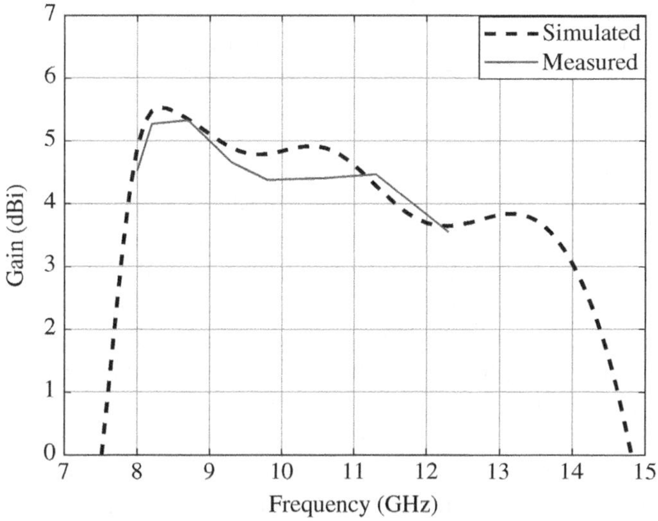

Figure 8.10 Simulated and measured gain values of the differential quasi-Yagi antenna.

Table 8.1 Differential and single-ended quasi-Yagi antennas.

Reference	Bandwidth (%)	Gain (dBi)	Efficiency (%)	X-pol (dB)	FBR (dB)
This work	73	4.4	94	−21	15
[8]	48	4.6	93	−12	12

reactive balun. It is a simple broadband structure and can excite the two terminals of the driver with signals of equal amplitude and reversed phases. However, it has a fundamental drawback, that is, the input port is matched but the output ports cannot be matched. The isolation between the two output ports is only 6 dB. This will affect the differential excitation quality if any loss of symmetry from the driver exists. It may also cause a parasitic resonance as the reflections occur between the driver and the balun junction. The parasitic resonance will not occur in the case of symmetry because the reflection from the driver will pass through the balun junction and be absorbed by the source resistance. The loss of symmetry will occur in the E-plane phased array because the two arms of the driver will see an incident oblique wave differently. The consequence of this broken symmetry could result in scan blindness at some frequencies. To avoid this potential issue, the solution is to use a resistive balun such as a Wilkinson power divider. However, the use of Wilkinson power divider will destroy the simplest of the original quasi-Yagi antenna. Our differential quasi-Yagi antenna has no such issue.

8.2.3.1 Mutual Coupling

The mutual coupling between elements needs to be considered especially in the design of a phased array because strong mutual coupling may cause scan blindness. The mutual coupling is determined by the transmission coefficient $|S_{21}|$ or $|S_{dd21}|$ of an array. Deal et al. determined the E-plane mutual coupling between two single-ended quasi-Yagi antennas implemented on the same substrate to be below −18 dB and, in most cases, below −20 dB for the center-to-center spacing equal to or greater than the half wavelength at 10 GHz. They also made effort to identify the source of mutual coupling and concluded from their measurements of the testing structures that, for 15-mm-array spacing, which corresponds to the half wavelength at the central frequency of 10 GHz, mutual coupling is almost solely due to coupling through the air [3].

To get a deeper insight into the coupling mechanism, we have conducted a simulated study of mutual coupling between two single-ended and two differential quasi-Yagi antennas, respectively. First, we keep the drivers, the substrate, and the truncated ground plane but remove all other building blocks. Figure 8.11a shows the simulated electric field distribution on the top surface of the substrate for the case with one driver fed by a lumped port and the other driver matched to a load at 10 GHz. It is evident from the figure that the surface wave of the TE_0 mode has been strongly excited and trapped in the ungrounded substrate region. It is more important to note that the grounded substrate region cuts off the surface wave of the TE_0 mode and forces it reflected to the ungrounded substrate region. The reflected surface wave of the TE_0 mode is an important source of mutual coupling. Then, we add the two directors in the model, and the simulated electric field distribution on the top surface of the substrate at 10 GHz is shown in Figure 8.11b. It is seen that there is a strong desirable coupling between the excited driver and its director due to the surface wave of the TE_0 mode for endfire radiation. There seems no effect by the directors on the coupling between the two drivers. Next, we add the CPS lines in the model and move the lumped port to the CPS input and the matched load to the other CPS input. Figure 8.11c shows the simulated electric field distribution on the top surface of the substrate at 10 GHz. Note that the CPS lines enhance the coupling between the two drivers. Finally, we add in the models the baluns to realize the single-ended quasi-Yagi array and tapered coupled microstrip lines (TCML) to realize the differential quasi-Yagi array, respectively. Figure 8.11d and e shows the simulated electric field distributions on the top surfaces of the substrates at 10 GHz for the cases of the single-ended

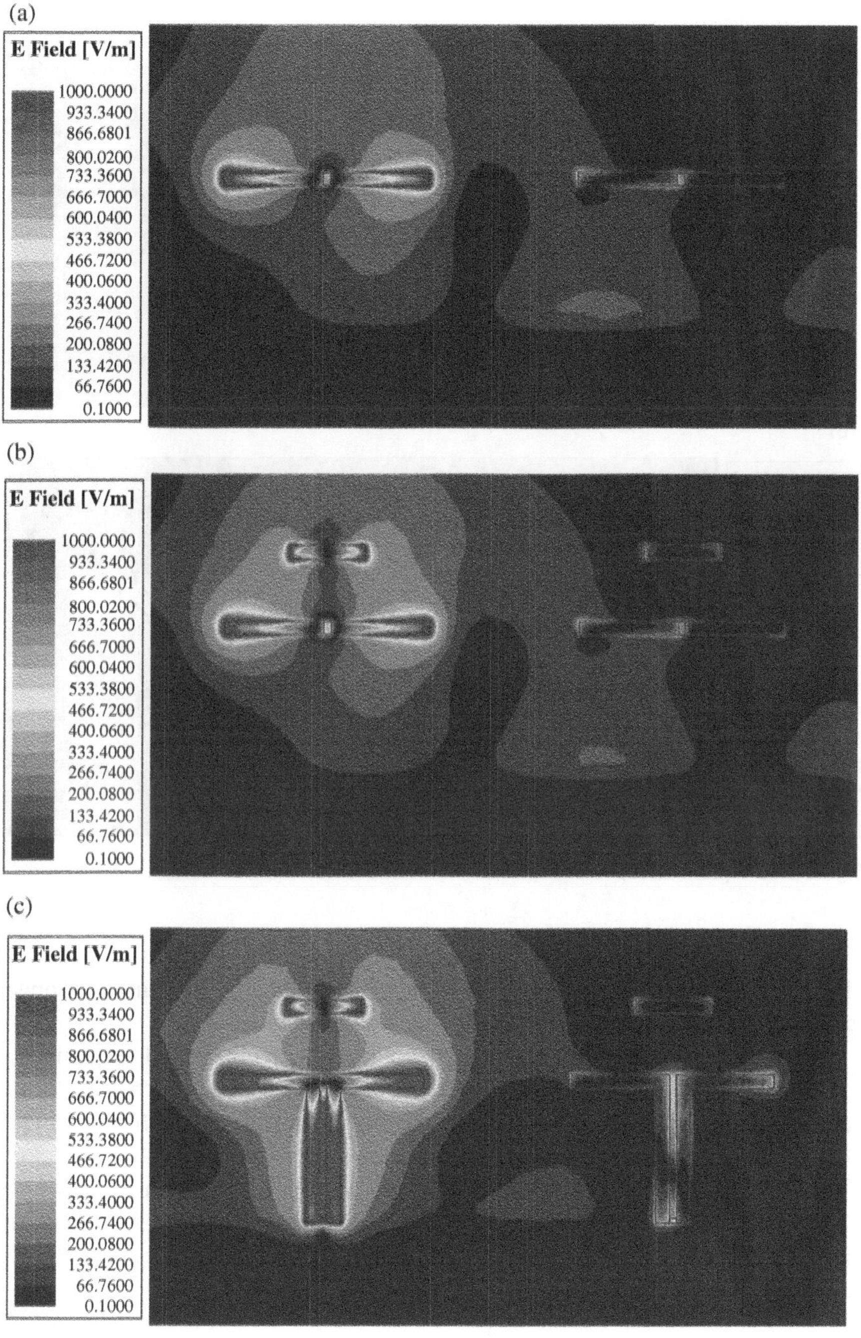

Figure 8.11 Simulated electric field distributions. (a) Only driver, (b) driver and director, (c) driver and director and CPS, (d) driver and director and CPS and balun, and (e) driver and director and CPS and TCML.

(d)

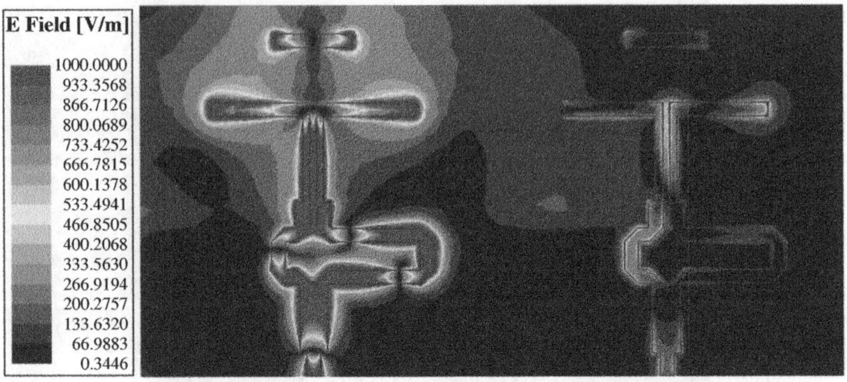

(e)

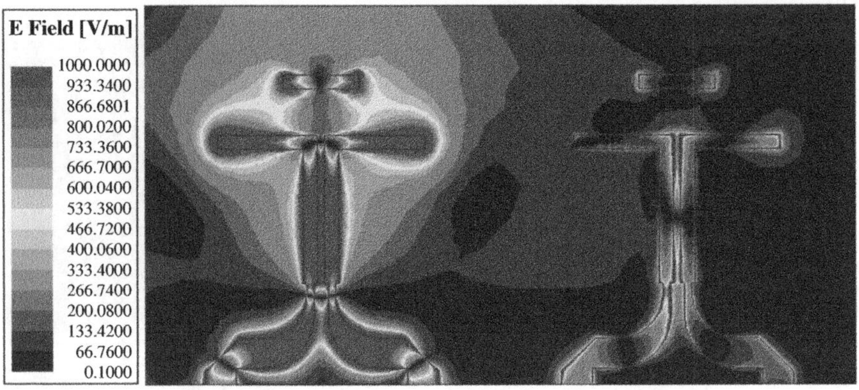

Figure 8.11 (Continued)

and differential quasi-Yagi arrays, respectively. Note that the coupling is stronger for the differential than for the single-ended quasi-Yagi array.

Figure 8.12 shows the simulated $|S_{dd11}|$ and $|S_{dd21}|$ as a function of frequency for the case of Figure 8.11a and b with a spacing of 15 mm between the two elements. As expected, the coupling level over the acceptable matching band from 9 to 11.5 GHz is almost the same between the two cases with and without the directors. Figure 8.13 shows the simulated $|S_{dd11}|$ and $|S_{dd21}|$ as a function of frequency for the case of Figure 8.11c with the spacing of 15 mm between the two elements. Note that the CPS extends the matching band to 15 GHz and reduces the coupling level over a wider bandwidth.

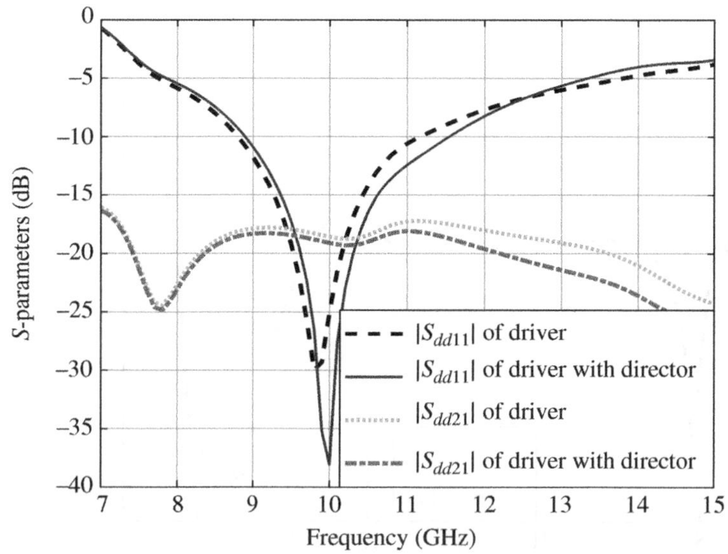

Figure 8.12 Simulated $|S_{dd11}|$ and $|S_{dd21}|$ as a function of frequency for the case of Figure 8.11a and b.

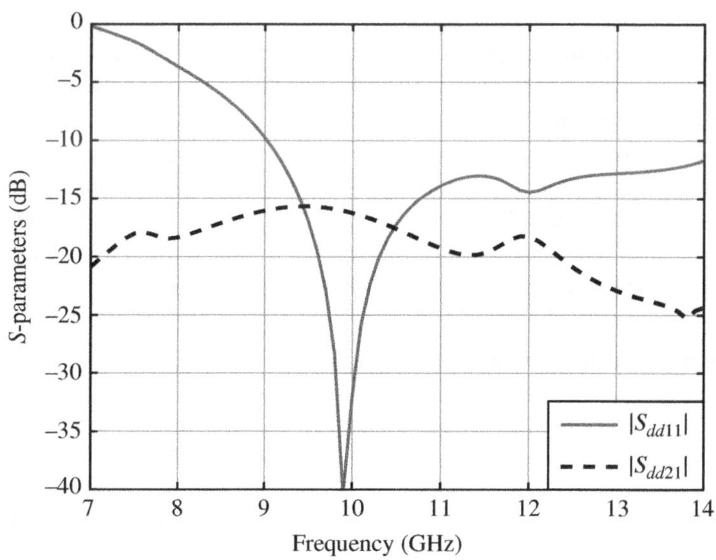

Figure 8.13 Simulated $|S_{dd11}|$ and $|S_{dd21}|$ as a function of frequency for the case of Figure 8.11c.

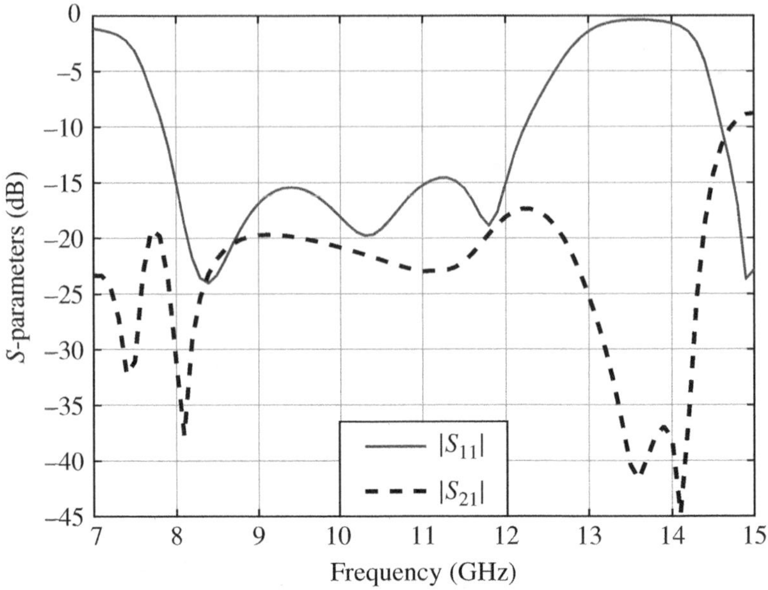

Figure 8.14 Simulated $|S_{11}|$ and $|S_{21}|$ as a function of frequency for the single-ended quasi-Yagi array with a spacing of 15 mm between the two elements.

Figure 8.14 shows the simulated $|S_{11}|$ and $|S_{21}|$ as a function of frequency for the single-ended quasi-Yagi array of Figure 8.11d with a spacing of 15 mm between the two elements. Figure 8.15 shows the simulated $|S_{dd11}|$ and $|S_{dd21}|$ as a function of frequency for the differential quasi-Yagi array of Figure 8.11e with a spacing of 15 mm between the two elements. Note that the matching band is from 7.9 to 12.2 GHz and from 8.0 to 14.0 GHz, respectively, for the single-ended and differential quasi-Yagi arrays. The maximum coupling level is almost the same for the two cases over the respective impedance bandwidths.

8.2.3.2 Decoupling Structures and Effects

It was seen from Figure 8.15 that the simulated mutual coupling level for the center-to-center spacing of 15 mm, which is equal to the half wavelength at 10 GHz is −18 dB. For a multiple-input and multiple-output (MIMO) array, the mutual coupling level of −25 dB is desirable. Hence, there is a need to further reduce the mutual coupling. A few decoupling structures such as neutralization lines [33], split-ring resonators [34], slit, and air holes have been attempted. The idea to add a meta-surface as a superstrate to reduce the mutual coupling has not been adopted in order to keep the low profile of the differential quasi-Yagi array [35].

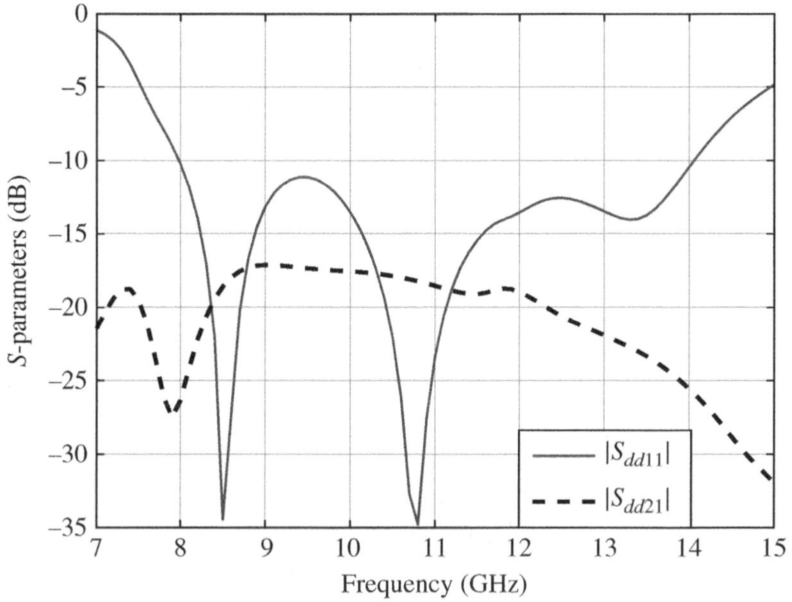

Figure 8.15 Simulated |S_{dd11}| and |S_{dd21}| as a function of frequency for the single-ended quasi-Yagi array with a spacing of 15 mm between the two elements.

The simulated decoupling effects of the above structures are summarized as follows. For the case of neutralization line, the coupling level is greatly reduced to −25 dB but the radiation patterns are distorted, and gain is reduced by 2 dB at 10 GHz. In the case of split-ring resonator, it fails to reduce the coupling level but increases the gain by 0.2–0.7 dB over the impedance bandwidth. For the cases of slit and air holes, they are very effective to reduce the mutual coupling to −25 dB and meanwhile do not affect the impedance bandwidth and radiation patterns. Hence, the *E*-plane linear arrays with the slit and air holes were fabricated and measured.

The two differential quasi-Yagi antenna elements are separated by 15 mm and fed with the SMA-connected coaxial cables. The slit is 2 mm wide cut in between the two elements and after the truncated ground plane. The air holes that have the same diameter of 0.5 mm are punched in between the two elements and after the truncated ground plane with an optimized pattern. It should be mentioned that the details of the SMAs and coaxial cables are unknown. The irregular solder joints and curved coaxial cables are hard to be modelled exactly. Hence, only measured results are discussed.

Figure 8.16 shows the measured $|S_{dd11}|$ and $|S_{dd21}|$ as a function of frequency for the two-element differential quasi-Yagi arrays without any decoupling structure, with the slit, and with the air holes, respectively. Measured results have confirmed that the simple decoupling structures are quite effective in reducing the mutual coupling level below −25 dB.

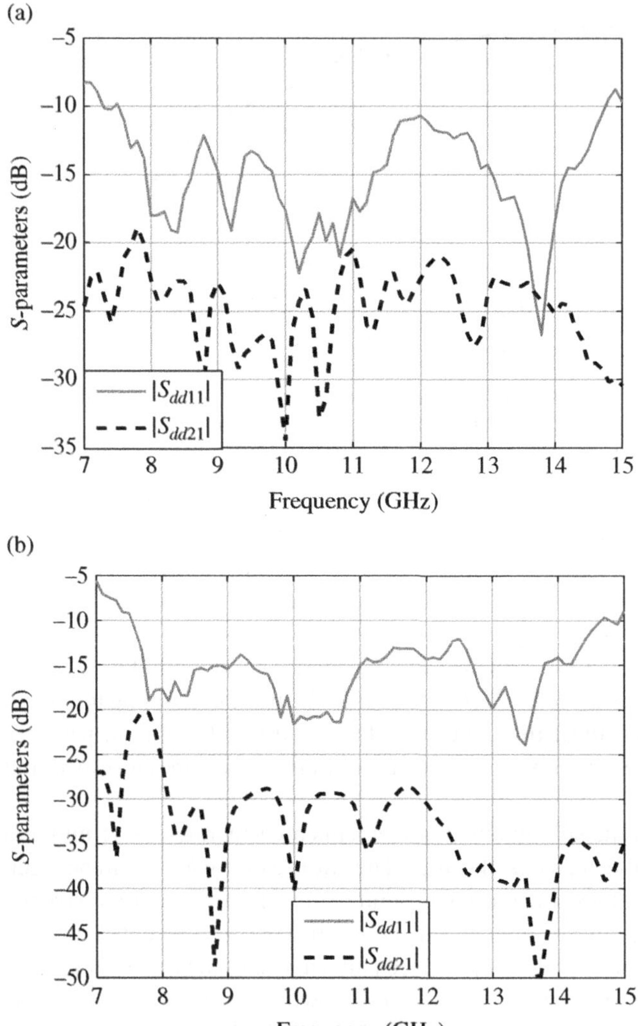

Figure 8.16 Measured $|S_{dd11}|$ and $|S_{dd21}|$ as a function of frequency for the differential quasi-Yagi antennas. (a) Solid, (b) slit, and (c) air holes.

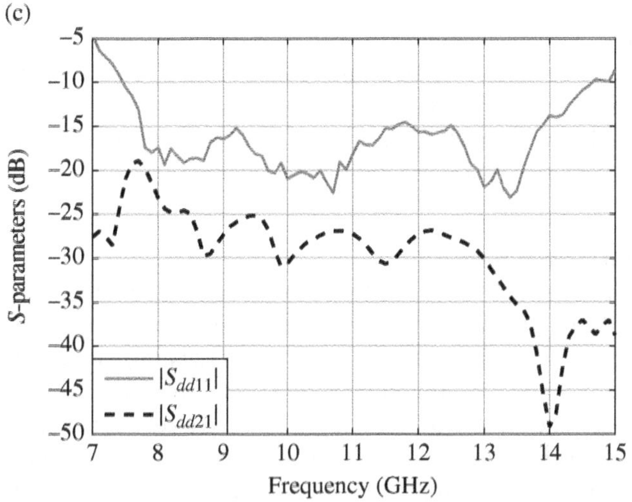

Figure 8.16 (Continued)

8.3 Fractal Antennas

A fractal exhibits a similar shape at an increasingly small scale. It has the property of self-similarity and is useful to design multiband and/or compact antennas [11]. There are a number of fractals. However, only a few are popular in antenna designs. They are Sierpinski gaskets and carpets, Koch snowflakes and islands, Hilbert curve, and Koch curves. Sierpinski gaskets and carpets are used in multiband antennas, Koch snowflakes and islands in miniaturized loop and microstrip patch antennas, and Koch curves in miniaturized dipole antennas [12]. This section only considers printed multiband antennas based on the Sierpinski gasket [13–16]. The interested reader in other fractal antennas is referred to the references in [11, 12]. Section 8.3.1 describes the Sierpinski gasket monopoles, including the original Sierpinski gasket monopole, the design formulas for a normal Sierpinski gasket monopole, and a Sierpinski gasket monopole fed by a microstrip line. Section 8.3.2 presents the Sierpinski gasket dipoles. Special emphasis is given to the analysis, design, and measurement of a Sierpinski gasket dipole fed by a differential CPS line.

8.3.1 The Sierpinski Gasket Monopoles

Since the introduction of the Sierpinski gasket monopole as a multiband antenna by Puente et al. in 1996 [13], there have been two important modifications to the original Sierpinski gasket monopole. One was made by Best who demonstrated that significant portions of self-similar gap structure could be removed from the Sierpinski gasket without adversely its multiband behavior in 2002 [14]. The other was done by Song et al. who modified the Sierpinski gasket monopole to have the ground plane that is planar with the Sierpinski gasket to realize the feeding with a microstrip line in 2003 [15]. The simple feeding and full planar structure will allow further flexibility in the application of these multiband antennas.

8.3.1.1 Original Sierpinski Gasket Monopole

Figure 8.17 shows the structure and coordinate system of the original Sierpinski gasket monopole [13]. Note that an equilateral triangular Sierpinski gasket is printed on a thin dielectric substrate ($\varepsilon_r = 2.17$, $h = 0.127$ mm), mounted upside-down over a square conductor ground plane of $80 \times 80\ \text{cm}^2$, and fed by a single coaxial probe through the ground plane. The Sierpinski gasket has triangles of five different scales. A factor of two exists between the two adjacent scales. Hence, the heights of the triangles are 10.9, 5.45, 2.72, 1.36, and 0.68 cm, respectively. Since the Sierpinski gasket has five different scales, it could be expected that the Sierpinski monopole would behave similarly to a triangular antenna but at five different bands. Measured results showed that the Sierpinski gasket monopole indeed had five different bands at 0.64, 1.62, 3.3, 6.4, and 12.6 GHz. The Sierpinski gasket monopole was not matched at the lowest band with a return loss of 5 dB. Also, the lowest band was slightly shifted from its expected frequency at 0.81 GHz. The poor matching and frequency shifting were most likely caused

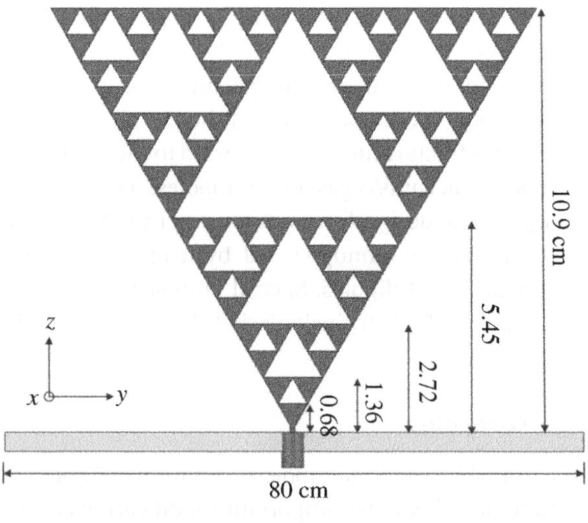

Figure 8.17 Structures of the original Sierpinski gasket monopole and coordinate system.

by the truncation effect. The Sierpinski gasket monopole was well-matched at the other bands with return losses of more than 15 dB.

8.3.1.2 Design Formulas

An improved formula for calculation of the resonant frequencies of a Sierpinski gasket monopole was given by Mishra et al. [16]. The formula was derived from the existing formulas in [15, 17] and can be expressed as:

$$f_r \approx \begin{cases} (0.15345 + 0.34px)\dfrac{c}{h_e}\delta^n & for\ n = 0 \\ 0.26\dfrac{c}{h_e}\delta^n & for\ n > 0 \end{cases} \tag{8.4}$$

where $p = \delta^{-1} - 0.230735$, c is the velocity of light in free space, h_e is the effective height of the largest gasket, and δ is the scale factor of the Sierpinski gasket geometry. The small n denotes the band number. For a conventional Sierpinski gasket geometry, the scale factor is taken to be two and the effective height can be estimated as:

$$h_e = \frac{\sqrt{3}}{2}\left[l + \frac{h}{\sqrt{\varepsilon_r}}\right]. \tag{8.5}$$

Table 8.2 compares the resonant frequencies of the Sierpinski gasket monopole predicted from the formula with measured values in [3, 5, 7]. It is evident from the table that the improved formula can accurately predict the resonant frequencies.

Table 8.2 Calculated and measured resonant frequencies in GHz.

Band	Measured [13] $\varepsilon_r = 2.17$ $h = 0.127$ mm $l = 125.86$ mm	Calculated	Measured [15] $\varepsilon_r = 4.4$ $h = 0.5$ mm $l = 46.19$ mm	Calculated	Measured [17] $\varepsilon_r = 2.5$ $h = 1.588$ mm $l = 102.5$ mm	Calculated
0	0.64	0.67	1.06	1.15	0.52	0.514
1	1.62	1.43	3.75	3.88	1.74	1.74
2	3.3	2.86	7.15	7.76	3.51	3.48
3	6.4	5.72	Not available	15.52	6.95	6.96
4	12.6	11.44	Not available	31.04	13.89	13.92

Given the resonant frequency and the substrate parameters, the formula for calculation of the side length of the generating equilateral triangle can be obtained from simple algebraic rearrangements of equations (8.4) and (8.5) as

$$l \approx \begin{cases} \dfrac{1}{\sqrt{3}}(0.3069 + 0.68px)\dfrac{c}{f_r}\delta^n - \dfrac{h}{\sqrt{\varepsilon_r}} & for\ n = 0 \\ \dfrac{0.52}{\sqrt{3}}\dfrac{c}{f_r}\delta^n - \dfrac{h}{\sqrt{\varepsilon_r}} & for\ n > 0 \end{cases} \quad (8.6)$$

8.3.1.3 A Sierpinski Gasket Monopole Fed by a Microstrip Line

Figure 8.18 shows the structure and coordinate system of the Sierpinski gasket monopole fed by a microstrip line [15]. Note that the whole structure consists of two portions. The upper portion has an equilateral triangular Sierpinski gasket printed on the top surface of a dielectric substrate ($\varepsilon_r = 2.2$, $h = 1.588$ mm). The lower portion has the microstrip line also printed on the top surface of the dielectric substrate. The ground plane is printed on the bottom surface of the lower portion of the dielectric substrate. The Sierpinski gasket has triangles of five different scales. A factor of two exists between the two adjacent scales. The heights of the triangles are 10.9, 5.45, 2.72, 1.36, and 0.68 cm, respectively.

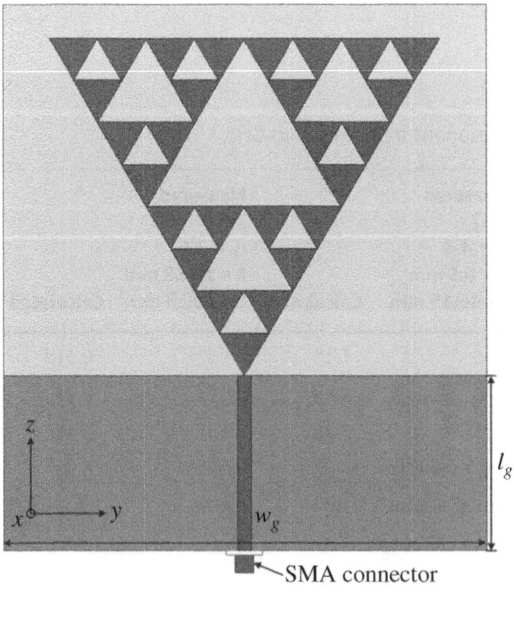

Figure 8.18 Structure of the Sierpinski gasket monopole fed by a microstrip line and coordinate system.

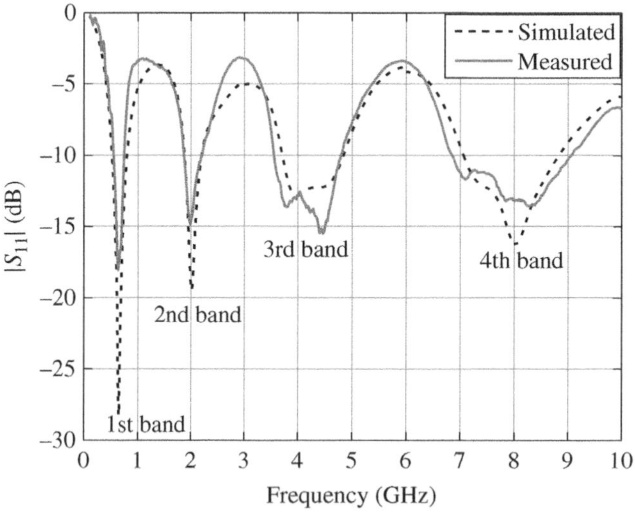

Figure 8.19 Measured and simulated $|S_{11}|$ of the Sierpinski gasket monopole fed by a microstrip line as a function of frequency.

Figure 8.19 shows the simulated and measured $|S_{11}|$ as a function of frequency. It can be seen that the Sierpinski gasket monopole has four different bands at 0.59, 2.00, 4.00, and 8.00 GHz, respectively. The impedance matching is good in all bands.

8.3.2 The Sierpinski Gasket Dipoles

A multiband dipole antenna based on the Sierpinski gasket was reported by Puente et al. in 1998 [17]. The dipole was made of two equilateral triangular gaskets facing each other at both apexes. The two equilateral triangular gaskets were printed on a dielectric substrate and fed by means of a coaxial tapered balun. Basically, the balun was built from a coaxial cable, which makes the multiband dipole antenna a nonplanar one.

Figure 8.20 shows the structure and coordinate system of the Sierpinski gasket dipole fed by a differential CPS line [13]. Note that the Sierpinski gasket dipole is symmetric and consists of two identical portions. The left portion has an equilateral triangular Sierpinski gasket, and a strip line printed on the left surface of a dielectric substrate ($\varepsilon_r = 2.2$, $h = 1.588$ mm). The right portion is printed on the right surface of the dielectric substrate. The Sierpinski gasket has triangles of four different scales. A factor of two exists between the two adjacent scales. The heights of the triangles are 7.707, 3.854, 1.927, and 0.963 cm, respectively.

Figure 8.21 shows the simulated and measured $|S_{d11}|$ of the Sierpinski gasket dipole as a function of frequency. Note that they agree well. The Sierpinski gasket

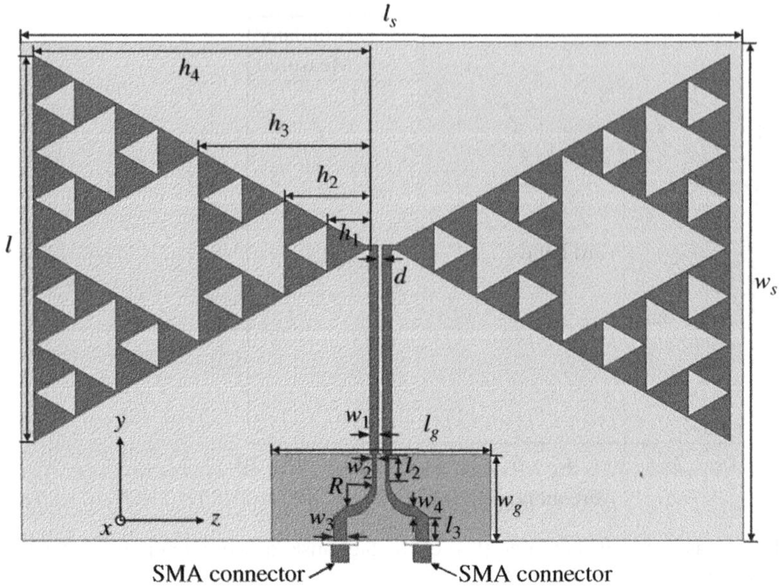

Figure 8.20 Structure of the Sierpinski gasket dipole fed by a CPS line and coordinate system.

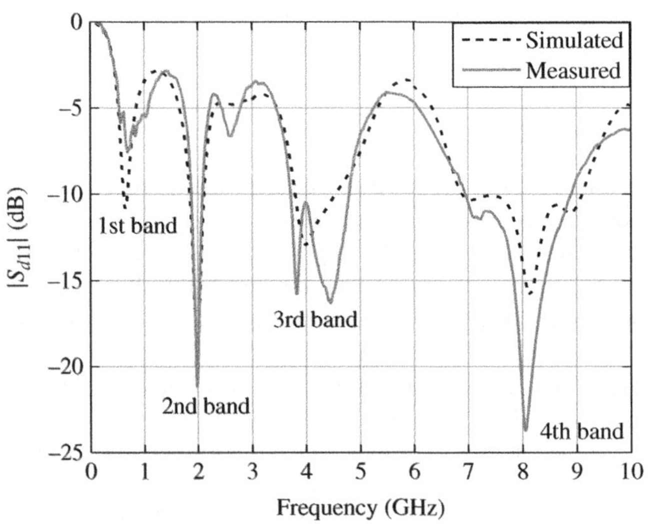

Figure 8.21 Measured and simulated $|S_{d11}|$ of the Sierpinski gasket dipole fed by a CPS line as a function of frequency.

dipole can operate in the four bands at 0.65, 2.00, 4.00, and 8.00 GHz, respectively. The impedance matching is good in all bands.

Figure 8.22 shows the simulated and measured radiation patterns at 2, 4, and 8 GHz in the *E* and the *H* planes. It is interesting to note the similarity between the patterns.

To get a deeper physical insight into the behavior of the Sierpinski dipole antenna, the current density distribution on the dipole was simulated. Figure 8.23 shows the simulated current distributions. The right column presents an expanded view of the region where most of the current is concentrated at each frequency. All of them are scaled by a factor of two, the same scale factor existing among bands. It is interesting to notice that the current density distributions on the right column are very similar among them; that is especially true if the effect of the smaller holes at the lower bands is neglected. This is a reasonable approximation since such holes are small compared to the wavelength and the current density is lower at the regions where the holes are located. Therefore, the similarity among the patterns shown in the previous section can now be explained: at each band, the current concentrates over a properly scaled substructure on the antenna which has the main contribution to the overall radiation pattern.

When an electromagnetic wave is fed through the tip of the antenna, it starts to propagate over the structure toward the flat end of the gasket. When the wave finds a cluster comparable in size to its wavelength, it becomes radiated; this way the power of the traveling wave is lost, and no current reaches the end of the antenna. The structure has many discontinuities that enhance radiation and emphasize such a process. If the wavelength is long, the small triangles of the shape contribute less to the overall radiation and the current can travel further over it; then a larger active region can be reached (which will have a similar shape to that of a shorter wavelength) and a similar radiation pattern for such a lower band will be obtained. Finally, the plots in Figure 8.23 can explain the truncation effect.

8.4 Spiral Antennas

Broadband antennas have always been desirable because of their stable characteristics of impedance, gain, and pattern with frequency. Among practical broadband antennas, planar spiral antennas are particularly attractive because they can be easily fabricated with PCB technology [18]. There are two popular types of printed spiral antennas. They are the Archimedean spiral antennas by Turner in 1953 [19]

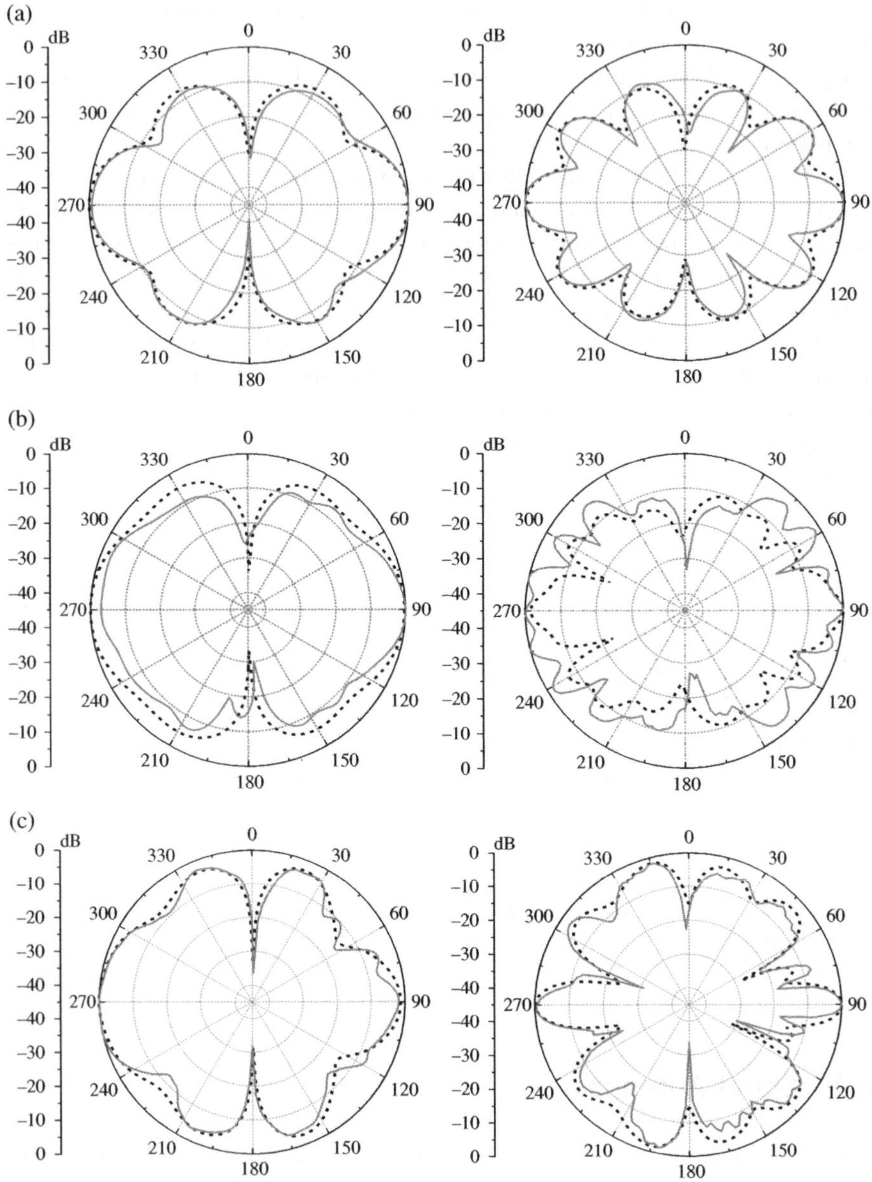

Figure 8.22 Measured and simulated radiation patterns of the Sierpinski gasket dipole fed by a CPS line at the resonant frequencies. (a) 2 GHz, (b) 4 GHz, and (c) 8 GHz.

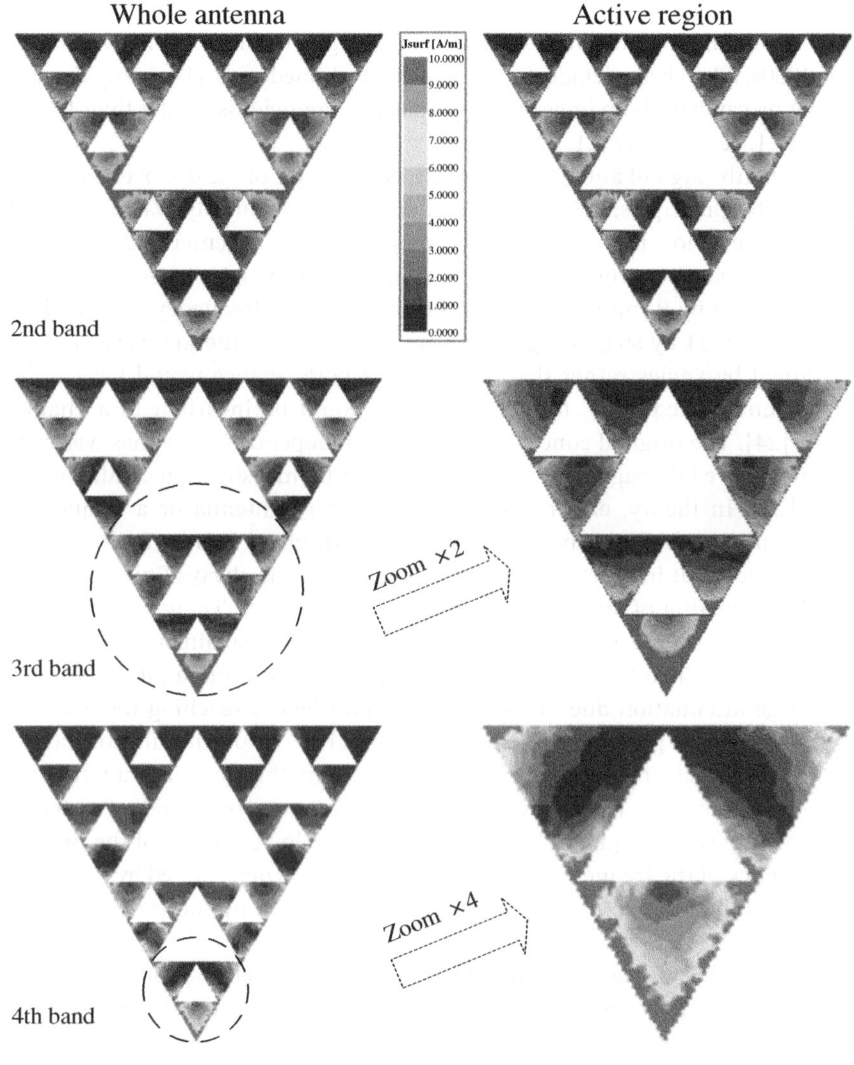

Figure 8.23 Simulated current density distributions of the Sierpinski gasket dipole fed by a CPS line at the resonant frequencies. The left column displays the whole antenna at the four upper bands while the right column shows a zoomed view of the active region at each band.

and the equiangular spiral antennas by Rumsey in 1954 [20]. Before the 1950s, an antenna bandwidth was generally not greater than 4 : 1. By the end of the 1950s, it had been demonstrated that the Archimedean and equiangular spiral antenna can have both impedance and pattern bandwidths greater than 10 to 1 and 20 to 1, respectively [21, 22].

The breakthrough of antenna bandwidth is the result of the discovery of three fundamental principles. The principle of self-complementarity was discovered by Mushiake who originated the self-complementary antenna and found its constant-impedance property in 1948 [23]. The principle of self-similarity was discovered by Rumsey who advanced the theory of frequency-independent antennas in 1954 by suggesting that if the shape of the antenna were entirely determined by angles rather than lengths, the performance would have to be independent of frequency, for such a shape would be invariant to a change of scale [24]. The original concept of frequency-independent antennas was subsequently proved through an extensive experimental investigation conducted by Dyson [25]. In theory, either a self-complementary antenna or a frequency-independent antenna must be extended from the origin to infinity for frequency independence. But in practice, the size of the antenna is always finite, which is not only a function of angle but also a function of length. Therefore, the principle of truncation must also be satisfied, which can be explained by the terminal effect [26]. When the antenna is excited at the feed terminal, if the current has a large attenuation due to effective radiation before reaching the end terminal, even if the part close to the end terminal is cut off, it will not have a significant impact on the antenna performance. In this case, the antenna of a finite size can have the characteristics of a frequency-independent antenna in a relatively wide frequency band. In general, the lower bound of the operating frequency of the frequency-independent antenna is determined by the cut of the end terminal, while the upper bound is determined by the cut of the feed terminal.

In this section, we consider the equiangular spiral antenna printed on a dielectric substrate. Readers interested in the Archimedean spiral antenna printed on a dielectric substrate can refer to [27]. A note on the difference between basic equiangular and Archimedean spiral antennas can be found in [28].

8.4.1 The Equiangular Spiral Antenna

Figure 8.24 shows the structure of a two-arm equiangular spiral antenna and coordinate system. It is printed on a substrate of dielectric constant ε_r, loss tangent δ, length l_s, width w_s, and thickness h. Note that each arm is tapered at its end to provide a better matching termination and is half-bow-tied at its input for better feeding.

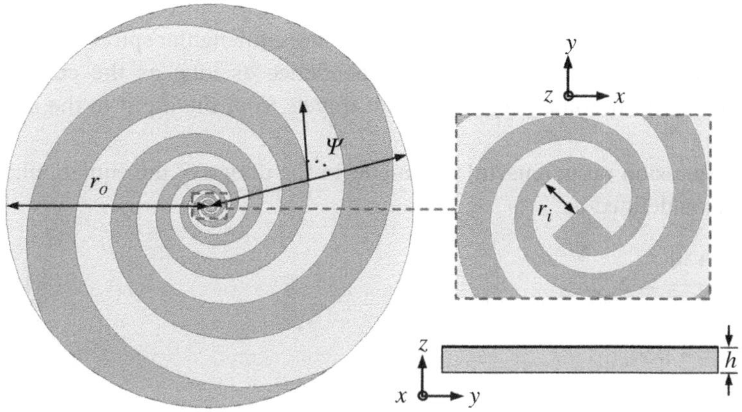

Figure 8.24 Structure of a truncated two-arm equiangular spiral antenna and coordinate system.

Early work on equiangular spiral antennas primarily focused on the operation of the antenna without a dielectric substrate. A few studies have been conducted in recent years to determine the effect of the substrate on the performance of the antenna [27, 29]. In the following, we focus on the frequency characteristics of the antenna in the presence of the substrate, examine the dependence of impedance, pattern, and gain on the dielectric constant and thickness, and explain the results according to the theory by Rumsey [24].

8.4.1.1 Frequency Independence

The shape of an arm of the equiangular spiral antenna is given by the formula in polar coordinates ρ and ϕ as

$$\rho = e^{\alpha(\phi+\phi_0)} = e^{\alpha\phi_0} e^{\alpha\phi} = \rho_0 e^{\alpha\phi} \tag{8.7}$$

where ρ_0 and α are constants. Let us divide both sides of (8.7) by an effective wavelength λ_e, we obtain

$$\frac{\rho}{\lambda_e} = \rho_0 e^{\alpha(\phi-\phi_1)} \tag{8.8}$$

where

$$\phi_1 = \frac{1}{\alpha}\ln(\lambda_e) = \frac{1}{\alpha}\ln\left(\frac{\lambda}{\sqrt{\varepsilon_e}}\right) = \frac{1}{\alpha}\ln\left(\frac{c\sqrt{\varepsilon_e}}{f}\right), \tag{8.9}$$

which shows that a change of wavelength λ or frequency f is equivalent to turning the antenna through the angle $\emptyset_1$. This implies that the equiangular spiral antenna has an inherent property of frequency independence as long as the current strength on the spiral becomes negligible as it travels from the input to the end terminal.

The spiral expansion ratio, a useful parameter in the design of the equiangular spiral antenna, is defined as

$$\zeta = \frac{e^{\alpha(\emptyset+2\pi)}}{e^{\alpha\emptyset}} = e^{2\pi\alpha}. \tag{8.10}$$

Typically, $\zeta \approx 4$ [29].

8.4.1.2 Current Distributions

The current flows along a spiral arm in a standing-wave pattern. The phase along arms 1 and 2 can be expressed as:

$$\psi^{1,2} = \psi_0^{1,2} - \beta_e s = \psi_0^{1,2} - \frac{2\pi}{\lambda_{eff}} s \tag{8.11}$$

where β_e is the effective phase constant, and s is the distance from the origin. Substituting for s into (8.11) yields

$$\psi^1 = \psi_0^1 - \frac{2\pi}{\lambda_e} \frac{e^{-\alpha\emptyset} - 1}{\alpha}, \tag{8.12}$$

$$\psi^2 = \psi_0^2 - \frac{2\pi}{\lambda_e} \frac{e^{-\alpha(\emptyset-\pi)} - 1}{\alpha} \tag{8.13}$$

where $(\emptyset - \pi)$ in (8.13) is due to the half-winding difference between the two arms.

For the differential excitation, $\psi_0^1 = 0$ and $\psi_0^2 = \pi$. The phase difference between the two arms should be

$$\psi^2 - \psi^1 = 2\,m\pi, \quad m = 0, \pm 1, \pm 2... \tag{8.14}$$

or

$$2\pi e^{\alpha\phi} \frac{1 - e^{-\alpha\pi}}{\alpha} = (2\,m - 1)\lambda_e, \quad m = 1, 2, 3, ... \tag{8.15}$$

for constructive radiation in the far field. When $m = 1$, the odd or the first spiral mode is obtained. The equiangular spiral antenna normally operates in this mode.

Now let us examine the current distributions on the equiangular spiral antenna printed on a Rogers RO3006 substrate of $\varepsilon_r = 6.15$, $\delta = 0.002$, $a = b = 32$ mm, and $h = 1.27$ mm. The antenna has the spiral parameters: $r_i = 3$ mm, $r_o = 11.4$ mm, and $\Psi = 79^o$. It is the same antenna as that reported in [29].

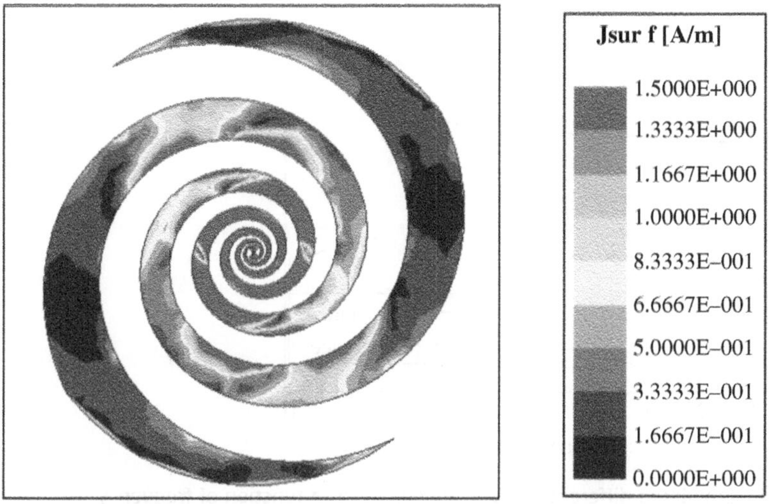

Figure 8.25 Simulated current distribution for the equiangular spiral antenna operating in the odd mode at 3 GHz.

Figure 8.25 shows the simulated current distribution for the equiangular spiral antenna operating in the odd mode at 3 GHz. It can be seen that the current has the same phase and flows in the same direction.

8.4.1.3 Input Impedance

Since the equiangular spiral antenna is self-complementary, its input impedance in free space is approximately 188 Ω. The input impedance would be reduced by a factor of $\sqrt{\varepsilon_r}$ if the antenna were in a homogeneous dielectric space. As the equiangular spiral antenna is in a nonhomogeneous space, the input impedance should be reduced by a factor of $\sqrt{\varepsilon_e}$ rather than the factor of $\sqrt{\varepsilon_r}$. The effective dielectric constant is smaller than the dielectric constant, due to the fact that the spiral arms face the air for the upper hemisphere. For the given substrate thickness, the effective dielectric constant depends on spiral arm width. Because the arm has different widths, it will have different dielectric constants. In the band of operation, the effective dielectric constant near the inner truncation appears to determine the impedance [29].

Figure 8.26 shows the simulated differential input impedance for the antenna operating in the odd mode. It is evident from the simulated result that the input impedance shows three distinct bands in the frequency domain. The lower and upper bands relate respectively to the outer and inner truncations that cause

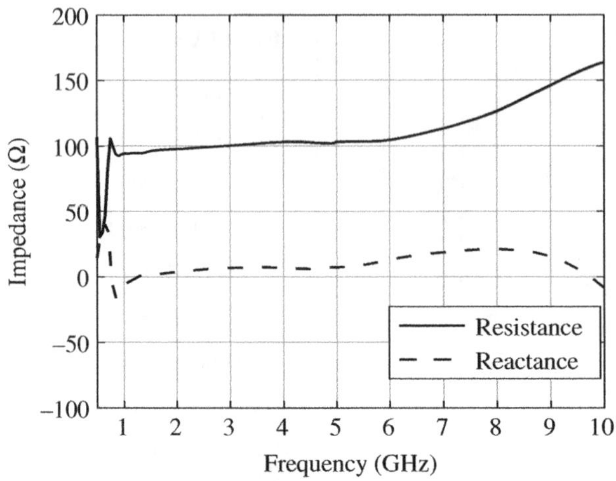

Figure 8.26 Simulated differential input impedance as a function of frequency.

reflections. The middle band from 1 to 5.5 GHz is the band of operation over which the antenna acts as a frequency-independent antenna.

8.4.1.4 Radiation Patterns

Figure 8.27 shows the simulated radiation patterns for the equiangular spiral antenna operating in the odd mode at 1.5, 3.5, and 5.5 GHz. Basically, the radiation patterns reach the peak in the boresight direction.

8.4.1.5 Gain

The equiangular spiral antenna should have a constant gain because it is self-similar. The gain can be estimated as $\pi/\sqrt{\varepsilon_e}$, which reveals that the gain of the equiangular spiral antenna decreases as the effective dielectric constant increases. Figure 8.28 shows the simulated gain values for the equiangular spiral antenna operating in the odd mode. Note that the gain values actually fluctuate. The fluctuation is related to the outer truncation.

8.5 Applications

The differential quasi-Yagi antenna has both the compactness of standing-wave antennas and broadband characteristics of traveling-wave antennas. It can be scaled linearly to any frequency band of interest. Hence, it has found wide

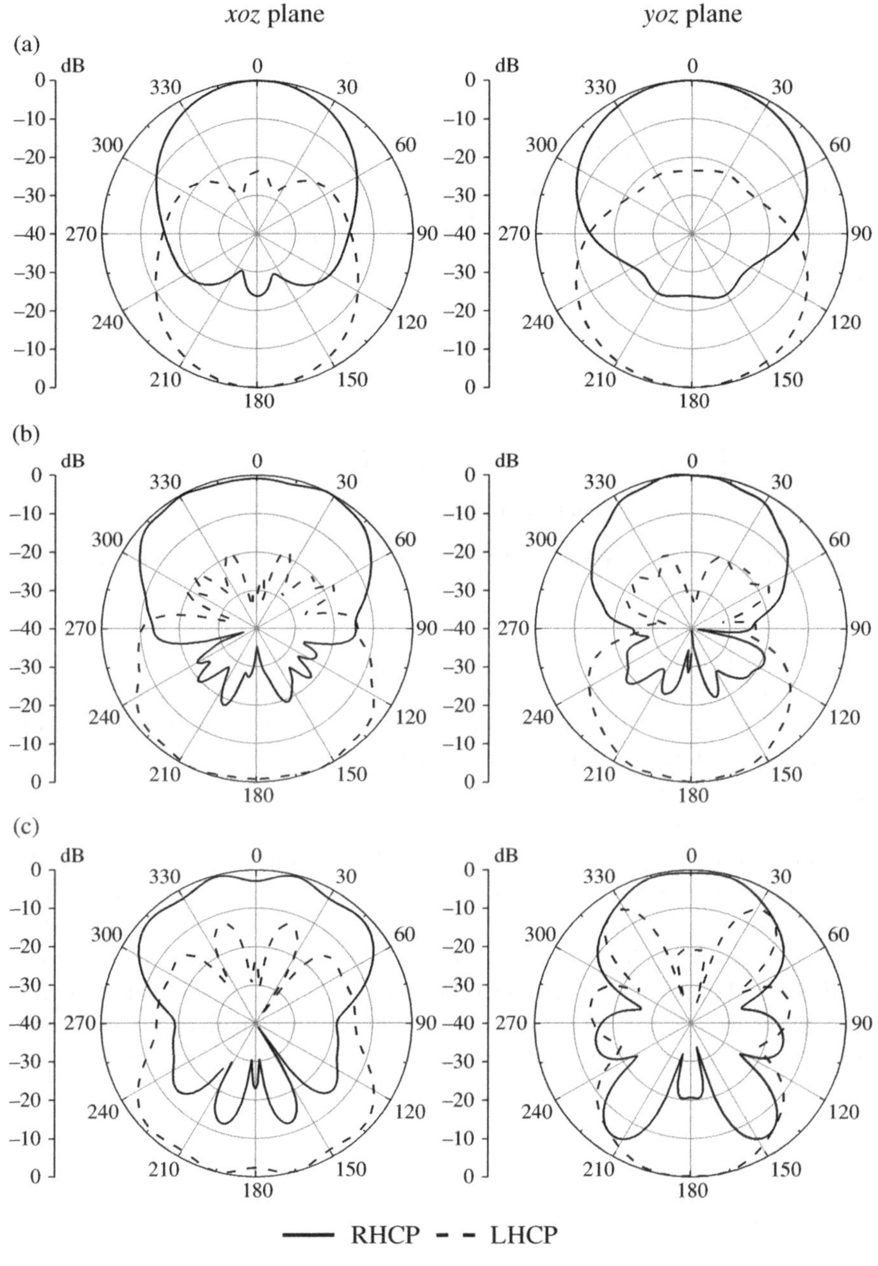

Figure 8.27 Simulated radiation patterns of the equiangular spiral antenna operating in the odd mode. (a) 1.5 GHz, (b) 3.5 GHz, and (c) 5.5 GHz.

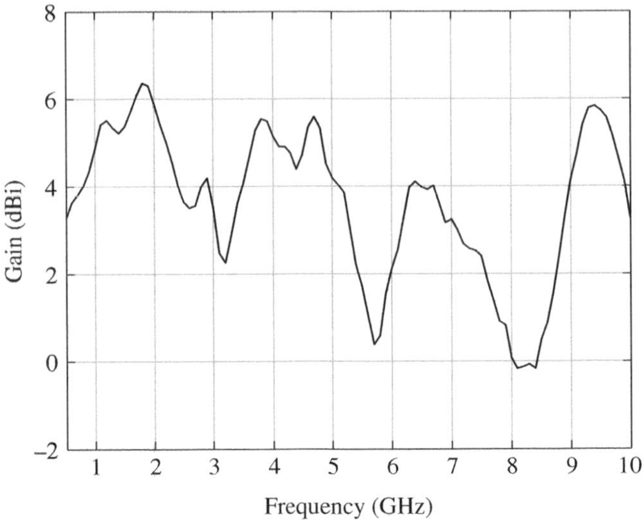

Figure 8.28 Simulated gain values of the equiangular spiral antenna operating in the odd mode.

applications in wireless communication systems [10]. Figure 8.29 shows the photos of the top and bottom views of 60-GHz CMOS radio SoC integrated with a four-element AiP that addresses the challenge of omnidirectional operation in the 60-GHz band using a switched-beam approach to enable broad angular coverage [10]. The AiP has a size of 11 mm × 11 mm × 0.5 mm that supports the differential transmit and single-ended receive paths in both broadside and endfire directions with four antennas. Both single-ended and differential patch antennas are for broadside direction, while both single-ended and differential quasi-Yagi antennas are for endfire direction. Note that the driver is a folded dipole rather than the ordinary dipole. This change facilitates the impedance matching between the driver and CPS from the manufacturable point of view. The simulated antenna peak gains are 4.3 and 5.1 dBi for the differential and single-ended quasi-Yagi antennas, respectively. The simulated impedance bandwidth is at least 4 GHz for both differential and single-ended quasi-Yagi antennas.

Fractal antennas have showed their advantages in real-life applications, such as RFIDs [30], and AiPs [31]. Spiral antennas have been used in various applications including direction finding, ground penetrating radar, global positioning systems, etc. Figure 8.30 shows the photo of the equiangular spiral antenna for UWB radios [32]. It has a size of 50 mm × 50 mm × 25 mm. The impedance bandwidth is from 2.4 to 5.8 GHz. The gain values are 3.5 dBi over the bandwidth.

(a) (b)

Figure 8.29 Photos of 60-GHz radio with quasi-Yagi antennas for portable terminals. (a) Top view and (b) bottom view. *Source:* Gu et al. [10]/with permission of IEEE.

Figure 8.30 Photo of the equiangular spiral antenna for UWB radios.

8.6 Concluding Remarks

The quasi-Yagi, fractal, and equiangular spiral antennas printed on dielectric substrates have been investigated. Particular emphasis has been given to understand the effects of the dielectric substrate on the antenna performance. For the equiangular spiral antennas, it has been found that the effect is a decrease in the impedance and bore-sight gain over the operating band, which becomes more severe as the dielectric constant and substrate thickness are increased.

References

1 Y. X. Qian, W. R. Deal, N. Kaneda, T. Itoh, "Microstrip-fed quasi-Yagi antenna with broadband characteristics," *Electron. Lett.*, vol. 34, no. 23, pp. 2194–2196, Nov.1998.

2 Y. X. Qian, W. R. Deal, N. Kaneda, T. Itoh, "A uniplanar quasi-Yagi antenna with wide bandwidth and low mutual coupling characteristics," in *Proc. IEEE Antenna Propagat. Symp.*, Orlando, FL, 11–16 Jul. 1999.

3 W. R. Deal, N. Kaneda, J. Sor, Y. X. Qian, T. Itoh, "A new quasi-Yagi antenna for planar active antenna arrays," *IEEE Trans. Microw. Theory Technol.*, vol. 48, no.6, pp. 910–918, Jun. 2000.

4 S. X. Ta, B. Kim, H. Choo, I. Park, "Slot-line-fed quasi-Yagi antenna," In *Proc. Int. Symp. Antennas Propagat. EM Theory*, Guangzhou, China, 29 Nov.–2 Dec. 2010.

5 H. D. Lu, L. M. Si, Y. Liu, "Compact planar microstrip-fed quasi-Yagi antenna," *Electron. Lett.*, vol. 48, no. 3, pp. 140–141, Feb. 2012.

6 M. Sun, et al., "Integration of Yagi antenna in LTCC package for differential 60-GHz radio," *IEEE Trans. Antennas Propagat.*, vol. 56, no. 8, pp. 2780–2783, August 2008.

7 N. G. Alexopoulos, P. B. Katehi, D. B. Rutledge, "Substrate optimization for integrated circuit antennas," *IEEE Trans. Microw. Theory Technol.*, vol. 31, no. 7, pp. 550–557, Jul. 1983.

8 M. K. H. Leong, T. Itoh, "Printed quasi-Yagi antennas," in *Printed Antenna for Wireless Communications*, R. Waterhouse, Ed. London: Wiley, 2007, ch. 3, pp. 69–102.

9 Z. H. Zhu, Y. P. Zhang, "Differential quasi-Yagi antenna and array," *ZTE Commun.*, vol. 23, no. 3, pp. 37–44, Sept. 2023.

10 X. X. Gu, D. X. Liu, C. Baks, B. Sadhu, A. Valdes-Garcia, "A multilayer organic package with four integrated 60GHz antennas enabling broadside and end-fire radiation for portable communication devices," in *Proc. IEEE Electronic Comp. and Technol. Conf.*, San Diego, CA, 26–29 May 2015.

11 J. P. Gianvittorio, Y. Rahmat-Samii, "Fractal antennas: A novel antenna miniaturization technique, and applications," *IEEE Antennas Propagat. Mag.*, vol. 44, no. 1, pp. 20–36, Feb. 2002.

12 D. H. Werner, S. Ganguly, "An overview of fractal antenna engineering research," *IEEE Antennas Propagat. Mag.*, vol. 45, no. 1, pp. 38–57, Feb. 2003.

13 C. Puente, J. Romeu, R. Pous, X. Garcia, F. Benitez, "Fractal multiband antenna based on the Sierpinski gasket," *Electron. Lett.*, vol. 32, no. 1, pp. 1–2, Jan. 1996

14 S. R. Best, "On the significance of self-similar fractal geometry in determining the multiband behavior of the Sierpinski gasket antenna," *IEEE Antennas Wireless Propagat. Lett.*, vol. 1, pp. 22–25, 2002.

15 C. T. P. Song, P. S. Hall, H. Ghafouri-Shiraz, "Perturbed Sierpinski multiband fractal antenna with improved feeding technique," *IEEE Trans. Antennas Propagat.*, vol. 51, no. 5, pp. 1011–1017, May 2003.

16 R. K. Mishra, R. Ghatak, D. R. Poddar, "Design formula for Sierpinski gasket pre-fractal planar-monopole antennas," *IEEE Antennas Propagat. Mag.*, vol. 50, no. 3, pp. 104–107, Jun. 2008.

17 C. Puente, J. Romeu, R. Pous, A. Cardama, "On the behavior of the Sierpinski multiband fractal antenna," *IEEE Trans. Antennas Propagat.*, vol. 46, no. 4, pp. 517–524, Apr. 1998.

18 M. McFadden, W. Scott, "Analysis of the equiangular spiral antenna on a dielectric substrate," *IEEE Trans. Antennas Propagat.*, vol. 55, no. 11, pp. 3163–3171, Nov. 2007.

19 E. M. Turner, Spiral Slot Antenna, Wright-Patterson AFB, Ohio, Technical Report. Note WGLR-55-8. WADC, 1955.

20 V. H. Rumsey, Frequency-independent antennas, *IRE Nat. Conv. Record* 5, Pt. 1, pp. 114–118, 1957.

21 J. A. Keiser, "The Archimedean two-wire spiral antenna," *IRE Trans. Antennas Propagat.*, vol. 8, no. 3, pp. 312–323, May 1960.

22 J. D. Dyson, "The equiangular spiral antenna," *IRE Trans. Antennas Propagat.*, vol. 7, pp. 181–187, April 1959.

23 Y. Mushiake, "The input impedance of a slit antenna," *Joint Convention Record of Tohoku Sections of IEE and IECE of Japan*, pp. 25–26, Jun. 1948.

24 V. H. Rumsey, *Frequency Independent Antennas*, Academic Press, Cambridge, MA, USA, 1966.

25 J. D. Dyson, "The unidirectional equiangular spiral antenna," *IRE Trans. Antennas Propagat.*, vol. 7, pp. 329–334, Oct. 1959.

26 S. A. Schelkunoff, *Electromagnetic Waves*, Van Nostrand, New York, 1943.

27 C. Fumeaux, D. Baumann, and R. Vahldieck, "Finite-volume time-domain analysis of a cavity-backed Archimedean spiral antenna," *IEEE Trans. Antennas Propagat.*, vol. 54, no. 3, pp. 844–851, Mar. 2006.

28 J. D. Dyson, R. Bawer, P. E. Mayes, J. I. Wolfe, "A note on the difference between equiangular and Archimedes spiral antennas (correspondence)," *IRE Trans. Microw. Theory Technol.*, vol. 9, no. 2, pp. 203–205, Mar. 1961.

29 M. McFadden, *Analysis of the Equiangular Spiral Antenna*, PhD Thesis, Georgia Institute of Technology, Dec. 2009.

30 L. Ukkonen, L. Sydanheimo, M. Kivikoski, "Read range performance comparison of compact reader antennas for a handheld UHF RFID reader," in *Proc. IEEE Int. Conf. on RFID*, Grapevine, TX, USA, 26–28 Mar. 2007, pp. 63–70.

31 J. S. Castany, C. Puente, J. M. Forn, Antenna package with reduced electromagnetic interaction with on chip elements, US Patent No: US 8330259 B2.

32 Y. P. Zhang, "Differential antennas: Fundamentals and applications," *Electromagn. Sci.*, vol. 1, no. 1, pp. 1–17, Mar. 2023.

9

Differential Antenna Measurements

9.1 Introduction

Antenna measurements ensure the antenna meets the specifications or is simply characterized [1]. Differential antenna measurements with a four-port mixed-mode vector network analyzer (VNA) are like single-ended antenna measurements with a two-port VNA. There should not be any issue with measuring differential antennas. However, since most antenna measurement systems are equipped with only two-port VNAs to measure single-ended antennas, baluns need to be added to the testing facilities to measure differential antennas. It is known that the nonideal nature of real baluns will cause measurement errors. Thus, how to remove the errors in a balun becomes essential in differential antenna measurements. This chapter discusses differential antenna measurements with an emphasis given to the balun calibration techniques. Section 9.2 describes the measurement of differential antenna impedance with the *S*-parameters, balun, and mixed-mode *S*-parameters methods, respectively. Section 9.3 presents the measurement of differential antenna efficiency with the Wheeler Cap and source-stirred chamber (SSC) methods. Section 9.4 examines the measurement of differential antenna radiation patterns and evaluates the effect of imperfect differential excitation on the radiation characteristics. Section 9.5 considers the measurement of differential antenna gain. Finally, Section 9.6 provides concluding remarks.

9.2 Impedance

The differential impedance is required in the design of the matching network between a differential antenna and a differential circuit [2–4]. Three methods for the measurement of the impedance of a differential antenna are the

Differential Antennas: Theory and Practice, First Edition. Yueping Zhang.

S-parameters [5], balun [6], and mixed-mode S-parameters methods [7]. It will be seen that the S-parameters method is probably the easiest method, and yields repeatable results, but cannot measure the radiation properties such as pattern and gain because no differential stimulus is applied to the differential antenna in the measurements. The balun method overcomes the drawbacks of the S-parameters method, and even can measure the impedance and radiation properties simultaneously, which is particularly useful in evaluating the alteration of the impedance as well as the radiation properties of the differential antenna, say, due to proximity effects. However, it will be seen that the removal of the effects of a balun is quite involved. Hence, reported impedance results are the characteristics of balun-antenna combination [8–10]. As the impedances of differential antennas are of prime concern, the effects of the balun must be calibrated out. The mixed-mode S-parameters method is certainly the most sophisticated method but it requires a more expensive and less accessible four-port mixed-mode VNA. Apparently, it is worthwhile making a comparison of the three methods.

9.2.1 The S-Parameters Method

The S-parameters method was reported by Meys and Janssens [5] and later improved by Palmer and Rooyen [11]. The S-parameters method uses a conventional two-port VNA and a custom-made test fixture. The test fixture facilitates the connection of the differential antenna under test (DUT) to the VNA. It was made of two short lengths of 50-Ω microstrip line, etched on printed-circuit boards and mounted back-to-back in [5], and two semi-rigid 50-Ω coaxial cables with their outer conductors soldered together in [11]. The test fixture is not always required. It depends on how the DUT is fed. In [5, 11], the DUTs are dipole antennas fed at the gap between the two arms by twin wires. For DUTs fed by two subminiature version A (SMA) connectors, no test fixture is thus required because the SMA connectors can be directly connected to the two-port VNA. Also, as the standard calibration kits can be used, the measurement can be made more accurately and conveniently with the following steps:

1) The two-port VNA is calibrated at the SMA connectors using the manufacturer's supplied standards and the short-open-load-through (SOLT) method.
2) The two-port S-parameters are measured with the differential antenna in place, as shown in Figure 9.1.
3) The impedance of the differential antenna is derived from the S-parameters as

$$Z_d = 2Z_0\frac{1 - S_{11}S_{22} + S_{12}S_{21} - S_{12} - S_{21}}{(1 - S_{11})(1 - S_{22}) - S_{21}S_{12}} \tag{9.1}$$

where Z_0 is reference impedance of 50 Ω.

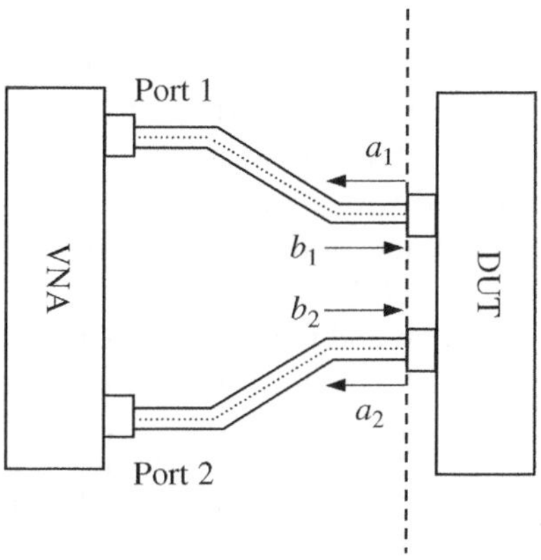

Figure 9.1 The measurement setup of the S-parameters method.

9.2.2 The Balun Method

A balun can be either a 180° power divider or a 180° hybrid coupler. The nonideal nature of the balun will cause measurements made with this method to be in error. Hence, the effects of the balun must be calibrated out. This subsection discusses how to remove the errors of the balun.

9.2.2.1 A 180° Power Divider as the Balun

Zhang and Tu developed the calibration technique to remove the errors of a 180° power divider as the balun [6]. Figure 9.2 shows the measurement setup. The calibration of this balun has four steps [6]. First, a full two-port calibration of the VNA to the end of the coaxial cables used to connect to the balun is performed. The VNA can now be considered ideal at the end of the coaxial cables. Then, the two-port S-parameters of the ten two-port standards with the VNA are measured

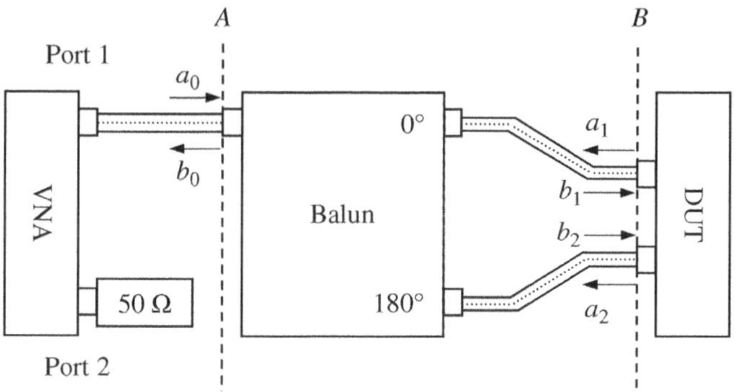

Figure 9.2 The measurement setup of the balun method using a 180° power divider.

and stored. The 10 two-port standards are short-open, short-load, short-short, match-open, match-match, match-short, open-open, open-match, open-short, and through, which can be easily realized by 2 VNA calibration kits. Next, one port of the VNA is connected to the input port of the balun, the other port of the VNA is terminated with a match load, and the one-port S-parameter from the VNA is measured and stored 10 times, each time with a different two-port standard on the output of the balun.

The balun and the output coaxial cables between reference planes A and B are represented by a three-port error network defined as a 3×3 S-parameters matrix

$$\begin{bmatrix} b_0 \\ b_1 \\ b_2 \end{bmatrix} = \begin{bmatrix} e_{00} & e_{01} & e_{02} \\ e_{10} & e_{11} & e_{12} \\ e_{20} & e_{21} & e_{22} \end{bmatrix} \begin{bmatrix} a_0 \\ a_1 \\ a_2 \end{bmatrix} \tag{9.2}$$

where a_i and b_i are the incident and reflected voltage waves at reference planes A and B. The DUT or the calibration standard is characterized at reference plane B by a 2×2 S-matrix

$$\begin{bmatrix} a_1 \\ a_2 \end{bmatrix} = [S_a] \begin{bmatrix} b_1 \\ b_2 \end{bmatrix} \tag{9.3}$$

where S_a is known for a two-port standard but is to be known for the DUT. One can rearrange (9.3) as

$$\begin{bmatrix} b_1 \\ b_2 \end{bmatrix} = [S_a]^{-1} \begin{bmatrix} a_1 \\ a_2 \end{bmatrix} = \begin{bmatrix} m_{11} & m_{21} \\ m_{12} & m_{22} \end{bmatrix} \begin{bmatrix} a_1 \\ a_2 \end{bmatrix}. \tag{9.4}$$

The measurement made at reference plane A yields

$$b_0 = S_m a_0 \tag{9.5}$$

where S_m denotes the measured one-port raw S-parameter data. Due to the reciprocity of the balun and DUT, one can get $e_{01} = e_{10}$, $e_{02} = e_{20}$, $e_{21} = e_{12}$ in (9.2) and $m_{12} = m_{21}$ in (9.4), which leads to

$$\begin{bmatrix} m_{11} & m_{12} \\ m_{12} & m_{22} \end{bmatrix} = \begin{bmatrix} e_{11} - \dfrac{e_{01}^2}{e_{00} - S_m} & e_{12} - \dfrac{e_{02}e_{10}}{e_{00} - S_m} \\ e_{12} - \dfrac{e_{02}e_{10}}{e_{00} - S_m} & e_{22} - \dfrac{e_{02}^2}{e_{00} - S_m} \end{bmatrix}. \tag{9.6}$$

The unknown e_{00}, e_{11}, e_{22}, $e_{01}{}^2$, $e_{02}{}^2$, e_{12}, and $e_{02}\, e_{10}$ can be solved from the above steps [5] through

$$S_m = e_{00} + \frac{\left(e_{01}^2 S_{a11} + e_{02}^2 S_{a22} + 2e_{01}e_{02}S_{a12}\right) + \left(2e_{01}e_{02}e_{12} - e_{02}^2 e_{11} - e_{01}^2 e_{22}\right)\left(S_{a11}S_{a22} - S_{a12}^2\right)}{1 - S_{a11}e_{11} - S_{a22}e_{22} - 2e_{01}S_{a12} + \left(e_{11}e_{22} - e_{12}^2\right)\left(S_{a11}S_{a22} - S_{a12}^2\right)}. \tag{9.7}$$

The S-parameters of the DUT can be obtained as

$$[S_a] = \begin{bmatrix} S_{a11} & S_{a12} \\ S_{a21} & S_{a22} \end{bmatrix} = \begin{bmatrix} e_{11} - \dfrac{e_{01}^2}{e_{00} - S_m} & e_{12} - \dfrac{e_{02}e_{10}}{e_{00} - S_m} \\ e_{12} - \dfrac{e_{02}e_{10}}{e_{00} - S_m} & e_{22} - \dfrac{e_{02}^2}{e_{00} - S_m} \end{bmatrix}^{-1}. \tag{9.8}$$

Finally, the input impedance of the DUT can be solved from

$$Z_{dif} = 2Z_0 \frac{1 - S_{a11}S_{a22} + S_{a12}S_{a21} - S_{a12} - S_{a21}}{(1 - S_{a11})(1 - S_{a22}) - S_{a21}S_{a12}}. \tag{9.9}$$

9.2.2.2 A 180° Hybrid Coupler as the Balun

Zhang and Tu applied the calibration technique developed by Curry [12] to remove the errors of the 180° hybrid coupler as the balun [6]. Figure 9.3 shows the measurement setup. The calibration procedure is as follows. The first step is to perform the full two-port calibration of the VNA to the end of the coaxial cables used to connect the balun. At the reference plane of the end of the coaxial cables, the VNA can then be considered ideal. The second step is to connect the two ports of the VNA to the Σ and Δ ports of the balun, and to measure the two-port S-parameters from the VNA five times, each time with a different standard on the balun's output. The S-parameters of the five standards are assumed to be exactly known. The third step is to determine the S-parameters of the actual balun. The S-parameters, $\boldsymbol{E}$, and T-parameters, $\boldsymbol{T}$ (transfer or chain-scattering matrix) of the balun are defined by

$$\begin{bmatrix} b_0 \\ b_3 \\ b_1 \\ b_2 \end{bmatrix} = E \begin{bmatrix} a_0 \\ a_3 \\ a_1 \\ a_2 \end{bmatrix} = \begin{bmatrix} E_1 & E_2 \\ E_3 & E_4 \end{bmatrix} \begin{bmatrix} a_0 \\ a_3 \\ a_1 \\ a_2 \end{bmatrix} = \begin{bmatrix} T_2T_4^{-1} & T_1 - T_2T_4^{-1}T_3 \\ T_4^{-1} & -T_4^{-1}T_3 \end{bmatrix} \begin{bmatrix} a_0 \\ a_3 \\ a_1 \\ a_2 \end{bmatrix}, \tag{9.10}$$

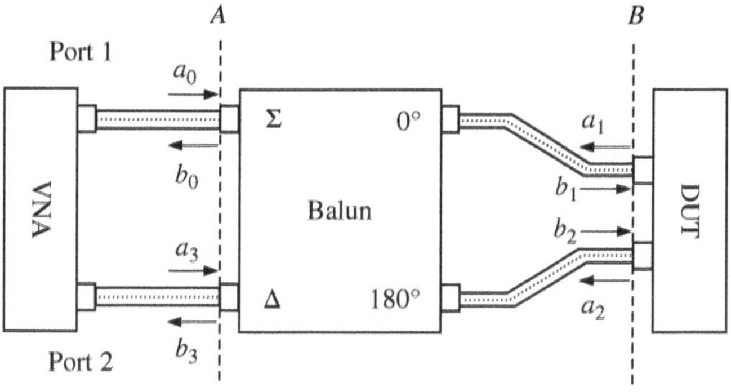

Figure 9.3 The measurement setup of the balun method using a 180° hybrid coupler.

$$\begin{bmatrix} b_0 \\ b_3 \\ a_0 \\ a_3 \end{bmatrix} = T \begin{bmatrix} a_1 \\ a_2 \\ b_1 \\ b_2 \end{bmatrix} = \begin{bmatrix} T_1 & T_2 \\ T_3 & T_4 \end{bmatrix} \begin{bmatrix} a_1 \\ a_2 \\ b_1 \\ b_2 \end{bmatrix} = \begin{bmatrix} E_2 - E_1 E_3^{-1} E_4 & E_1 E_3^{-1} \\ -E_3^{-1} E_4 & -E_3^{-1} \end{bmatrix} \begin{bmatrix} a_1 \\ a_2 \\ b_1 \\ b_2 \end{bmatrix} \tag{9.11}$$

where $\boldsymbol{E_i}$ and $\boldsymbol{T_i}$ are 2×2 submatrices [13].

The determination of the *S*-parameters of the actual balun requires solving the calibration equation, given by

$$T_1 S_a + T_2 = S_m (T_3 S_a + T_4) \tag{9.12}$$

where the S_a are the actual *S*-parameters of the calibration standard or of the DUT, and the S_m are the measured (uncorrected) *S*-parameters of the calibration standard or of the DUT. Each measurement of a standard results in four equations. Theoretically, measurements with 4 different standards would provide 16 equations in 16 unknowns to determine submatrices T_1 to T_4. However, it was found from simulation that the equations are singular for any four standards. Measurements with five different standards are hence strictly needed [13]. Measurements of five different standards result in a system with more equations than unknowns. A unique solution to the system is unlikely. The singular-value decomposition needs to be used to find a least-squares solution [12]. Once the *T* matrix is known, the *S*-parameters, *E*, of the actual balun, can be found from (9.10).

Having calibrated the VNA and the balun, we now proceeded to find the impedance of the DUT. The embedded *S*-parameters of the DUT (S_m) were first measured, and then the de-embedding was performed using the following equation

$$S_a = (T_1 - S_m T_3)^{-1} (S_m T_4 - T_2). \tag{9.13}$$

Finally, the impedance of the DUT is found from

$$Z_d = 2Z_0 \frac{1 - S_{a11} S_{a22} + S_{a12} S_{a21} - S_{a12} - S_{a21}}{(1 - S_{a11})(1 - S_{a22}) - S_{a21} S_{a12}} \tag{9.14}$$

where Z_0 is reference impedance of 50 Ω.

9.2.3 The Mixed-Mode *S*-Parameters Method

The theory of the mixed-mode *S*-parameters was developed by Bockelman and Eisenstadt for general microwave differential-circuit characterization [14, 15]. The mixed-mode *S*-parameters given in (1.21) can be measured by either a four-port mixed-mode VNA or a four-port VNA. It was found that the four-port mixed-mode VNA has a higher accuracy than the four-port VNA when measuring a differential device. However, it was also shown that when a differential device is

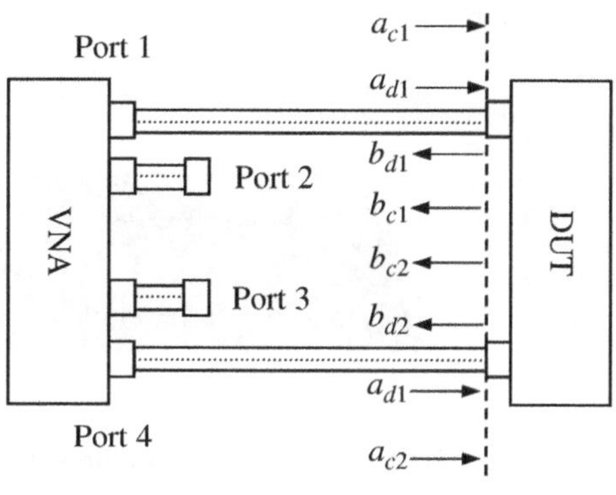

Figure 9.4 The measurement setup of the mixed-mode S-parameters method.

measured by a four-port mixed-mode VNA, the mixed-mode S-parameters (S_{dd} and S_{cc}) have about equal-magnitude error of the corresponding parameters of the transformed four-port S-parameters measured by a four-port VNA; while the mode conversion terms (S_{dc} and S_{cd}), as measured by a four-port mixed-mode VNA, can have substantially lower error than those measured by a four-port VNA [14, 15]. The measurement setup of the mixed-mode S-parameters method is shown in Figure 9.4. The measurement involved the following steps:

1) The four-port mixed-mode VNA is calibrated at the SMA connectors using the manufacturer's supplied standards.
2) The unused two ports are terminated with matched loads or left open.
3) The mixed-mode S-parameters are measured with the DUT in place, and under true differential stimulus signals of identical amplitude and with 180° phase shift, directly at the DUT.
4) The impedance of the DUT is derived from the mixed-mode S-parameters as

$$Z_d = 2Z_0 \frac{1 + S_{d11}}{1 - S_{d11}} \tag{9.15}$$

where Z_0 is reference impedance of 50 Ω.

9.2.4 Comparison of the Methods

An R&S four-port ZVB8 network analyzer was used in the measurements. It could measure the mixed-mode S-parameters of a differential device and the S-parameters of a single-ended device from 300 kHz to 8 GHz. A hybrid coupler J-212-180 from ET Industries was used as the balun.

The broadband differential load was made of two single-ended loads. It had a known differential impedance of 100 Ω from DC to 3 GHz. The narrowband

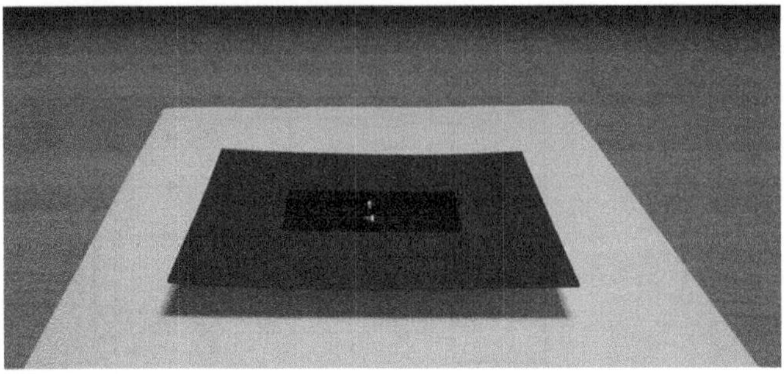

Figure 9.5 Photo of the fabricated differential rectangular microstrip patch antenna.

differential microstrip patch antenna was constructed on an RT/Duriod 5880 substrate with a thickness of 0.787 mm. To have sufficient physical separation to make it possible to feed the antennas with 3.5 mm SMA connectors, we designed it to operate at 2.4 GHz, and determined the patch dimensions of width 63 mm and length 40 mm on the middle of the substrate using the cavity model. Figure 9.5 shows a photo of the fabricated narrowband differential microstrip patch antenna. Note that the existing warpage was so irregular that it made it almost impossible to accurately simulate the structure.

Figure 9.6 shows a photo of one-port male and female short, open, and match standards, together with a two-port thru from a 3.5 mm calibration kit. They were employed to realize two-port standards for the balun method. Due to sexed connectors in the VNA and balun, the two-port standards had to be female. One-port male short, open, and match standards were transformed into one-port female standards with female-to-female adaptors and denoted as short(T), open(T), and match(T), respectively.

The short-short(T) represented a two-port standard for which the one-port female short standard was connected to Port 1 of the VNA or Port 0° of the balun, the transformed one-port female short standard was connected to Port 2 of the

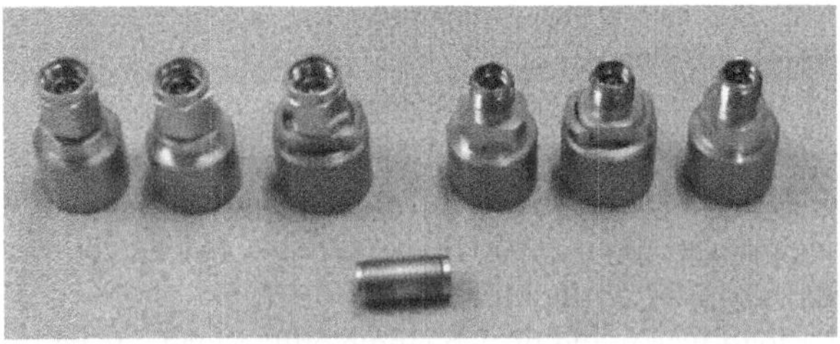

Figure 9.6 Photo of the male and female standards.

VNA or Port 180° of the balun, and vice versa. The same description was also true for the open-open(T) and match-match(T) two-port standards.

Based on our extensive measurements, we found that one standard must be the thru. Two standards can be any two pairs of open(T)-open, short(T)-short, and match(T)-match. The other two standards can be any two pairs of match-open, match-short, short-match, and open-match. Two-port standards, such as short-open or open-short, cannot be used. Alternatively, the first four standards can be thru, match(T)- match, short(T)-short, open(T)-open; the last standards can be any.

Figure 9.7 shows the measured results of the broadband balanced load from 1.5 to 7.5 GHz. The three combinations in Table 9.1 follow the above guidelines. Note that the three combinations can work over the whole frequency range. After the removal of the balun effects, respectively, with the three combinations, the differential impedance was indeed very close to the known value of 100 Ω.

Figure 9.8 compares the measured differential impedances of the broadband balanced load from 1.5 to 7.5 GHz obtained with the three methods. It is seen that the measured differential impedances are almost the same from the three methods. Table 9.2 shows the uncertainties of using each of the three methods, which were calculated from the variations in data obtained from several measurements over the frequency range from 1.5 to 3 GHz, where the true differential impedance value was known. The largest uncertainties of the balun method can be attributed to the calibration standards, which had errors at each frequency.

Little is known about the measured impedances of differential microstrip patch antennas from the balun method. Most reported results were the characteristics of a balun-antenna combination [8–10]. Figure 9.9 compares the measured impedances of the narrowband differential rectangular microstrip patch antenna from 2.0 to 3.0 GHz obtained from the three methods. It is evident that the impedance through the balun by the VNA, the raw data, was not correct. The calibrated impedance of the balun method is almost the same as the values obtained from the other two methods.

The three methods were first validated by measuring the known 100 Ω impedance of the broadband differential load, and then employed to measure the unknown impedance of the narrowband differential rectangular microstrip patch antenna. The measured impedances were almost the same as those obtained by the three methods. On the whole, the S-parameters and mixed-mode S-parameters methods were easy to implement. The difference was that the S-parameters method used a common two-port VNA, while the mixed-mode S-parameters method required a more expensive and less accessible four-port mixed-mode VNA. Although the calibration was more involved, the balun method made it possible to not only measure the impedance but also measure the radiation patterns and gain of the differential antenna using a common two-port VNA.

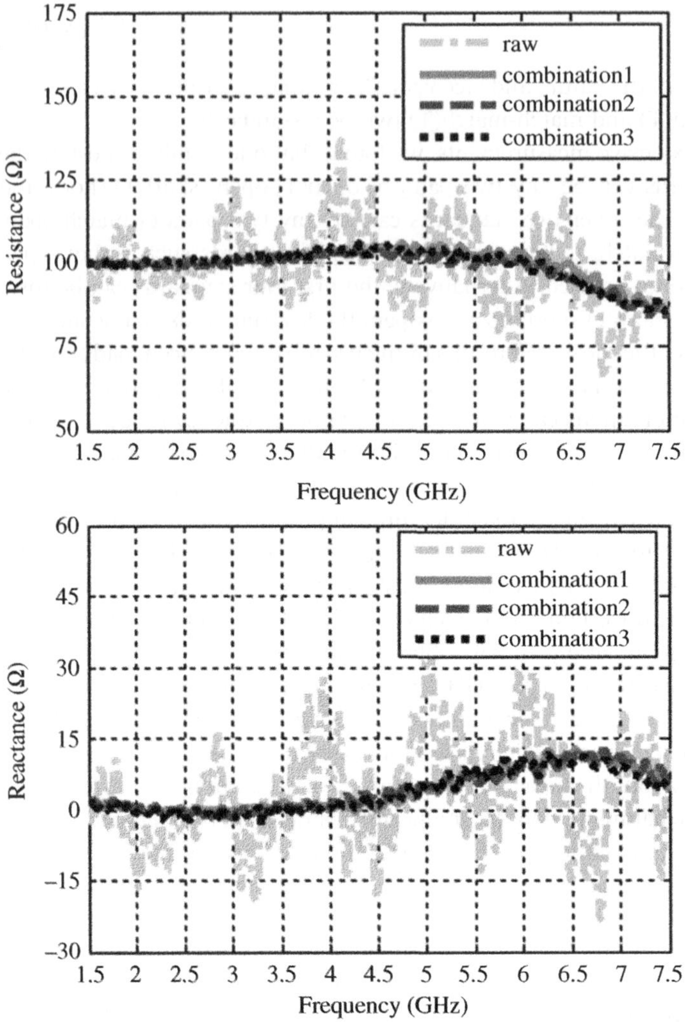

Figure 9.7 The calibration effects of three appropriate combinations of five standards on the measured impedances of the broadband differential load.

Table 9.1 Three appropriate combinations of the five standards.

Combination 1	Combination 2	Combination 3
Thru	Thru	Thru
Short(T)-short	Match(T)-match	Match(T)-match
Open(T)-open	Open(T)-open	Short(T)-short
Short-match	Match-short	Open(T)-open
Open-match	Open-match	Open-short

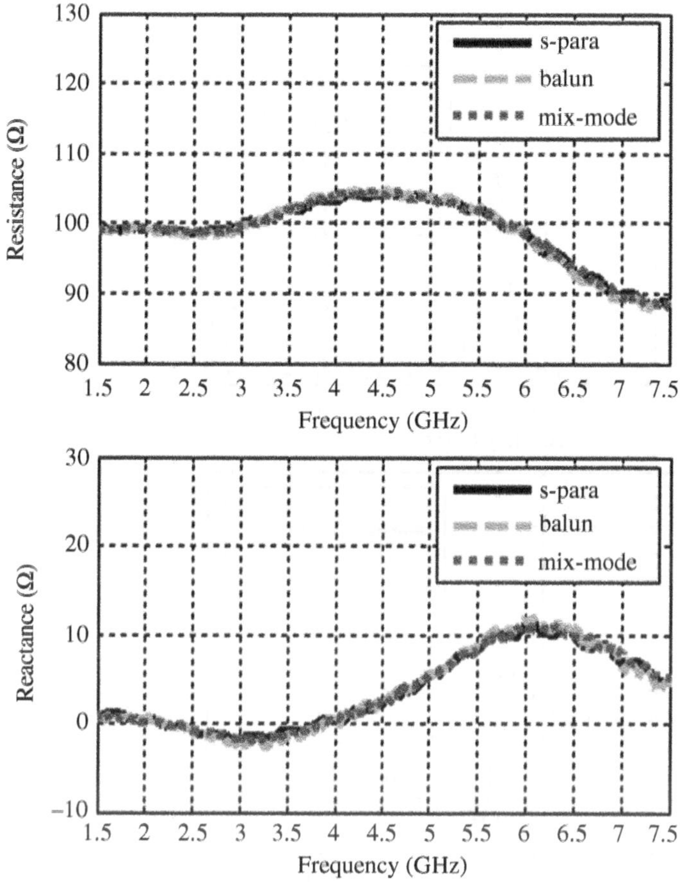

Figure 9.8 The measured differential impedances of the broadband balanced load.

Table 9.2 Measured differential impedances and uncertainties.

	Measurement methods		
Balanced load impedance	***S*-parameter**	**Balun**	**Mix-mode**
Resistance (Ω)	100.0 ± 1.5	100.0 ± 2.4	100.0 ± 1.6
Reactance (Ω)	0.0 ± 1.1	0.0 ± 1.7	0.0 ± 1.1

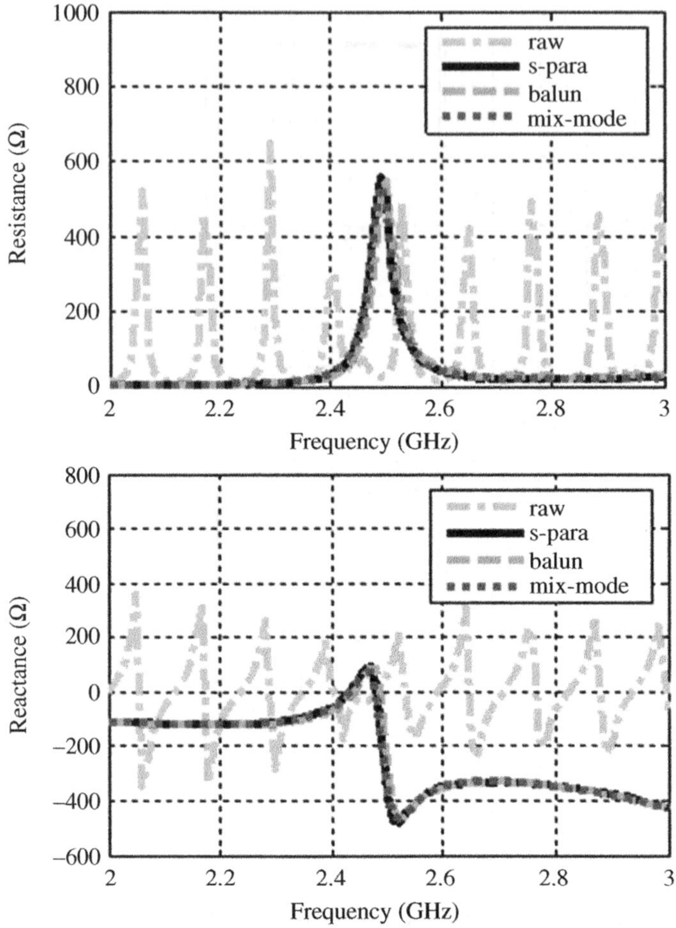

Figure 9.9 The measured impedances of the narrowband differential rectangular microstrip patch antenna.

9.3 Efficiency

The efficiency, an important antenna parameter, is defined as the ratio of the power radiated by an antenna to the power delivered to the antenna. Pozar and Kaufman argued that measurement is often the only way to reliably determine antenna efficiency because factors such as surface roughness, tolerance effects, and spurious radiation are difficult to account for in a theoretical calculation and can have a drastic effect on the actual antenna efficiency. Thus, they evaluated three methods for the measurement of printed antenna efficiency. They showed that the Wheeler Cap method is probably the easiest to implement and gives the most repeatable results. The radiometric method is probably the most complex but is capable of giving good results if high-quality components are used carefully. The directivity/gain method is probably the most common method but is highly subject to inaccuracies and uncertainties [16].

9.3.1 The Wheeler Cap Method

According to the Wheeler Cap method [16–18], if a differential antenna can be modelled by a series *RLC* circuit, its efficiency can then be calculated as

$$\eta = \frac{Re\left(Z_d^{fs}\right) - Re\left(Z_d^{wc}\right)}{Re\left(Z_d^{fs}\right)}, \tag{9.16}$$

while if the differential antenna can be modelled by a parallel *RLC* circuit, its efficiency needs to be calculated as

$$\eta = \frac{Re\left(Z_d^{wc}\right) - Re\left(Z_d^{fs}\right)}{Re\left(Z_d^{wc}\right)} \tag{9.17}$$

where $Re\left(Z_d^{fs}\right)$ and $Re\left(Z_d^{wc}\right)$ are the real parts of the impedances of the differential antenna at the resonant frequency in free space and in a Wheeler Cap, respectively [19].

The R&S four-port network analyzer ZVB8 was used in the measurements. Three probe-fed differential microstrip antennas were designed and fabricated as the test antennas. The first antenna was a rectangular differential microstrip antenna on a lossy FR4 substrate for operating at 2 GHz ($\lambda = 150$ mm). The patch dimensions were 74.4 mm wide and 35 mm long on the grounded substrate of size 150 mm × 150 mm × 1.6 mm. The second antenna was a circular differential microstrip antenna on a low-loss RT/duriod 5880 substrate for operating at 2.4 GHz ($\lambda = 125$ mm). The patch had a radius of 23 mm on the grounded substrate of size 125 mm × 125 mm × 0.787 mm. The third antenna was a rectangular differential microstrip antenna on a low-loss RT/duriod 5880 substrate for operating at 3.8 GHz ($\lambda = 79$ mm). The patch dimensions were 40 mm wide and 25 mm long on the grounded substrate of size 79 mm × 79 mm × 0.787 mm. SMA connectors were soldered to the three antennas for testing. The ground planes of the three antennas were short-circuited to a large aluminum ground plane, respectively, for mounting the aluminum cubic cap of size 230 mm × 230 mm × 20 mm. The guidelines in [16] such as making good electrical contact between the cap and the ground plane and centering the cap over the differential microstrip antenna were followed in our measurements.

The $|S_{d11}|$ of the three antennas were measured in free space. The first antenna had a bandwidth ($|S_{d11}| \leq -10$ dB) from 1.96 to 2.01 GHz with the smallest $|S_{d11}|$ of −28 dB at 1.9875 GHz. The second antenna had a bandwidth from 2.44 to 2.46 GHz with the smallest $|S_{d11}|$ of −19 dB at 2.4537 GHz. The third antenna had a bandwidth from 3.82 to 3.88 GHz with the smallest $|S_{d11}|$ of −15.56 dB at 3.8512 GHz.

The impedances of the three differential antennas were measured with the mixed-mode *S*-parameters method. It was found that the first antenna did not resonate as its impedance did not show zero reactance, while the second and third antennas resonated because their impedances showed zero reactance. The resonances occurred at 2.4350 GHz as the parallel and at 2.4650 GHz as the series resonances, respectively, for the second antenna. Similarly, the resonances occurred at 3.7425 GHz as the parallel and at 3.87 GHz as the series resonances, respectively, for the third antenna. It was also found that the three antennas resonated in the cap. However, the resonant frequencies in the cap were slightly different from those in free space. The shift of resonant frequencies was attributed to the presence of the remaining cavity modes in [18] and here to the perturbation of the current distributions of the antennas by the cap, indicating the Wheeler method may not accurately yield the efficiency of microstrip antennas.

It is well known that a microstrip antenna operates at the fundamental mode, that is, the TM_{01} mode for a rectangular patch and TM_{11} for a circular patch. The resonant frequency of the fundamental mode is different from the parallel and series resonant frequencies of the microstrip antenna. The resonant frequency of the fundamental mode is the frequency at which the resistance reaches the peak value, and the reactance is equal to the average of sum of its maximum and minimum values of the microstrip antenna. Ideally, the reactance of the fundamental mode at the resonant frequency should be zero. The positive reactance of the microstrip antenna at the resonant frequency of the fundamental mode is due to the feed inductance and existence of excited higher-order modes. If the feed inductance and excited higher-order modes were negligible, the resonance of the fundamental mode would occur at the same frequency as the parallel resonance of the microstrip antenna and the series resonance of the microstrip antenna would not occur. In other words, the microstrip antenna should be modelled by a parallel *RLC* circuit; the efficiency (rigorously speaking, the radiation efficiency) should be calculated according to (9.16) at the resonant frequency of the fundamental mode rather than at the parallel or series resonant frequencies of the microstrip antenna.

Based on the above elucidation and the peak resistances in the cap being larger than those in free space, we can conclude that it is the parallel *RLC* circuit model that should be applied to the three differential microstrip antennas to calculate their radiation efficiencies (or simply speaking, efficiencies). It should be also clear

Table 9.3 Measured and simulated efficiencies for differential microstrip patch antennas.

Antennas	Measured	Simulated
1st	49.03%	41.9%
2nd	75.46%	82.55%
3rd	88.77%	89.82%

that the operation efficiencies of the three differential microstrip antennas at their operating frequencies where the $|S_{d11}|$ go to the minimum values should be smaller than the measured radiation efficiencies.

Table 9.3 lists the measured and simulated efficiencies for the three differential microstrip patch antennas. The simulated efficiencies were obtained with the HFSS using the nominal substrate loss tangent values. The measured uncertainty is consistent with well under ±0.08% for the first antenna, ±0.25% for the second antenna, and ±0.36% for the third antenna. The uncertainties were estimated from the variations in measured data from one measurement to the next; each of the measurements was made several times [16]. The simulated efficiency is lower than the measured one by 7% for the first antenna, larger than the measured one by 7% for the second antenna, and larger than the measured one by 1% for the third antenna. The difference between the simulated and measured efficiencies is similar to those reported in [18].

Table 9.4 lists the measured and simulated efficiencies for the three single-ended antennas, that is, one feed connected to a 50-Ω single-ended source, and the other feed left open. Comparing with the data given in Table 9.3, one can see that there is a significant difference between the efficiencies of the first antenna driven differentially with dual probe feeds and "single-ended-ly" with one probe feed. This can be explained by the patch dimensions of the first antenna, which were determined properly for a differential excitation but improperly for a single-ended excitation.

Table 9.4 Measured and simulated efficiencies for single-ended microstrip patch antennas.

Antennas	Measured	Simulated
1st	18.71%	17.82%
2nd	75.26%	83.50%
3rd	88.67%	91.12%

The patch width is about two times the patch length. The undesirable TM_{20} mode was suppressed by the differential excitation but energized by the single-ended excitation in the band. As a result, the efficiency of the first antenna-driven "single-ended-ly" with one probe feed is much lower than that driven differentially with dual probe feeds. The patch dimensions for the second and third antennas were properly determined for both excitations. Hence, the efficiencies of the second and third antennas driven differentially with dual probe feeds and "single-ended-ly" with one probe feed are quite similar. The measured uncertainty is consistent with well under ±0.25% for the first antenna, ±0.02% for the second antenna, and ±0.20% for the third antenna.

The application of the Wheeler method for the measurement of the efficiency of differential microstrip antennas was made. It was shown that the Wheeler method yields efficiencies repeatable within ±0.40%. It was also shown that although additional parasitic inductance and capacitance are introduced from the dual feeds, the probe-fed differential microstrip antennas may exhibit nonresonance (the first antenna) or parallel and series resonances (the second and third antennas) within the band of operation; the parallel *RLC* model is the valid model to use for them. More importantly, it was found that a significant difference may exist between the efficiencies of a microstrip antenna driven differentially with dual probe feeds or "single-ended-ly" with one probe feed, depending on the choice of patch dimensions.

9.3.2 The Source-Stirred Chamber Method

The Wheeler Cap method is generally suitable for narrowband small antenna measurements. Huang et al. proposed an SSC method for a wide range of antennas (electrically small and large, narrow, and wideband antennas). They demonstrated that the SSC method is accurate, cost-effective, convenient, and efficient [20]. Hence, one can use it to measure differential antenna efficiency. The measurement procedure has four steps. The first step is to measure the S_{d11} of the antenna in free space S_{d11fs} over the frequency range of interest. The second step is to measure the S_{d11} of the antenna inside the SSC at various positions and at different (three orthogonal) orientations to obtain as many S_{d11ssc} as possible. The third step is to calculate the radiation efficiency by

$$\eta = \sqrt{\frac{|S_{d11ssc}|^2 - |S_{d11fs}|^2}{1 - 2|S_{d11fs}|^2 + |S_{d11ssc}|^2 |S_{d11fs}|^2}} \tag{9.18}$$

The fourth step is to plot all radiation efficiency values obtained on the same plot as a function of the frequency and link the peaks to finally yield the radiation efficiency of the differential antenna.

9.4 Radiation Pattern

A real balun is imperfect. Its amplitude and phase imbalances can be up to 1 dB and 10°, respectively. The effects of such imbalances on the radiation patterns of the differential microstrip patch antenna were discussed in [21]. It was shown that they degrade the polarization purity in the *H*-plane. The degradation by the phase imbalance of 10° is more severe than that by the amplitude imbalance of 1 dB. The degradation can be understood from the fact that they decrease the cancellation extent. The higher-order modes, in particular the TM_{20} mode, cannot be sufficiently suppressed. Consequently, the polarization purity in the *H*-plane is degraded. In addition, they do not alter the radiation mechanism of the microstrip patch antenna, the shapes of radiation patterns, especially in the *E*-plane, are therefore not changed.

Recently, Hou et al. simulated the effects of imperfect differential excitation on the radiation patterns of the differential laminated resonator antenna for four cases [22]. The first case is the perfect case for which amplitude and phase imbalances were 0 dB and 0°. The other three cases are imperfect for which amplitude and phase imbalances are 1 dB and 0°, 0 dB and 10°, and 1 dB and 10°, respectively. Figure 9.10 shows the simulated radiation patterns at 5.65 GHz. It is seen that the co-polarized radiation patterns hardly change in the *E* and *H* planes with the imperfect differential excitations. However, the cross-polarized radiation pattern in the *H* plane degrades with the imperfect differential excitations. The phase imbalance of 10° causes more degradation than the amplitude imbalance of 1 dB.

Figure 9.11 compares the simulated and measured *E*-plane and *H*-plane radiation patterns of the differential laminated resonator antenna at 4.35, 5, and 5.65 GHz, respectively. It is evident that the simulated and measured co-polarized radiation patterns agree very well for most directions. There are differences between the cross-polarized radiation patterns, which are quite weaker and hard to measure accurately. The large discrepancies in the co-polarized radiation patterns near the backsight direction are due to the limitation of the antenna anechoic chamber.

9.5 Gain

There are two basic methods that can be used to measure the gain of a differential antenna: absolute gain and gain comparison techniques. The absolute gain method requires no a priori knowledge of the transmitting or receiving antenna gain. If the receiving and transmitting antennas are identical, one measurement and use of the transmission formula is sufficient to determine the gain.

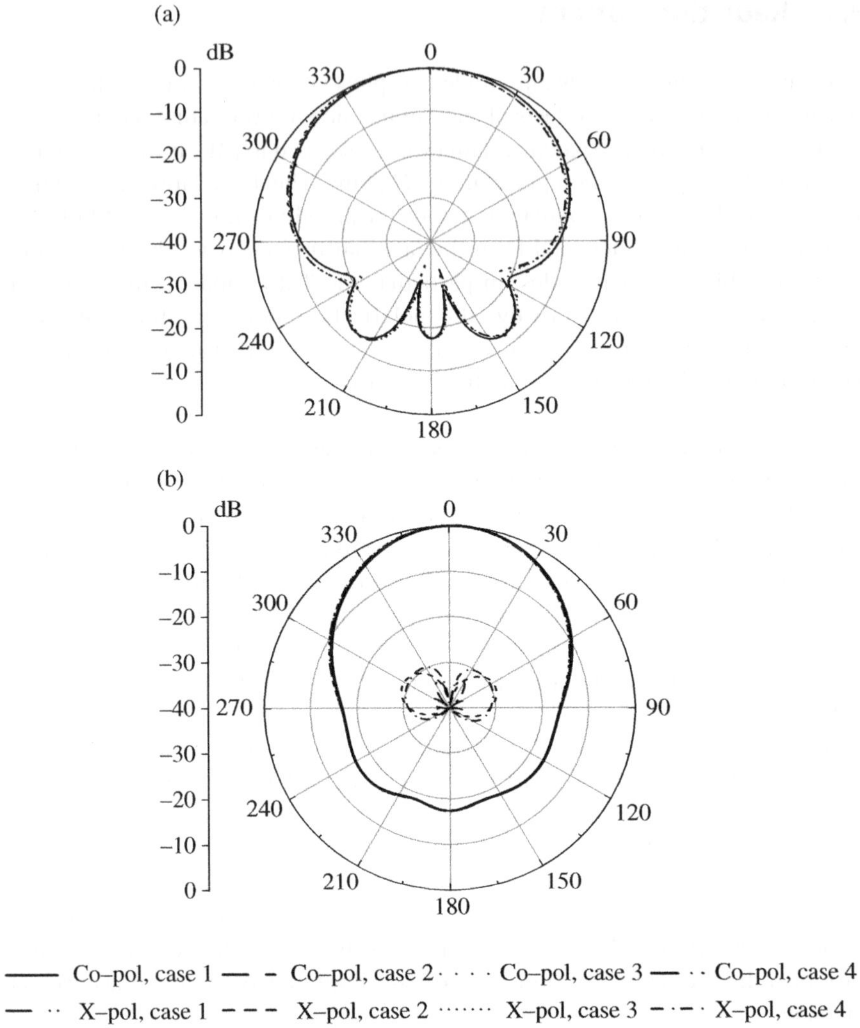

Figure 9.10 Simulated radiation patterns of the differential laminated resonator antenna at 5.65 GHz for four cases. (a) The *E* plane and (b) the *H* plane.

Figure 9.12 shows the measured gain values of the differential microstrip grid array antenna using the absolute gain method with an Agilent four-port VNA. Two identical differential microstrip grid array antennas were used. The measured highest gain was 21.0 dBi at 24.28 GHz and the measured 3-dB gain bandwidth was from 23.09 to 25.27 GHz (or 9.03% at 24.15 GHz), the simulated highest gain was

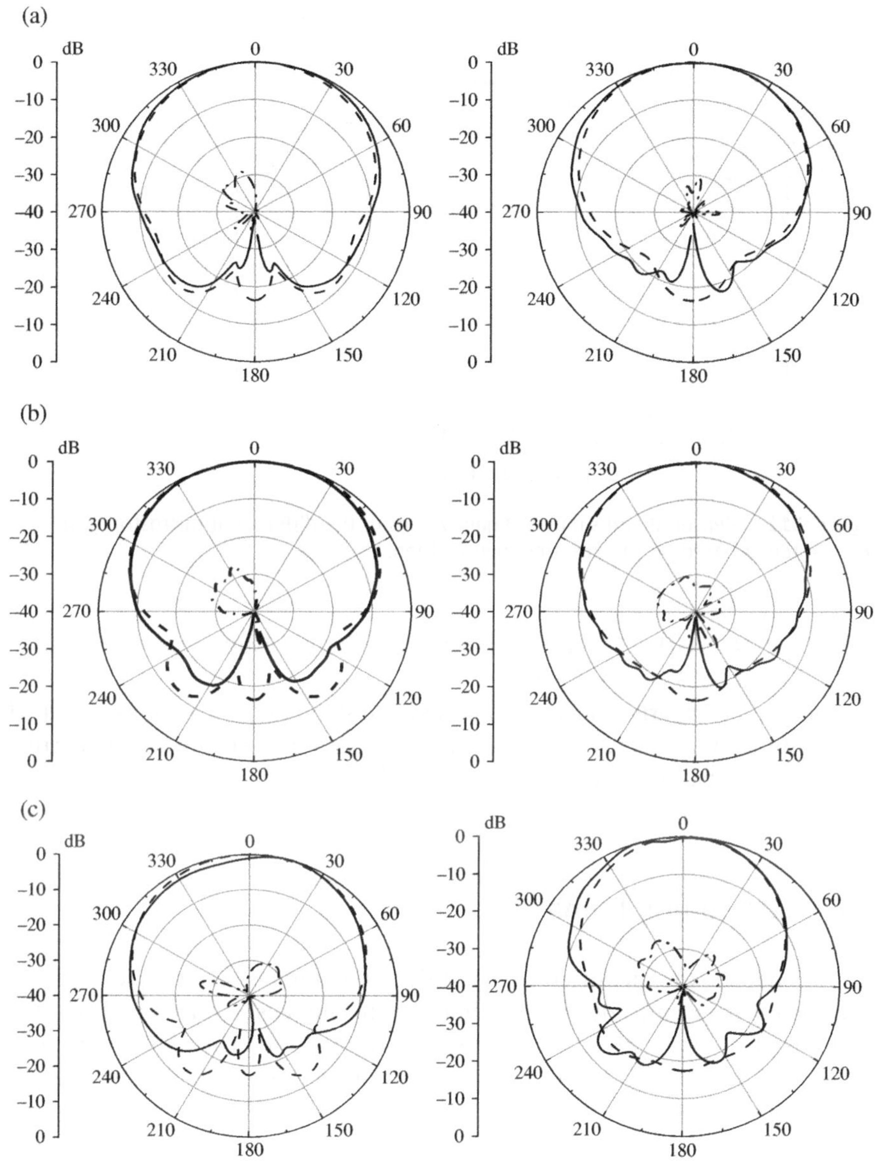

Figure 9.11 Simulated and measured *E*-plane and *H*-plane patterns of the differential laminated resonator antenna. (a) 4.35 GHz. (b) 5 GHz. (c) 5.65 GHz.

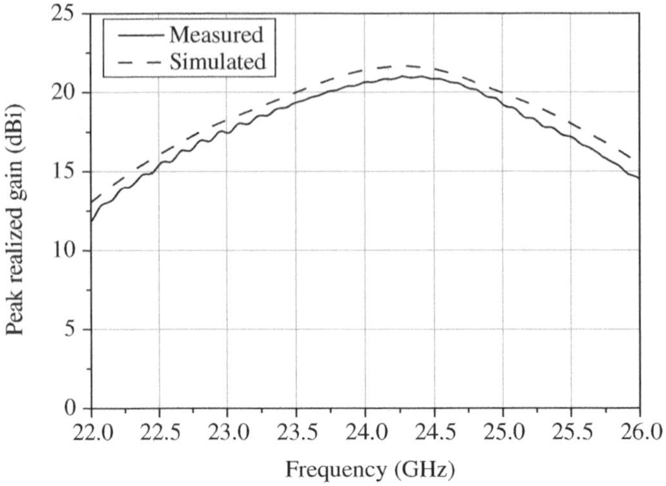

Figure 9.12 Measured and simulated gain values of the differential microstrip grid array antenna based on the absolute gain method.

21.6 dBi at 24.25 GHz and the simulated 3-dB gain bandwidth was from 23.12 to 25.35 GHz (or 9.23% at 24.15 GHz).

In the gain comparison method, precalibrated standard gain antennas are used to determine the absolute gain of the DUT. The losses of the balun and connecting cables need to be calibrated out.

9.6 Concluding Remarks

In this chapter, the measurements of differential antennas were discussed. An emphasis was given to the balun calibration techniques. It was found that the removal of the errors of the balun was quite involved. The calibration technique to remove the errors of a 180° power divider as the balun and the calibration technique to remove the errors of the 180° hybrid coupler as the balun was developed for the measurements of the impedances of differential antennas. The amplitude and phase imbalances of a real balun can be up to 1 dB and 10°, respectively. The effects of such imbalances on the measurements of the radiation patterns of differential antennae were investigated. It was shown that they degrade the polarization purity. The degradation by the phase imbalance of 10° was more severe than that by the amplitude imbalance of 1 dB.

References

1 IEEE Recommended Practice for Antenna Measurements, IEEE 149-2021, 2022.

2 T. Brauner, R. Vogt, W. Bachtold, "A differential active patch antenna element for array applications," *IEEE Microw. Wirel. Comp. Lett.*, vol. 13, no. 4, pp. 161–163, Apr. 2003.

3 Y. P. Zhang, J. J. Wang, Q. Li, X. J. Li, "Antenna and transmit/receive switch for single-chip radio transceivers of differential architecture," *IEEE Trans. Circuits Systems I*, vol. 55, no. 11, pp. 3564–3570, Dec. 2008.

4 S. V. Hum, H. Y. Xiong, "Analysis and design of a differentially-fed frequency agile microstrip patch antenna," *IEEE Trans. Antennas Propagat.*, vol. 58, no.10, pp. 3122–3130, Oct. 2010.

5 R. Meys, F. Janssens, "Measuring the impedance of balanced antennas by an S-parameter method," *IEEE Antennas Propagat. Mag.*, vol. 40, no. 6, pp. 62–65, Dec. 1998.

6 Y. P. Zhang, Z. H. Tu, "Measuring the impedance and efficiency of differentially driven microstrip antenna by two balun methods," *IEEE Trans. Antennas Propagat.*, vol. 62, no.3, pp. 1246–1252, Mar. 2014.

7 Z. H. Tu, Y. P. Zhang, "Comparison of three methods for the measurement of the differential impedance of a balanced device," *IEEE Antennas Propagat. Mag.*, vol. 55, no. 5, pp. 142–151, Oct. 2013.

8 W. H. Hsu, K. L. Wong, "A dual capacitively fed broadband patch antenna with reduced cross-polarization radiation," *Microw. Opt. Technol. Lett.*, vol. 26, pp. 169–171, Aug. 5, 2000.

9 C. H. K. Chin, Q. Xue, H. Wang, "Broadband patch antenna with a folded plane pair as a differential feeding scheme," *IEEE Trans. Antennas Propagat.*, vol. 55, no. 9, pp. 2461–2467, Sep. 2007.

10 Y. C. Ou, G. M. Rebeiz, "Differential microstrip and slot-ring antennas for millimeter-wave silicon systems," *IEEE Trans. Antennas Propagat.*, vol. 60, no. 6, pp. 2611–2619, Jun. 2012.

11 K. D. Palmer, M. W. Rooyen, "Simple broadband measurement of balanced loads using a network analyzer," *IEEE Trans. Instrument. Meas.*, vol. 55, no. 1, pp. 266–272, Jan. 2006.

12 C. R. Curry, "How to calibrate through balun transformers to accurately measure balanced systems," *IEEE Trans. Microw. Theory Technol.*, vol. 51, no. 3, pp. 961–965, Mar. 2003.

13 K. Silvonen, "LMR 16 – a self-calibration procedure for a leaky network analyzer," *IEEE Trans. Microw. Theory Technol.*, vol. 45, no. 7, pp. 1041–1049, Jul. 1997.

14 D. E. Bockelman, W. R. Eisenstadt, "Pure-mode network analyzer for on-wafer measurements of mixed-mode S-parameters of differential circuits," *IEEE Trans. Microw. Theory Technol.*, vol. 45, no. 7, pp. 1071–1077, Jul. 1997.

15 D. E. Bockelman, W. R. Eisenstadt, R. Stengel "Accuracy estimation of mixed-mode S-parameter measurements," *IEEE Trans. Microw. Theory Technol.*, vol. 47, no. 1, pp. 102–105, Jan. 1999.

16 D. M. Pozar and B. Kaufman, "Comparison of three methods for the measurement of printed antenna efficiency," *IEEE Trans. Antennas Propagat.*, vol. 36, no. 1, pp. 136–139, Jan. 1988.

17 W. E. McKinzie III, "A modified Wheeler cap method for measuring antenna efficiency," in *Proc. IEEE AP-S Int. Symp.*, Montreal, Canada, 13-18 Jul. 1997, pp. 542-545.

18 H. Choo, R. Rogers, H. Ling, "On the Wheeler cap measurement of the efficiency of microstrip antennas," *IEEE Trans. Antennas Propagat.*, vol. 53, no. 7, pp. 2328–2332, Jul. 2005.

19 Y. P. Zhang, Z. H. Chen, "The Wheeler method for the measurement of the efficiency of differentially-driven microstrip antennas," *IEEE Trans. Antennas Propagat.*, vol. 62, no. 6, pp. 3436–3439, Jun. 2014.

20 Y. Huang, Y. Lu, S. Boyes, T. H. Loh, "Source-stirred chamber/cap method for antenna radiation efficiency measurements," in *Proc. the 5th European Conf. Antennas Propagat.*, 11–15 Apr. 2011, Rome Italy, pp. 164-168.

21 Y. P. Zhang, "Design and experiment on differentially-driven microstrip antennas," *IEEE Trans. Antennas Propagat.*, vol. 55, no. 10, pp. 2701–2708, Oct. 2007.

22 Y. W. Hou, Z. J. Shao, Y. P. Zhang, J. F. Mao, "A wideband differentially fed dual-polarized laminated resonator antenna," *IEEE Trans. Antennas Propagat.*, vol. 69, no. 7, pp. 4148–4153, Jul. 2021.

Index

Printed and bound by CPI Group (UK) Ltd, Croydon, CR0 4YY

17/09/2026

14964332-0003